Advances in 21st Century Human Settlements

Indexed by SCOPUS

This Series focuses on the entire spectrum of human settlements – from rural to urban, in different regions of the world, with questions such as: What factors cause and guide the process of change in human settlements from rural to urban in character, from hamlets and villages to towns, cities and megacities? Is this process different across time and space, how and why? Is there a future for rural life? Is it possible or not to have industrial development in rural settlements, and how? Why does 'urban shrinkage' occur? Are the rural areas urbanizing or is that urban areas are undergoing 'ruralisation' (in form of underserviced slums)? What are the challenges faced by 'mega urban regions', and how they can be/are being addressed? What drives economic dynamism in human settlements? Is the urban-based economic growth paradigm the only answer to the quest for sustainable development, or is there an urgent need to balance between economic growth on one hand and ecosystem restoration and conservation on the other – for the future sustainability of human habitats? How and what new technology is helping to achieve sustainable development in human settlements? What sort of changes in the current planning, management and governance of human settlements are needed to face the changing environment including the climate and increasing disaster risks? What is the uniqueness of the new 'socio-cultural spaces' that emerge in human settlements, and how they change over time? As rural settlements become urban, are the new 'urban spaces' resulting in the loss of rural life and 'socio-cultural spaces'? What is leading the preservation of rural 'socio-cultural spaces' within the urbanizing world, and how? What is the emerging nature of the rural-urban interface, and what factors influence it? What are the emerging perspectives that help understand the human-environment-culture complex through the study of human settlements and the related ecosystems, and how do they transform our understanding of cultural landscapes and 'waterscapes' in the 21st Century? What else is and/or likely to be new vis-à-vis human settlements – now and in the future? The Series, therefore, welcomes contributions with fresh cognitive perspectives to understand the new and emerging realities of the 21st Century human settlements. Such perspectives will include a multidisciplinary analysis, constituting of the demographic, spatio-economic, environmental, technological, and planning, management and governance lenses.

If you are interested in submitting a proposal for this series, please contact the Series Editor, or the Publishing Editor:

Bharat Dahiya (bharatdahiya@gmail.com) or
Loyola D'Silva (loyola.dsilva@springer.com)

Bharat Dahiya · Francesco de Pascale ·
Orlando De Pietro · Piero Farabollini ·
Francesca Romana Lugeri · Leonardo Mercatanti
Editors

Disaster Resilience and Human Settlements

Emerging Perspectives in the Anthropocene

Springer

Editors
Bharat Dahiya
School of Global Studies
Thammasat University
Bangkok, Thailand

School of Public Leadership
Stellenbosch University
Stellenbosch, Western Cape, South Africa

Orlando De Pietro
Department of Mathematics
and ComputerScience
University of Calabria
Rende (CS), Italy

Francesca Romana Lugeri
University of Camerino
Camerino, Italy

Rome, Italy

Francesco de Pascale
Department of Culture and Society
University of Palermo
Palermo, Italy

Piero Farabollini
School of Sciences
and Technologies—Geology Division
University of Camerino
Camerino, Italy

Leonardo Mercatanti
Department of Culture and Society
University of Palermo
Palermo, Italy

ISSN 2198-2546 ISSN 2198-2554 (electronic)
Advances in 21st Century Human Settlements
ISBN 978-981-99-2247-5 ISBN 978-981-99-2248-2 (eBook)
https://doi.org/10.1007/978-981-99-2248-2

This Springer imprint is published by the registered company Springer Nature Singapore Pte Ltd.
The registered company address is: 152 Beach Road, #21-01/04 Gateway East, Singapore 189721, Singapore

To the social scientists, natural scientists, humanists, urban and regional planners, policymakers, and professionals studying and building disaster resilience for humansettlements in the Anthropocene epoch

Foreword by Prof. Michele Cometa

Since the year 2000, at the very beginning of the new century, chemist P. J. Crutzen and biologist E. F. Stoermer proposed the term "Anthropocene" to designate the "new" epoch in which we live, international cultural studies, i.e., all disciplines concerned with human culture, have been well aware that a confrontation with the natural sciences is now unavoidable. Sciences seemingly far removed from the so-called "humanities," such as geology, climatology, physics, and chemistry. Within a few years the writings of a historian like Dipesh Chakrabarty, who was able to make a connection between climatic events and capitalist economy, would have made this confrontation even more inescapable.

Today, we know that this convergence has a much older origin. As early as the late nineteenth century, an Italian geologist, Antonio Stoppani, realized that the era in which we live is the era of geological and climatic transformation of the planet, an era in which even the most technical observations on the geology of the earth cannot disregard the analysis of the modifications made to the planet by human cultures. Stoppani spoke prophetically of the "Antropozoic" era.

The era of the Anthropocene has forced all humanities disciplines to revise their DNA, especially those that have always walked the difficult ridge of "nature" and "culture" such as geography, anthropology, psychology, and cognitive neuroscience.

Today, Stoppani, forgotten by the history of Italian science, is unanimously recognized as one of the forerunners of the Anthropocene theory, as Etienne Turpin and Valeria Federighi, among others, have made well clear. Stoppani, in a chapter of his *Corso di geologia* (1873—Bernardoni and Brigola publishing, Milan), devoted to the *First Stone Age or Archeolithic Period. Early Period of the Anthropozoic Era*, discusses the "influence of man on telluric nature," imagines a post-apocalyptic scenario in which a geologist from Mars visits a land in which man has become extinct after exerting his "telluric force" for millions of years. A being that has moved mountains, relocated plants, and revolutionized the planet geologically as well. For the Martian geologist, therefore, no other geology can be a given than one that also studies the cultures that have inhabited the planet:

> A new era has thus begun with man. Let us admit, eccentric though it might be, the supposition that a strange intelligence should come to study the Earth in a day when human progeny, such as populated ancient worlds, has disappeared completely. Could he study our epoch's geology on the basis of which the splendid edifice of gone worlds' science was built? Could he, from the pattern of floods, from the distribution of animals and plants, from the traces left by the free forces of nature, deduct the true, natural conditions of the world? Maybe he could but always and only by putting in all his calculations this new element: human spirit. At this condition, as we, for instance, explain the mounds of terrestrial animals' bones in the deeps of the sea, he, too, could explain the mounds of sea shells that savage prehistoric men built on the coasts that they inhabited. But if current geology, to understand finished epochs, has to study nature irrespective of man, future geology, to understand our own epoch, should study man irrespective of nature. So that future geologist, wishing to study our epoch's geology, would end up narrating the history of human intelligence. That is why I believe the epoch of man should be given the dignity of a separate new era. (Stoppani, 1873: 739).

This volume deals with the question of the Anthropocene and that of so-called "natural" disasters, showing, precisely, that there is nothing natural about disasters and human beings have always "collaborated" in the natural history of disasters. It is no coincidence that literature and philosophy, at least since the time of the Lisbon earthquake, has been asking about "man's place in the world." A question that must now be posed from a profoundly anti-anthropocentric perspective, since along with human beings' disruptive intervention, the extinction of the human species—and thus a survival of life on the planet that does not include *Homo sapiens*—should perhaps also be thought of as a possible outcome.

Stoppani could still believe that the "Anthropozoic" era was an era full of future because the human seed is still too young for it to become extinct. However, the twenty-first century, with its nuclear tests, and more recent with its climate catastrophes, make the vision of a post-apocalyptic age ever more vivid and certainly more realistic.

That is why this volume's call for a new culture of resilience where, beyond the fashionable term, the focus is on a pragmatic education to prevention, is as urgent as ever.

If I may be allowed a personal note, I will conclude with a sentence I read in the essay of a young German high school student, early in my career when I was not an intellectual, then, but a very young student: "Disasters are the product of human imprudence."

I think this volume could adopt this "naïve" statement as an exergue.

This work is, moreover, the first result of an institutional agreement between the School of Global Studies of Thammasat University and the Department of Cultures and Societies of the University of Palermo and is certainly intended to be discussed by our students, from whom we, the faculty, have much to learn.

Prof. Michele Cometa
Director
Department of Culture and Society
University of Palermo
Palermo, Italy

Michele Cometa teaches Cultural History and Visual Culture at the University of Palermo. He has been a fellow of the CNR, DAAD, of the Sterling and Francine Clark Art Institute (Williamstown, MA) and Research Associate Professor at the Italian Academy (Columbia University, NY). He has held courses and seminars in various Italian and foreign universities and research institutes. He founded and coordinated the international Ph.D. in *European Cultural Studies/Europäische Kultustudien.*

In 1999 he received the Ervino-Pocar Prize of the City of Gorizia for literary translation. In 2017 he was awarded the Pozzale Luigi Russo Prize (Empoli) for the book "Why do stories help us to live". In 2019 he was awarded the Winckelmann-Medaille of the City of Stendal (Germany).

His most recent publications include: *La scrittura delle immagini* (Cortina, 2012); *Archeologie del dispositivo* (Pellegrini, 2016); *Perché le storie aiutano a vivere* (Cortina, 2017); *Il Trionfo della morte di Palermo. An allegory of modernity* (Quodlibet, 2017); *Literature and Darwinism. Introduction to biopoetics* (Carocci, 2018); *How to study culture. Practices, tactics, forms of writing* (Palermo UP, 2019); *Visual culture. A genealogy* (Cortina, 2020).

Foreword by Asst. Prof. Dr. Prapaporn Tivayanond Mongkhonvanit

It is my pleasure to write a Foreword to this significant volume, *Disaster Resilience and Human Settlements: Emerging Perspectives in the Anthropocene*, co-edited by Bharat Dahiya, Francesco De Pascale, Orlando De Pietro, Piero Farabollini, Francesca Romana Lugeri, and Leonardo Mercatanti. It explores an increasingly important area of global concern that has been emphasized by academics, professionals, and policymakers.

In recent years, the world has witnessed a number of hydro-meteorological and geophysical disasters. The impact of climate change is evident in the increasing number and intensity of extreme weather events, such as Jeanne hurricane (2004), Katrina hurricane (2005), Durian typhoon (2006), Sidr cyclone (2007), Nargis cyclone (2009), Haiyan typhoon (2013), and Maria hurricane (2017). The Gujarat earthquake (2001), Bam earthquake (2003), Indian Ocean earthquake and tsunami (2004), Kashmir earthquake (2005), Sichuan earthquake (2008), Haiti earthquake (2010), Tohoku earthquake and tsunami (2011), Nepal earthquake (2015), earthquake in central Italy (2016), and Sulawesi earthquake and tsunami (2018) have shown the destructive power of geophysical disasters. These hydro-meteorological and geophysical disasters have caused major loss of lives and livelihoods and damage to properties in urban and rural human settlements around the world.

In view of this, strengthening disaster resilience for human settlements has become one of the key priority areas in the global sustainable and resilient development agendas. The academic community, including those engaged in the study and research in the emerging field of Global Studies, has been playing its part in conducting research and studies on the various aspects of disaster resilience and human settlements in the current Anthropocene era.

This book project has brought together scholars based on three continents—Asia, Australia, and Europe, and their cutting-edge research on the emerging perspectives of disaster resilience and human settlements in the Anthropocene. In the first part, the book begins with a thoughtful introduction written by the Bharat Dahiya and his co-editors. Four chapters in the second part examine flood risks in urban areas in relation to disaster resilience assessment and governance. In the third part, four chapters study the perceptions and representations of disasters and climate change in the Anthropocene era. And five chapters in the forth part focus on postdisaster management and recovery with regard to urban landscape, pandemic, and community resilience.

I would like to congratulate the co-editors for their dedication, perseverance, and hard work in accomplishing this collaborative book project. It is never easy to complete such successful projects when its partners are based on different continents. This becomes even more difficult when a pandemic causes major lockdowns around the world, as happened in the case of COVID-19 during the years 2020–2022.

In March 2022, the School of Global Studies, Thammasat University signed a Memorandum of Understanding (MOU) with the Department of Culture and Society, University of Palermo, Italy. I am delighted to note that several scholars from these two institutions have played an important role in putting together this co-edited volume.

Taking this opportunity, I would like to thank the co-editors for their kind invitation to write this Foreword. I extend my best wishes to all scholars involved in the preparation of this rich collection of studies on disaster resilience and human settlements in the Anthropocene era.

Asst. Prof. Dr. Prapaporn Tivayanond Mongkhonvanit
Dean
School of Global Studies
Thammasat University
Bangkok, Thailand

Asst. Prof. Prapaporn Tivayanond Mongkhonvanit is the Dean of the School of Global Studies at Thammasat University. She was a Board Member of the National Council on Social Welfare of Thailand and the Founding Director of Social Policy and Development (SPD) International Programme at Thammasat University. Prapaporn is currently leading the SDG Lab at the School of Global Studies, and is working on promoting a vital connection between localization and globalization under the Sustainable Development Goals (SDGs).

Prapaporn lectures and conducts research in areas of developmental welfare and social protection in Asia. She has researched the connection between Thailand's welfare scheme and the informal sector, paying particular attention to OTOP producers and Thailand's social entrepreneurship. Prapaporn has been a consultant at the United Nations Children's Fund (UNICEF) Regional Office for East Asia and the Pacific and the Association of Southeast Asian Nations (ASEAN). She has also conducted studies in Thailand's early childhood care and education for UN Women. Her study for United Nations Research Institute for Social Development (UNRISD) was on Thailand's universal health scheme. She had led training initiatives for the 'Girls in ICT' for the United Nations International Telecommunication Union (ICT). She has also worked on fair trade issues in Thailand, and is now researching on shared values, social bonds, social policy innovation, and the impact of the fourth industrial revolution on the marginalized groups.

Prapaporn is the recipient of Asia Society's Asia 21 Young Leaders Award. She holds an MA in International Relations from Victoria University of Wellington, New Zealand, and a DPhil in Social Policy from the University of Oxford.

Preface

For decades, Disaster studies researchers have questioned the use of the expression "natural disasters" (Wisner et al., 2012; Kelman, 2020; Alexander et al., 2021). In fact, disasters cannot be defined as natural, since they are processes born from the relationship of mutual complication between nature and society (De Pascale and Dattilo, 2021), resulting from the interaction between a destructive natural agent (such as, indeed, an earthquake, a tsunami, a hurricane, a volcanic eruption, a flood) and the territory on which it impacts. Often, in fact, the territory is made more vulnerable by social, cultural, economic, environmental, and institutional factors, making it possible for the effects of the extreme natural event to be catastrophic.

Disaster studies from a multidisciplinary perspective began in the 1960s by a group of sociologists—E. L. Quarantelli in collaboration with R. R. Dynes and E. Haas—who founded the Disaster Research Center (DRC) at Ohio State University. These researchers, using qualitative research methods, engaged in the collection of data relating to the preparation of communities in dealing with and managing mass emergency situations (Castellani, 2018).

A growing intensity and frequency of disasters have been observed in the recent decades which affected human settlements, whether mega cities, metropolitan cities, or small- and medium sized towns. The geological epoch, the Anthropocene, proposed by Crutzen and Stoermer (2000), highlights the intensity and pervasiveness of human activity in the dynamics of the Earth system.

The Anthropocene, unlike other previous eras, is commonly recognized as the era in which the impact of human beings is most incisive in determining the dynamics of the climate and the planet in which we live. The Nobel Prize in Chemistry Paul Crutzen (2005) argued, for example, that, from a geological point of view, the era in which we live presents modifications linked to human action that can persist for millions of years, giving an irreversible imprint on the transformations of the terrestrial environment. Unlike the Pleistocene, Holocene, and all previous eras, therefore, it is characterized above all by the impact of human beings on the environment (De Pascale and Dattilo, 2015). "The new force—says Paul Crutzen—of which the extraterrestrial observer distinguished the action is us, capable of moving more matter than the volcanoes and the wind combined, of making entire continents degrade, of altering the cycle of water, nitrogen, and carbon and to produce the most abrupt and marked increase in the amount of greenhouse gases in the atmosphere in the last 15 million years" (Crutzen, 2005, pp. 25–26). "Certain geological periods—continues Crutzen—are characterized by the fossil remains of disappeared species; the Anthropocene is characterized by the species which has suddenly become decisive for the balance of the Earth and the climate" (Crutzen, 2005, p. 26).

According to Paul Crutzen, we are only at the beginning of the Anthropocene, and it is not at all clear what the impact on the climate of the greenhouse gases we are pouring into the atmosphere will be. From paleoclimatology studies, we know that changes, even the most abrupt ones, occur on the scale of thousands of years. The definitive effects of today's anomaly could be seen in many centuries (De Pascale and Dattilo, 2015).

According to Clive Hamilton, the Anthropocene cannot be defined simply as the increase of the impact of humans on the environment, which would simply extend what humans have been doing for millennia (Hamilton, 2016). The Anthropocene Working Group, which includes Crutzen, initially leaned toward the idea that the beginning of the Anthropocene could date back to the industrial revolution, with other scholars who have interpreted the Anthropocene as a continuing impact of individuals on the biosphere (Monastersky, 2015b, p. 145; Hamilton, 2016). As Hamilton points out, changes in landscape and vegetation can bear the hallmarks of humans, but these cannot have a sufficient impact on the planet to bring about a new geological era. Other scholars argue that the start date of the Anthropocene should be closely linked to the time when human societies began to play a decisive role in shaping the Earth's ecosystems (Monastersky, 2015b; Hamilton, 2016). Furthermore, according to Hamilton, the Anthropocene began when humans caused a significant impact to the point of changing the functioning of the Earth system (Hamilton, 2016).

Some scholars include archeology in the debate, dating the beginning of the Anthropocene to the expansion of agriculture, more than 5000 years ago, or to a surge in mining activity, more than 3000 years ago (Ellis, 2011). Yet other scholars insist that the Anthropocene is the most recent phase of a process that began 50,000 years ago with human geographical expansion (Dattilo and De Pascale, 2019). Despite these academic speculations, members of the Anthropocene Working Group have proposed 1945 as the unequivocal point at which individuals produced a significant change in the functioning of the Earth system (Zalasiewicz et al. 2015).

The first A-bomb test in 1945 contributed to the first stratigraphic presence of radioactive elements on Earth and the much more intense nuclear detonations, which took place over the next half century, contributed significantly to that process (Monastersky, 2015a). Other factors that have contributed to this profound change concern, for example, the accumulation of carbon dioxide in the atmosphere, the number of motor vehicles increased from 40 to 800 million, the world production of plastic which reached 300 million tons in 2015 (McNeill and Engelke, 2018). Furthermore, in this period known as the "Great Acceleration," there have been increasingly significant migrations from rural areas to urban centers, fuelling the growth of megacities (Monastersky, 2015a). These changes have exerted enormous pressure on the environment and the biosphere and are the basis of the concept of the Anthropocene, literally known as the "Age of Man."

Starting from this reference framework, this book deals with the emerging perspectives on disaster resilience and human settlements in the larger context of the Anthropocene. The chapters explore urban and rural perspectives focusing on the current and emerging perspectives on disaster resilience through a holistic approach, involving scientists, humanists, planners, policymakers, and professionals in the global debate.

Bangkok, Thailand — Bharat Dahiya
Palermo, Italy — Francesco de Pascale
Rende, Italy — Orlando De Pietro
Camerino, Italy — Piero Farabollini
Rome, Italy — Francesca Romana Lugeri
Palermo, Italy — Leonardo Mercatanti

References

Alexander, D, Gaillard JC, Kelman I, Marincioni F, Penning-Rowsell E, van Niekerk D et al (2021) Academic publishing in disaster risk reduction: past, present, and future. Disasters 45(1):5–18

Castellani S (2018) Disaster resilience e spazi pubblici. Strade, piazze e aree verdi nella storia dei terremoti aquilani (1315–2009), Tesi di dottorato, Università degli Studi di Padova, Padova

Crutzen, Paul J (2005) Benvenuti nell'antropocene! Mondadori, Milano

Dattilo V, De Pascale F (2019) Geomorphology and philosophy: a STEAM survey about the anthropocene. In: De La Garza A, Travis C (eds) The STEAM revolution. Transdisciplinary Approaches to Science, Technology, Engineering, Arts and Mathematics. Springer, Cham, p 203–219

De Pascale F, Dattilo V (2015) La semiosi dell'Antropocene. Riflessioni tra geoetica e semiotica a partire dal triangolo di Peirce. Filosofi(e)Semiotiche 2(1):90–98

De Pascale, F, Dattilo, V (2021) The geoethical semiosis of the anthropocene: the peircean triad for a reconceptualization of the relationship between human beings and environment. Ann Am Assoc Geogr 111(3):647–654, https://doi.org/10.1080/24694452.2020.1843994

Ellis E (2011) Anthropogenic transformation of the terrestrial biosphere. Phil Trans R Soc A 369:1010–1035. https://doi.org/10.1098/rsta.2010.0331

Hamilton C (2016) Define the anthropocene in terms of the whole earth. Nature 536, 251, http://www.nature.com/news/define-the-anthropocene-in-terms-of-the-wholeearth-1.20427. https://doi.org/10.1038/536251

Kelman I (2020) Disaster by choice: how our actions turn natural hazards into catastrophes. Oxford University Press, Oxford, UK

Mcneill JR, Engelke P (2018) La Grande accelerazione. Una storia ambientale dell'Antropocene dopo il 1945. Einaudi, Torino

Monastersky R (2015a) First atomic blast proposed as start of anthropocene. Nature, http://www.nature.com/news/first-atomic-blast-proposed-as-start-of-anthropocene-1.16739. https://doi.org/10.1038/nature.2015.16739

Monastersky R (2015b) Anthropocene: the human age. Nature 519(7542):144–147

Wisner B, Gaillard JC, Kelman I (eds) (2012) Handbook of hazards and disaster risk reduction. Routledge, Abingdon, UK

Zalasiewicz J et al (2015) When did the anthropocene begin? A mid-twentieth century boundary level is stratigraphically optimal. Quat Int 383:196–203

Acknowledgements

No major work can be completed without the collaboration and cooperation of many people. Likewise, the preparation and publication of this anthology, *Disaster Resilience and Human Settlements: Emerging Perspectives in the Anthropocene*, has benefitted from the contribution and support of many individuals whom the Editors would like to acknowledge.

First of all, the Editors would like to thank the contributing authors of the various chapters included in this volume: Iftekhar Ahmed, Fabrizio Bendia, Cristina Casareale, Alessandra Colocci, Gianluca Esposito, Massimiliano Fazzini, Eleonora Gioia, Teresa Graziano, Miquel Grimalt, Zeeshan Ibrar, Vania Mancinelli, Noemi Marchetti, Fausto Marincioni, Vincenzo Marsala, Giovanni Messina, Enrico Miccadei, Giorgio Paglia, Joan Rosselló, Charles Travis, Gaetano Sabato, and Vinita Yadav. They conducted cutting-edge research in varied contexts ranging from Europe to Asia and to the Pacific and documented it systematically for the benefit of readers.

Second, the Editors are grateful to Prof. Michele Cometa, Director, Department Culture and Society, University of Palermo Palermo, Italy, and Asst. Prof. Dr. Prapaporn Tivayanond Mongkhonvanit, Dean, School of Global Studies, Thammasat University, Pathum Thani, Thailand, for writing thought provoking Forewords to this co-edited book.

Third, the Editors would like to thank Dr. Loyola D'Silva, Executive Editor; Ms. Shalini Monica Clement Selvam; Ms. Manopriya Saravanan, together with Springer's entire team, for their support and cooperation throughout the preparation and publication of this co-edited volume.

Contents

Editors and Contributors

About the Editors

Bharat Dahiya is Director of Research Center for Sustainable Development and Innovation at the School of Global Studies, Thammasat University, Thailand. He is an Extraordinary Professor at the School of Public Leadership, Stellenbosch University, Western Cape, South Africa, and an Adjunct Faculty at the Asian Institute of Technology, Thailand.

An award-winning urbanist, Bharat combines cutting-edge research, policy analysis and development practice aimed at examining and tackling socio-economic, environmental and governance issues in the global urban context of sustainable development. Since the early-1990s, his research and professional work has focused on sustainable cities and urbanization, strategic urban planning and development, urban infrastructure, urban environment, climate change and urban resilience, post-disaster recovery and reconstruction, and cultural heritage and landscapes.

Working with the World Bank, UN-HABITAT, Asian Development Bank, UNDP, United Nations University, and Metropolis, he has initiated, led and/or contributed to international development projects and programmes on sustainable urban development around the world. Some of the projects that Bharat led include the preparation of the first-ever 'Citywide Pro-poor Ger-area Upgrading Strategy and Investment Plan for Ulaanbaatar', and 'Community-Led Ger Area Upgrading in Ulaanbaatar City' (Mongolia); 'Earthquake-resistant and Energy-efficient School Reconstruction in Sichuan Province', and 'Getting Children Back to School: Providing Prefabricated Classroom Units to the Earthquake-Affected Children of Gansu Province' (China); and 'City Development Strategy for Thanh Hoa City in a Regional Development Context' (Viet Nam).

Throughout his career, Bharat has been involved in the preparation and dissemination of cutting-edge knowledge products. At the World Bank headquarters in Washington DC, USA, he conducted the first-ever systematic review of the Bank's investments for improving urban liveability, published as a co-authored book, Urban Environment and Infrastructure: Toward Livable Cities (2004). At UN-HABITAT, he led, conceptualized, and coordinated the preparation of United Nations' inaugural report on The State of Asian Cities 2010/11 (2010). In 2018, he co-authored Partnering for Sustainable Development: Guidelines for Multi–stakeholder Partnerships to Implement the 2030 Agenda in Asia and the Pacific (UNU-IAS, and UN-ESCAP). He was the lead author of Metropolis's first-ever Asian Metropolitan Report (2021). He has co-edited two academic anthologies, New Urban Agenda for Asia-Pacific: Governance for Sustainable and Inclusive Cities (Springer, 2020) and Practising Cultural Geographies: Essays in Honour of Rana PB Singh (Springer, 2022).

Bharat advises several professional, academic, private sector, and non-profit organisations around the world. Since 2019, he has been a Member of the International Advisory Board for the UN-HABITAT's World Cities Report (Nairobi, Kenya). Since early-2014, he has been the Series Editor for the SCOPUS-indexed Springer book series, Advances in 21st Century Human Settlements, which has a growing collection of over 40 monographs and edited volumes. He serves as a member of the editorial board of several international journals as well as the International Scientific Board of the book series, 'Geographies of the Anthropocene', published by Il Sileno Edizioni, Italy.

He holds a Ph.D. in Urban Governance, Planning and Environment from the University of Cambridge, UK. He read for his M.A. in Geography from Jawaharlal Nehru University, and Master of Planning from School of Planning and Architecture, both based in New Delhi, India.

Francesco De Pascale, Ph.D., is Research Fellow at the Faculty of Communication Sciences of the University of Teramo (Italy) and Teaching Assistant in Geography of Cultural Heritage at the University of Palermo (Italy). He mainly deals with the geography of risk and the social perception of the risks associated with natural hazards and climate change in the context of Disaster Studies. He is the Editor-In-Chief of the scientific series "Geographies of the Anthropocene" (Il Sileno Edizioni) and Associate Editor of the journals AIMS Geosciences, Stratigrafie del Paesaggio, Filosofi(e)Semiotiche. He is Editor of the Topical Collection "Ethics in Geosciences", in MDPI Geosciences, with Piero Farabollini, Francesca Romana Lugeri and Francesco Muto. He is the author of over eighty scientific publications with publishers of national and international relevance. Among his publications: F. De Pascale, V. Dattilo (2021), The geoethical semiosis of the Anthropocene: the Peircean triad for a reconceptualization of the relationship between human beings and environment, Annals of the Association of American Geographers, vol. 111, issue 3, 647–654; F. De Pascale, C. Travis, The Covid-19 Testimonies Map: Representing Italian "Pandemic Space" Perceptions with Neogeography Technologies. In: Travis C. et al. (Eds.), Routledge Handbook of the Digital Environmental Humanities, Routledge, London, 2022.

Orlando De Pietro is Associate Professor (SSD M-PED/04—Experimental Pedagogy. SC 11/D2 - Didactics, Special Pedagogy and Educational Research) at the Department of Mathematics and Computer Science, University of Calabria. He is a professor of: Educational Research Methodology, Theory and Methods of Evaluation, Educational Technologies at the Degree Course in Primary teacher Education. He is scientific coordinator of several research projects focused on the study, design and implementation of virtual environments for online learning supported by intelligent agents and e-Learning 2.0 platforms.

Studies and scientific research activities concern the applications of Virtual Reality and Augmented Reality to educational-training processes and, in recent years, the methodology of educational research. The research, in particular, concerns the use of educational technologies for the realization of innovative educational-training environments, didactic design and assessment for competencies. The design learning-teaching processes for the acquisition of Soft Skills, fundamental in work contexts. The research interest is focused, therefore, toward those educational and innovative strategies (Service Learning, Digital Storytelling, Debate, STEAM, etc.) that enable subjects to mature Soft Skills.

The activity described is supported by more than 100 publications in journals and conferences, national and international, and monographs, of which we cite: De Pietro O., De Pascale F. (2020). STEAM educational paths to fight educational poverty and reduce the disaster risk: an experimental activity in a primary school. Q-TIMES WEBMAGAZINE, vol. no. 4, p. 167–191; Nuzzaci A., De Pietro O., Altomari N., Valenti A. (2020). The self-perceived skills of teachers in initial training: validation of a scale to measure methodological skills. LLL, vol. Vol. 16, p. 338–353, ISSN: 2279-9001; De Pietro O., Altomari N. (2019). A Tool to Measure Teachers' Soft Skills: Results of a Pilot Study. Advances in Social Science and Culture, vol. No. 1, p. 245–257.

Piero Farabollini is currently an Associate Professor in Physical Geography and Geomorphology at the University of Camerino (Italy). He received his degree in Geology from the University of Camerino in 1987 and a Ph.D. in Earth Sciences from the University of Perugia in 1995.

Piero is the Head of the Marche Regional Council of Geologist, a Member of List of Experts to the MIUR (Ministry of Education and Scientific Research), and a Member of List of Experts in Research, Development, Innovation and Technology Transfer at the Marche Region (Decree of Regional Council n. 4/S11 of 27.01.2009).

Most of his past research has focused on the reconstruction of quaternary geomorphological evolution of the umbro-marchean Apennines. His scientific activity is directed to the Applied Geomorphology and to the Engineering Geomorphology with particular attention to the evaluation, prevention and prediction of natural hazards; in fact in this sector was responsible and co-coordinator of several projects sponsored by various organizations and companies (Marche Region,

Province of Macerata, IRPI-CNR, Marche Regional Basin Authority, Authority of Inter-basin of the river Tronto, Aquater, GNDT_CNR, RFI spA) on hydrogeological and seismic risk assessment and other activities related to the environmental planning and laws.

Part of his activity is aimed at geoethics in the geologist profession, at the communication and dissemination of Earth Sciences and at the dissemination and awareness of natural hazards and related risks. As part of the activity carried out as Commissioner of the 2016 Earthquake Reconstruction, great commitment was placed in the dissemination of the activities to be activated for the purpose of a conscious and adequate post-emergency reconstruction also for the purposes of effective resilience.

He is the author of some hundreds of papers and has been the tutor of about 100 experimental degree theses.

Francesca Romana Lugeri is an Italian geologist (degree in Earth Sciences 1984). Further studies led her to a B.Sc. in Social Sciences (2004), a M.Sc. in Geography (2006), and a Ph.D. in Environmental Science and Public Health (2016). She worked as an Associate Researcher at the University of Camerino, and as a Technologist Researcher at Istituto Superiore per la Protezione e la Ricerca Ambientale (ISPRA), both based in Italy. Her research activity covers topics related to science communication, sustainable development, natural and cultural landscape, risk prevention, education, teaching and learning. She has designed and implemented unconventional science communication projects and is the author of a number of articles published in peer-reviewed journals.

Leonardo Mercatanti (Catania, 1972) is graduated in Urban Geography at the Faculty of Economics (University of Catania). He has an international Master's degree (with thesis) in "Economics and Valorisation of Cultural Heritage" from University of Catania, Pisa Normale School and David Hume Institute (Edinburgh). Leonardo Mercatanti has an international Ph.D. in "Development politics and territorial management" from University of Trieste. From 2005 to 2017 he was Researcher in Geography and since 2017 he is Associate Professor in Geography at the Department of Culture and society, University of Palermo. He is lecturer in Geography of tourism and Geography of cultural heritage. Since 2004 he has been involved in national research projects (PRIN) concerning the urban development, politics of urban sustainability and cultural tourism in Sicily. He was member of the scientific board of the International Congress "Commerce spaces in the urban change processes" (Catania, October 25–26, 2007) organized by "Department of Economy and Territory" (University of Catania); "Peoples and Cultures of the World" first and second edition (Palermo, 2019 and 2022); "Cinema and the City: Interdisciplinary Perspectives" (Palermo, 2019). He speaks English and Spanish fluently and has written many specialized articles (in Italian and in English) about commerce, population, cultural heritage, tourism and ethnic conflicts. He has contributed with papers and posters to national and international congresses. He is Director of the series "Geografia, culture e società", Nuova Trauben-Turin; he is a member of the Editorial Scientific Committee "Geographic narrations of the territory", Aracne publisher-Rome. He was guest editor for the special issues on "Sustainability and Risk Perception: multidisciplinary approaches" (2020) and "Digital Education, Geography and Multidisciplinarity: Themes, Methods and Critical Issues" (2022) of the international journal "AIMS Geosciences."

Contributors

Iftekhar Ahmed School of Architecture and Built Environment, University of Newcastle, Callaghan, Australia

Joan Bauzà-Llinàs Climaris Research Group, Geography Department, Universitat de Les Illes Balears, Palma, Spain

Fabrizio Bendia School of Science and Technology, University of Camerino, Camerino, Italy

Cristina Casareale Università Politecnica Delle Marche, Ancona, Italy

Alessandra Colocci Università Politecnica Delle Marche, Ancona, Italy

Bharat Dahiya School of Global Studies, Research Center for Sustainable Development and Innovation, Thammasat University, Bangkok, Thailand

Francesco De Pascale Department of Culture and Society, University of Palermo, Palermo, Italy

Orlando De Pietro Department of Mathematics and Computer Science, University of Calabria, Rende, Italy

Gianluca Esposito Laboratory of Environmental Geomorphology and GIS, Department of Engineering and Geology, Università degli Studi "G. d'Annunzio" Chieti-Pescara, Chieti Scalo, CH, Italy

Piero Farabollini School of Sciences and Technologies—Geology Division, Camerino University, Camerino (MC), Italy

Massimiliano Fazzini Laboratory of Environmental Geomorphology and GIS, Department of Engineering and Geology, Università degli Studi "G. d'Annunzio" Chieti-Pescara, Chieti Scalo, CH, Italy

Eleonora Gioia Università Politecnica delle Marche, Ancona, Italy

Teresa Graziano University of Catania, Catania, Italy

Miquel Grimalt-Gelabert Climaris Research Group, Geography Department, Universitat de Les Illes Balears, Palma, Spain

Zeeshan Ibrar School of Planning and Architecture, Department of Regional Planning, New Delhi, India

Francesca Romana Lugeri Rome, Italy

Vania Mancinelli Laboratory of Environmental Geomorphology and GIS, Department of Engineering and Geology, Università degli Studi "G. d'Annunzio" Chieti-Pescara, Chieti Scalo, CH, Italy

Noemi Marchetti Università Politecnica Delle Marche, Ancona, Italy

Fausto Marincioni Università Politecnica delle Marche, Ancona, Italy

Vincenzo Marsala SGI Studio Galli Ingeneria S.r.l., Sarmeola di Rubano, PD, Italy

Leonardo Mercatanti Department of Culture and Society, University of Palermo, Palermo, Italy

Giovanni Messina University of Messina, Messina, Italy

Enrico Miccadei Laboratory of Environmental Geomorphology and GIS, Department of Engineering and Geology, Università degli Studi "G. d'Annunzio" Chieti-Pescara, Chieti Scalo, CH, Italy

Giorgio Paglia Laboratory of Environmental Geomorphology and GIS, Department of Engineering and Geology, Università degli Studi "G. d'Annunzio" Chieti-Pescara, Chieti Scalo, CH, Italy

Joan Rosselló-Geli Climaris Research Group, Universitat de Les Illes Balears, Palma de Mallorca, Spain

Gaetano Sabato University of Palermo, Palermo, Italy

Charles Travis Department of History, University of Texas, Arlington, USA; Trinity Centre for Environmental Humanities, Trinity College Dublin, Dublin, Ireland

Vinita Yadav School of Planning and Architecture, Department of Regional Planning, New Delhi, India

Introduction

Disaster Resilience and Human Settlements in the Anthropocene

Bharat Dahiya, Francesco De Pascale, Orlando De Pietro, Piero Farabollini, Francesca Romana Lugeri, and Leonardo Mercatanti

Abstract This chapter provides an overview of the book's theme, 'Disaster Resilience and Human Settlements: Emerging Perspectives in the Anthropocene'. First, the chapter sets the context starting from the emerging perspectives on disaster resilience and human settlements in the larger context of the proposed new geological era of Anthropocene. It explores the impact of disasters on the human settlements, giving examples and illustrating the theoretical reference framework regarding the birth of the idea of the Anthropocene. Second, the chapter focuses on the interdisciplinary nature of the 'disaster resilience and human settlements' theme in relation to the various global development agendas, including the Expanded Brown Agenda, the Hyogo Framework of Action, the Sendai Framework for Disaster Risk Reduction, the 2030 Agenda for Sustainable Development, the Sustainable Development Goals, the Paris Agreement, and the New Urban Agenda. In doing so, it dwells in particular on the theme of social vulnerability as an element capable of transforming an extreme

B. Dahiya (✉)
School of Global Studies, Research Center for Sustainable Development and Innovation, Thammasat University, Bangkok, Thailand
e-mail: bharatdahiya.tu@gmail.com

F. De Pascale · L. Mercatanti
Department of Culture and Society, University of Palermo, Palermo, Italy
e-mail: francesco.depascale@unipa.it

L. Mercatanti
e-mail: leonardo.mercatanti@unipa.it

O. De Pietro
Department of Mathematics and Computer Science, University of Calabria, Rende, Italy
e-mail: orlando.depietro@unical.it

P. Farabollini
School of Sciences and Technologies – Geology Division, Camerino University, Camerino (MC), Italy
e-mail: piero.farabollini@unicam.it

F. R. Lugeri
Rome, Italy
e-mail: francesca.lugeri@unicam.it

B. Dahiya et al. (eds.), *Disaster Resilience and Human Settlements*,
Advances in 21st Century Human Settlements,
https://doi.org/10.1007/978-981-99-2248-2_1

natural event into disaster. From the various international agendas, it emerges, that the current and emerging perspectives on disaster resilience via-a-vis must be studied and addressed through a holistic and integrated approach. Such an approach ought to involve natural and social scientists, humanists, planners, policymakers, and professionals in the global debate. Finally, the chapter introduces the structure of the book including its various parts and the outlines of the papers presented therein.

Keywords Anthropocene · Disaster resilience · Human settlements · SDGs · Social vulnerability · New Urban Agenda · The Paris Agreement

1 The Current Context

Natural hazards often cause severe impacts within and beyond national borders. Earthquakes and tsunamis, volcanic eruptions, floods, landslides, hurricanes can impact several states and even become a global disaster (for example, the COVID-19 pandemic). The 'Global Assessment Report on Disaster Risk Reduction' for 2022 notes that the global annual average economic loss associated with geophysical, and climate- and weather-related disasters during the past decade (2011–2020) was estimated at US $170 billion (UNDRR 2022, p. 32). The impacts of extreme natural events highlight the interconnectedness of global society, even though the events that triggered extreme impacts usually have local origins (Ismail-Zadeh 2022). In various areas of the planet, the intense urbanization, which took place without taking due account of the areas in which, for example, geo-hydrological and hydraulic events could have occurred, has led to a considerable increase in exposed and vulnerable elements and, therefore, of the risk (De Pascale et al. 2019).

Human settlements, whether mega cities, metropolitan cities, or small- and medium-sized towns, are affected severely by disasters. The world has seen the terrible impacts of disasters on human settlements of all size and shapes caused by the Odisha cyclone in India (1999), heatwaves in Europe (2003), Aceh tsunami in Indonesia (2004), hurricane Katrina in the United States (2005), cyclone Nargis in Myanmar, and earthquake in China (2008), earthquake in Haiti (2010), Tohoku earthquake and tsunami in Japan (2011), hurricane Sandy in the United States (2012), typhoon Haiyan in the Philippines (2013), earthquake in Nepal (2015), hurricane Maria in Dominica, St Croix, and Puerto Rico (2017), bushfires in Australia (2019–2020), floods in South Asia (2020), and fires and floods in Europe, and floods in Pakistan (2022). When human settlements are impacted by disasters, there is a tremendous loss of lives, livelihoods and property (UN-ESCAP 2017; UNDRR 2019; UN-HABITAT 2020). Specifically, the disasters in the early twenty-first century' have caused an alarm for governments and communities around the world. Governments and the various stakeholders have started to take preventive measures to save people, properties and infrastructures, and to avoid direct and indirect losses. However, the physical and social vulnerabilities to natural and human-induced hazards are still increasing due to the number of vulnerable objects and the clustering of populations and infrastructures in the exposed areas to extreme natural events (Ismail-Zadeh 2018).

The growing intensity and frequency of disasters has been observed in the recent decades, as noted above. The geological epoch, the Anthropocene, proposed by Crutzen and Stoermer (2000), highlights the intensity and pervasiveness of human activity in the dynamics of the Earth system. While in geology the recognition of a new geological epoch requires some procedures and the identification of markers (golden spike) to establish the start of an era, in the human and social sciences, the Anthropocene has already initiated an intense debate focusing intellectual attention on the various processes and phenomena that link human activity with nature and the global environment.

Unlike the Pleistocene, the Holocene and all previous eras, the Anthropocene is characterized above all by the impact of human beings on the environment. "The new force", says Paul Crutzen (1933–2021), "of which the extraterrestrial observer distinguished the action is us, capable of moving more matter than the volcanoes and the wind combined, of making entire continents degrade, of altering the cycle of water, nitrogen and carbon and to produce the most abrupt and marked increase in the amount of greenhouse gases in the atmosphere in the last 15 million years" (Crutzen 2005, pp. 25–26). According to Crutzen, "certain geological periods are characterized by the fossil remains of disappeared species; the Anthropocene is characterized by the species that has suddenly become decisive for the balance of the Earth and the climate" (Crutzen 2005, p. 26). The idea was born by chance, during a meeting of the scientific committee of the International Geosphere-Biosphere Program, which was held on the morning of February 22, 2000, in Cuernavaca, Mexico. The chairman of the meeting was talking about human activity in the Holocene, when Crutzen interrupted him to observe that the Holocene had set, and we were now in the Anthropocene. The Anthropocene, therefore, stands for "the geological epoch of man"; just as anthropic means "relative to man" and anthropogenic means "man-made" (Crutzen 2005, p. 27).

According to Paul Crutzen, we are only at the beginning of the Anthropocene, and it is not clear what the impact of greenhouse gases—which we are pouring into the atmosphere, will be on the climate. From paleoclimatology studies we know that changes, even the most abrupt ones, occur on the scale of thousands of years. The definitive effects of today's anomaly could be seen in many centuries. Earth is supposed to continue warming for many decades to come, sea levels will rise, and sudden climatic events will occur. Crutzen (2005) observes that it is very unlikely, even if theoretically plausible, that we will face a new ice age; the uncertainty is due to our very limited forecasting abilities and the unpredictability that characterizes our civilization.

There is only one certainty: the human impact on the environment will grow unless we all work together to tackle the problem as fast as possible. Barring unforeseen disasters, the world population will increase further, and its agricultural and industrial activities will occupy ever larger areas. In the Anthropocene, human beings are the single and important factor that most affects the change in climate and the Earth's surface. We cannot go back, but we can, however, study the transformation process underway, learn to control it and try to manage it.

2 Disaster Resilience, Human Settlements, and Global Agendas

2.1 Global Agendas and Policy Responses

Disasters are the phenomena that arise at the intersection between the nature and human society. Disasters have always marked the human and natural history of the planet and conditioned human settlements.

2.1.1 'Expanded Brown Agenda'

Since the 1992 Earth Summit in Rio de Janeiro, the urban environmental issues have been known and addressed as the *brown agenda.* Until the late-1990s, the brown agenda comprised of two components. First were the "traditional environmental health issues including the limited availability of land suitable for low-income groups and the lack of acceptable shelter and services to prevent the transmission of pathogens in the environment"; the second set of problems owed "to rapid industrialisation such as 'toxic/hazardous wastes, water and air pollution, noise pollution and industrial accidents" (Williams 1997, pp. 18–19). Based on their levels of development, cities around the world had had to deal with either or both these problems.

During 2002–2003, the World Bank conducted a global review "assessing the quantity and quality of ongoing investments aimed at improving the urban environment, the role of the various sectors involved, and the policy implications for the Bank's future work" (Bigio and Dahiya 2004, p. ix). The World Bank's investment projects, which include the objectives related to enhancing the quality of life in cities and towns, originate in multiple sectors including urban development, environment, energy, transport, social protection, rural development, health, education, public sector governance, and financial and private sector development (Bigio and Dahiya 2003). Following extensive consultations organized by the Urban Environment Thematic Group at the World Bank, and building on the 'brown agenda', Bigio and Dahiya (2004) outlined the 'expanded brown agenda' which included four urban environmental goals (p. 8):

Goal 1: Protect and enhance environmental health in urban areas.
Goal 2: Protect water, soil, and air quality in urban areas from contamination and pollution.

Goal 3: Minimize the urban impact on natural resources at the regional and global scales.

Goal 4: Prevent and mitigate the urban impacts of disasters and climate change.

"These goals constitute the expanded brown agenda for the first part of the twenty-first century, framing the Bank's purpose in this area" (Bigio and Dahiya 2003, p. 8). Thus, the 'expanded urban agenda' brought into focus the prevention and mitigation of the urban impacts of disasters and climate change. "Urban environmental challenges should be tackled holistically" (Bigio and Dahiya 2003, p. 4) for achieving meaningful results towards improving urban liveability.

2.1.2 Hyogo Framework for Action

The international efforts towards strengthening policy support and capacity development for disaster risk reduction (DRR) received an impetus when the United Nations declared the decade of 1990s as the 'International Decade for Natural Disaster Reduction' with the aim of contributing to technical and scientific buy-in and to make DRR agenda imperative. The first United Nations World Conference on Disaster Reduction held in Japan in 1994 adopted the 'Yokohama Strategy and Plan of Action'. In 1999, the United Nations made the decision to establish the secretariat of the 'United Nations International Strategy for Disaster Reduction' (UNISDR). Late in the year 2004, the world saw the massive destruction caused by the Indian Ocean tsunami. Soon after, the second United Nations World Conference on Disaster Reduction in January 2005 in the Japanese city of Kobe in Hyogo Prefecture adopted the 'Hyogo Framework for Action 2005–2015: Building the Resilience of Nations and Communities to Disasters' (or HFA). The HFA adopted five 'priorities for action' (UNISDR 2005, p. 6):

1. Ensure that disaster risk reduction is a national and a local priority with a strong institutional basis for implementation.
2. Identify, assess and monitor disaster risks and enhance early warning.
3. Use knowledge, innovation and education to build a culture of safety and resilience at all levels.
4. Reduce the underlying risk factors.
5. Strengthen disaster preparedness for effective response at all levels.

In relation to the HFA's five 'priorities for action', the Kyoto University and UNISDR (2010) outlined 20 DRR tasks that were to be implemented by local and city-level stakeholders (Table 1).

Table 1 Local/city-level disaster risk reduction tasks related to five priorities of the hyogo framework of action

Local/city governance (HFA Priority 1 related)
Task 1. Engage in multi-stakeholder dialogue to establish foundations for disaster risk reduction
Task 2. Create or strengthen mechanisms for systematic coordination for DRR
Task 3. Assess and develop the institutional basis for disaster risk reduction. Task 4. Prioritize disaster risk reduction and allocate appropriate resources
Risk assessment and early warning (HFA Priory 2 related)
Task 5. Establish an initiative for community risk assessment to combine with country assessments
Task 6. Review the availability of risk-related information and the capacities for data collection and use
Task 7. Assess capacities and strengthen early warning systems
Task 8. Develop communication and dissemination mechanisms for disaster risk information and early warning
Knowledge management (HFA Priority 3 related)
Task 9. Raise awareness of disaster risk reduction and develop education program on DRR in schools and local communities
Task 10. Develop or utilize DRR training for key sectors based on identified priorities
Task 11. Enhance the compilation, dissemination and use of disaster risk reduction information
Vulnerability reduction (HFA Priority 4 related)
Task 12. Environment: Incorporate DRR in environmental management
Task 13. Social needs: Establish mechanisms for increasing resilience of the poor and the most vulnerable
Task 14. Physical planning: Establish measures to incorporate disaster risk reduction in urban and land-use planning
Task 15. Structure: Strengthen mechanisms for improved building safety and protection of critical facilities
Task 16. Economic development: Stimulate DRR activities in production and service sectors
Task 17. Financial/economic instruments: Create opportunities for private sector involvement in DRR
Task 18. Emergency and public safety; disaster recovery: Develop a recovery planning process that incorporates DRR
Disaster preparedness (HFA Priority 5 related)
Task 19. Review disaster preparedness capacities and mechanisms, and develop a common understanding
Task 20. Strengthen planning and programming for disaster preparedness

Source Kyoto University and UNISDR (2010, p. v)

2.1.3 Sendai Framework for Disaster Risk Reduction

In March 2011, the Sendai region was struck by the 'Great East Japan Earthquake' that included the triple disasters of a 9.0–9.1 magnitude earthquake, a massive tsunami, and the Fukushima Daiichi nuclear disaster. The triple disaster resulted in almost 20,000 causalities (Wikipedia 2023; Fig. 1). The third United Nations World Conference on Disaster Risk Reduction, held in Sendai, Japan, in March 2015, considered the lessons learned, identified the gaps, and discussed the future challenges. This

Fig. 1 Extensive destruction of the lower area of Rikuzentakata city in Iwate Prefecture, Japan, caused by the tsunami, March 2011. *Source* Mitsukuni Sato—Flickr: IMG_0888, CC BY 2.0, https://commons.wikimedia.org/w/index.php?curid=14824547

resulted in the adoption of the 'Sendai Framework for Disaster Risk Reduction 2015–2030', which provides the global roadmap for developing safe and resilient communities around the world (UNISDR 2015). The Sendai Framework adopted four priority areas (UNISDR 2015, p. 14):

1. Priority 1: Understanding disaster risk.
2. Priority 2: Strengthening disaster risk governance to manage disaster risk.
3. Priority 3: Investing in disaster risk reduction for resilience.
4. Priority 4: Enhancing disaster preparedness for effective response and to "Build Back Better" in recovery, rehabilitation and reconstruction.

According to United Nations Disaster Risk Reduction (UNDRR), whilst some hazards are natural and unavoidable, disasters are often made by human actions and decisions (De Pascale and Dattilo 2021). Further, according to the definition attributed by UNDRR (Mizutori 2020), disasters result when a natural or human-induced hazard affects human settlements that lack disaster preparedness, and whose population is vulnerable because of poverty, exclusion or other social disadvantages.

In Europe, through international research programs for DG ECHO (Directorate-General for European Civil Protection and Humanitarian Aid Operations), following a bottom-up approach, good practices in civil protection and risk reduction have been compiled, which could be shared through bulletins and periodic publications with decision makers and professionals of the sector (Tagliacozzo et al. 2022). From this work, a vision paper was derived in which the results of thematic works were elaborated and put into a system to outline the actions to be promoted in order to move from knowledge to the tools and procedures necessary for risk management policies and for use by decision makers (Di Bucci et al. 2022).

2.1.4 The 2030 Agenda for Sustainable Development and the Sustainable Development Goals

The DRR processes are closely linked to the reduction of social vulnerability. The achievement of DRR priorities could be complemented by implementing the goals of the 2030 Agenda for Sustainable Development (United Nations 2015), signed by the governments of 193 member countries of the United Nations. The 2030 Agenda considers social issues an integral part of sustainable development and outlines the 17 Sustainable Development Goals or SDGs. These include fighting poverty, eradicating hunger, reducing inequalities, quality education globally, achieving greater social justice and peace and gender equality. SDG11 aims at making cities and human settlements inclusive, safe, resilient and sustainable (United Nations, 2022). Building resilience to disasters is key to achieving the goal of eradicating extreme poverty. Being one of the key factors in the disaster risks, given the way it generates economic and social vulnerability, poverty has contributed significantly to the increase in the risk conditions of the population which hinder the progress of sustainable development the most. The impacts of disasters undermine hard-earned progress in both developing and developed countries, further exacerbating the poverty conditions of the most vulnerable and already poor.

2.1.5 The Paris Agreement

The Paris Agreement was signed by 196 parties in December 2015 and entered into force in November 2016. Its 'long term temperature goal' aims at limiting the global temperature rise to well below 2 °C whilst pursing to limit the increase to 1.5 °C (UNFCCC 2016). Further, the Paris Agreement also focuses on taking appropriate actions with regard to global peaking and 'climate neutrality'; mitigation; adaptation; loss and damage; sinks and reservoirs; voluntary cooperation/market- and non-market-based approaches; finance, technology and capacity-building support; climate change education, training, public awareness, public participation and public access to information; transparency, implementation and compliance; and global stocktake (UNFCCC 2016).

One of the key agenda items that is part of the Paris Agreement is the setting up of a 'Loss and Damage' fund (as noted above). In November 2022, the United Nations Climate Change Conference, or Conference of the Parties (COP27), held in Sharm el-Sheikh, Egypt, resulted in an agreement to establish a global 'Loss and

Damage' fund (UNFCCC 2022a). The transitional committee, which is responsible for the operationalization of the 'Loss and Damage' fund (UNFCCC 2022b), "must assure that the most vulnerable and affected groups can access this critical financial support" (Dahiya and Okitasari 2022, p. 1285). As the impacts of climate change become increasingly evident, there are growing calls for enhanced climate action around the world.

2.1.6 New Urban Agenda

In Habitat III or the United Nations Conference on Housing and Sustainable Urban Development, held in Quito, Ecuador, in October 2016, the New Urban Agenda (United Nations 2017) was adopted. The New Urban Agenda has made three transformative commitments for sustainable urban development: (i) Sustainable urban development for social inclusion and ending poverty; (ii) Sustainable and inclusive urban prosperity and opportunities for all; and (iii) Environmentally sustainable and resilient urban development (United Nations 2017). The 'Effective Implementation' will be achieved through 'building the urban governance structure: establishing a supportive framework', and 'planning and managing urban spatial development'. The 'mobilization of financial resources', 'capacity development', and 'innovation and technology' have been identified as the 'means of implementation'. The New Urban Agenda envisaged that cities and human settlements would "[a]dopt and implement disaster risk reduction and management, reduce vulnerability, build resilience and responsiveness to natural and human-made hazards and foster mitigation of and adaptation to climate change" (United Nations 2017, p. 7).

2.2 *Understanding the Current Challenges and Responses*

Today, more than ever, the need to understand the current challenges and responses seems timely and necessary. This is for several reasons:

1. It is now clear that global climate change is affecting the lives of everyone. It has triggered frequent and disastrous events linked to natural risk. These events are followed by victims and destruction of homes, infrastructures, cultural assets.
2. Sometimes these destructions leave only rubble on the ground and the restoration of buildings with historical and cultural value becomes impossible: the damage is irreversible.
3. Furthermore, a territory afflicted by disasters related to natural hazards and climate change remains isolated, marginalized for a long period of time, and it is necessary to rebuild from scratch all the networks that allow the survival and development of a city: cultural networks, infrastructural ones, administrative ones.

4. Territories risk being forgotten, as is the fate of its surviving citizens who have lost loved ones or their settlements.
5. The structure of cultural identity is partially modified, marked forever by disastrous events.
6. Sometimes the disastrous event also marks the history of future generations. The economic sector is heavily involved, as a place can suddenly stop being touristy. A tourist destination can go into decline without respecting Richard Butler's well-known TALC-Tourism Area Life Cycle Model (Butler 1980).

It is therefore of fundamental importance to study shared resilience systems. It is important to share good practices. An important role in this sense is certainly entrusted to governance actors and scholars, who must work in synergy by offering their diversified skills.

All of this leads towards the need for an in-depth analysis of disaster resilience (Tiernan et al. 2019) for human settlements. The complex human ecosystems show a lot of different ways in responding to the various kinds of disasters, depending mostly on their geographic, environmental, social, and economic factors. In other words, the responses to contemporary disasters at the individual, community, and societal levels are deeply linked to the environmental settings as well as to the local cultures.

In this regard, the following thematic research questions emerge:

Question 1. In the Anthropocene, what are the underlying factors that intensify the impacts of disasters in the different geographical contexts of cities, towns, and regions (with urban and/or rural perspectives)?

Question 2. In the Anthropocene, what are the triggers that activate the strengthening of disaster resilience and/or responses in the different geographical contexts of cities, towns, and regions (with urban and/or rural perspectives)?

Question 3. In the Anthropocene, what are the different pathways to improve the planning and governance to address the challenges posed by increasing number and intensity of disasters in cities, towns, and regions (with urban and/or rural perspectives)?

Question 4. In the Anthropocene, how does the role of various stakeholders change and/or need to change in tackling/responding to the increasing number and intensity of disasters in cities, towns, and regions (with urban and/or rural perspectives)?

Question 5. In the Anthropocene, what sort of financial instruments are needed and/or how the current financial instruments need to be improved to tackle/respond to the increasing number and intensity of disasters in cities, towns, and regions (with urban and/or rural perspectives)?

3 Collection of Essays

3.1 *Thematic Focus*

Starting from this reference framework, the chapters included in this volume focus on the emerging perspectives on disaster resilience and human settlements in the larger context of the Anthropocene. The chapters explore urban and rural perspectives focusing on the current and emerging perspectives on disaster resilience through a holistic approach, involving scientists, humanists, planners, policymakers, and professionals in the global debate. Particularly, the chapters respond the abovementioned thematic research questions.

The chapters included in this volume feature on one or more of the following sub-themes:

1. Urban and rural narratives and geographies of twenty-first century human settlements against disasters: empirical and theoretical studies following human, social and natural sciences approaches.
2. Presentation of scenarios and case studies representing the Anthropocene.
3. Proposals about disaster resilience to natural hazards and adapting to climate change in the multidisciplinary context of human settlements: early warning systems, and new technologies, reducing mortality, strengthening resilience of vulnerable groups, risk governance and institutional strengthening, livelihoods, risk communication, geo-education and perception.
4. Proposals about disaster preparedness for effective response reconstruction in the urban and rural areas of human settlements: emergency response and disaster preparedness, reconstruction practices, sustainable and community-based approaches.
5. Socio-economic systems, globalization, pandemic, and the crisis of twenty-first century.

3.2 *Structure of the Book*

This volume includes fourteen chapters that are organized into four parts. In Part I, this introductory chapter provides an overview of the book's theme, disaster resilience and human settlements in the Anthropocene. Part II includes a collection of four papers that focus on flood risks in urban areas with regard to disaster resilience assessment and governance. Four chapters in Part III explore perceptions and representations of disasters and climate change in the Anthropocene. Part IV includes five papers that dwell on post-disaster management and recovery in relation to urban landscape, pandemic, and community resilience.

Part II: Flood Risks in Urban Areas—Disaster Resilience Assessment and Governance

This Part consists of four contributions that focus on risk assessment in urban contexts and on the related practices implemented to reduce the risk of damage to property and, above all, to the people. All cases were studied with a multidisciplinary approach.

Eleonora Gioia and Fausto Marincioni's contribution, titled "Leaving nothing to chance: reducing flood risk by evaluating simulation exercises in urban contexts", moves exactly in this direction. The two authors propose the case study relating to simulations and exercises to mitigate the flood risk carried out in three particularly fragile Italian regions: Emilia-Romagna, Marche and Abruzzo. These regions are known to be highly exposed to natural hazards, with visible consequences due to landscape modifications and geomorphological processes. The extent of adverse phenomena is nowadays accentuated by the effects of global climate change. The chapter adopts a methodology that takes into account three dimensions: the management of structures affected by risk, the networks between subjects involved in prevention and rescue and finally the so-called proactivity (sensemaking) of the various organizations. Through targeted interviews, also was made an attempt to evaluate the reactions to the possibility of a catastrophic event. Research has highlighted the need for more effective training and more realistic scenarios. It is certainly an effective model that can also be applied to other similar territorial contexts.

Miquel Grimalt and Joan Rosselló's study examines "Urban growth and increasing flood impact in the city of Palma: a loss of resilience capacity". It shows that on the one hand, urban expansion, largely without control during the final decades of the twentieth century, is the main cause of flood-related impacts and damages. On the other hand, the lack of a public and private responses only enhances the resilience failure.

Iftekhar Ahmed's chapter focuses on the process of building resilience to floods of housing in informal settlements in Honiara, the capital and largest city of the Solomon Islands. Much of the existing research on this topic suggests a contextual approach, and in this study, "Addressing the Impacts of Inland Floods on Informal Housing in Honiara, Solomon Islands", such an approach has been followed. Globally, urban informal settlements present a critical urban policy challenge, and the Pacific region is no exception. Although cities in Pacific countries are smaller than in the wider Asian region, they are growing rapidly; in cities such as Honiara, an estimated 40 percent of people live in informal settlements. This is a topic that has been widely explored for a long time and diverse approaches have been implemented around the world, yet the challenge persists. In many cases, the problem of housing in informal settlements has amplified with the complexities ushered in by climate change and urbanisation in the Anthropocene epoch. Through an analysis of the local challenges and opportunities, this chapter posits a framework for housing improvement in informal settlements of Honiara, which will have relevance for the wider Pacific region, bearing connections to key global frameworks including the Sendai Framework for Disaster Risk Reduction and the SDGs.

Vinita Yadav and Zeeshan Ibrar examine the adaptive capacity to flood risks in the state of Bihar in India. Their chapter, "Adaptive Capacity Analysis of Flood Prone Regions in Bihar, India", focuses on quantifying the adaptive capacity of community development blocks. Under the districts, community development blocks are administrative units important for implementing planning and development administration, including for building resilience. In Bihar, 74 percent of the geographical area is flood prone. In 2017, the flood affected population of 0.97 million resided in 663 villages in the Kishanganj district. Hence, Kishanganj district was chosen as a case study. Using an integrated approach, the chapter analyses the role of six aspects, i.e., education, economy, infrastructure, technology, livelihood, and institutions. The community development blocks of Kishanganj district are ranked using a composite index. The 'educational commitment' and 'institutional setups' are the outliers and play a crucial role in the flood adaptation process. The flood risk policies ought to develop the social and educational institutions in order to strengthen the adaptive capacity of the region.

Part III: Perceptions and Representations of Disasters and Climate Change in the Anthropocene

This Part III includes four case studies that are useful for defining different perspectives and types of environmental risks.

Gaetano Sabato, in his chapter "Cruise tourism, risk perception and public narratives in Syracuse, Italy", studies the perception of risk and the issue of safety in the tourism sector. In more detail, attention is focused on the relationship between residents and tourists who use cruise ships, intended as carriers of an environmental risk. Could tourism therefore represent an environmental threat or is it a development opportunity? It is a central question of cultural geography: the contribution examines the journalistic literature and the related narration relating to the permanence of two cruise ships for a long period in the city of Syracuse-Sicily. Teresa Graziano also moves along this line of research, with the contribution entitled "Social media in risk perception and disaster management: a geographical perspective". This time the scale is global and refers to the epochal change in the relations between territories and communication. The excess of data production, also accentuated by bottom-up participation, has consequences on the perception of risk and on the management of disasters. It is the so-called paradigm of *e-participation*, investigated with reference to Italy, which however has advantages, such as increasing awareness, and limits, such as the overlapping of skills and functions.

Charles Travis, with the chapter "Environment as a Weapon: History, Gaia, and the Geohazards of War", studies the individual-nature relationship in a two-way way: on the one hand, individuals have always used the territories and environmental resources indiscriminately during countless wars; on the other hand, today nature is putting up a strong resistance and has even entered, so to speak, into war with humanity, after centuries of exploitation. The author uses the results offered by the practices of literary, cultural and historical geography.

Leonardo Mercatanti, in the chapter titled "Living on Mount Etna between risk, beauty and need: a field survey on villages struck by 2018 earthquake" carries out field research shortly after a devastating earthquake occurred in some municipalities of Etna-Sicily. It is an area subject to earthquakes of considerable intensity after a few decades. Through the collection of information, the inspection and the results of interviews and questionnaires, the author reflects on the dynamics and reasons that see the population and the actors of governance insisting, despite the strong environmental risk, on staying in a notoriously dangerous territory.

Part IV: Post-disaster Management and Recovery—Urban Landscape, Pandemic and Community Resilience

De Pascale's chapter, titled "Narratives of urban resilience and sustainability in southern Italy: the case studies of Matera (Basilicata) and Filadelfia (Calabria)", focuses on the concept of urban resilience and on some thematic declinations of the term, by comparing two case studies in southern Italy: Matera (Basilicata) and Filadelfia (Calabria). The paper, through the technique of storytelling, demonstrates how the concept of urban resilience is closely linked to socio-cultural structures.

On the Abruzzo region wrote Giorgio Paglia, Massimiliano Fazzini, Gianluca Esposito, Vania Mancinelli, Vincenzo Marsala and Enrico Miccadei, with an article titled "Disaster resilience assessment for drainage network and urban landscape after heavy meteorological events: examples from the middle Adriatic coastal area (Abruzzo Region, Central Italy)". The case study shows that, despite the numerous mitigation activities carried out in recent decades, the coastal area of the Abruzzo region has been severely affected by flood events which have confirmed the high degree of vulnerability to natural hazards in the area. In any case, the path taken seems to be the correct one: it is necessary to insist on training and simulations. A fundamental role is played by in-depth knowledge of the territory, in its various facets. In this sense, the cartographic and photographic apparatus and the quantitative analyses proposed in the contribution, in addition to their interpretation, provide the solid foundations of a model applicable in other territorial contexts.

Piero Farabollini, Fabrizio Bendia and Francesca Romana Lugeri wrote the chapter titled "Revitalizing the wounded territory: the 'geo-hiking's' potential", They reflected extensively on the vital spaces of individuals, subject to dynamics and balances that today more than ever require serious and constant investigations. These are real dynamic and complex ecosystems that require the definition of an effective geomorphological cartography that over time needs to be compared to follow their evolution. Similarly, it is essential to study the evolution of the practices implemented in the area to mitigate the various environmental risks. Among these, the seismic risk is the most difficult to manage, given its almost total unpredictability. The authors' reflection brings into play the concept of landscape and the fundamental role of sharing information related to risk.

Cristina Casareale, Noemi Marchetti and Alessandra Colocci, in their chapter "Sense of belonging and response to climate change: how the relation with local territories influences cli-mate resilience", analyse data on the resilience of the various communities of the Veneto Region under various social and territorial profiles. What factors affect the sense of place of individuals and different communities? To this end, over 1200 questionnaires were administered, the results of which, discussed in the contribution, highlight a greater awareness of the environmental risk by young people, who are more likely to get involved even by individuals. The level of integration in the territory (or perception of it) offers very interesting starting points for reflections. The survey aims to report any differences also at a territorial level (internal coast).

Giovanni Messina is the author of the chapter titled "From Disasters to the Pandemic. A Study on the EU Solidarity Fund". He examines the intervention framework of the European Union Solidarity Fund (EUSF) and, with reference to Italy, reflects on the type of resources allocated, outlining a geography of the areas of intervention.

References

Bigio AG, Dahiya B (2003) World bank investments for the urban environment. Environment Strategy Notes, No. 8. Environment Department, World Bank Group, Washington, DC. http://documents.worldbank.org/curated/en/366001468151475459/World-Bank-investments-for-the-urban-environment

Bigio AG, Dahiya B (2004) Urban environment and infrastructure: toward livable cities. Directions in development series. The World Bank, Washington, DC. https://openknowledge.worldbank.org/handle/10986/15018

Butler RW (1980) The concept of the tourist area life-cycle of evolution: implications for management of Resources. Can Geogr 24(1):5–12

Crutzen PJ (2005) Benvenuti nell'Antropocene!, Ed. A. Parlangeli, Mondadori, Milano

Crutzen PJ, Stoermer EF (2000) The anthropocene. IGBP Global Change Newslett 41:17–18. http://www.igbp.net/download/18.316f18321323470177580001401/1376383088452/NL41.pdf

Dahiya B, Okitasari M (2022) Accessing the loss and damage climate fund. Science 378(6626):1285. https://doi.org/10.1126/science.adf9670

De Pascale F, Dattilo V (2021) The geoethical semiosis of the Anthropocene: the Peircean triad for a reconceptualization of the relationship between human beings and environment. Ann Assoc Am Geogr 111(3):647–654

De Pascale F, Antronico L, Coscarelli R (2019) La percezione del rischio idrogeologico in Calabria: il caso studio della Costa degli Dei. Archivio Di Studi Urbani e Regionali, XLX 124(2019):171–199

Di Bucci D, Casarotti C, Morsut C, Polese M, Rebora N, Giuliani F, Rodrigues T, Viegas DX, Fagà G, Centoducati C, De Siervo G, Dolce M, Gotti A, Hamidova E, Kruke BI, Prota A, Ribeiro LM, Capone F, Farabollini P, Petrenj B, Rizzoli V, Tagliacozzo S (2022) Shared basic elements for a 'European Doctrine on Disaster Risk and Crisis Management'—The RODMAP vision paper. ROADMAP Project, 32 pp. https://doi.org/10.57580/VPDOI

Ismail-Zadeh A (2018) Geohazard analysis for disaster risk reduction and sustainability. In: Beer T, Li J, Alverson K (eds) Global change and future earth. Cambridge University Press, Cambridge, pp 349–363

Ismail-Zadeh A (2022) Natural hazards and climate change are not drivers of disasters. Nat Hazards 111:2147–2154. https://doi.org/10.1007/s11069-021-05100-1

Kyoto University and UNISDR (2010) A guide for implementing the Hyogo framework for action by local stakeholders. https://www.unisdr.org/files/13101_ImplementingtheHFA.pdf Accessed 05 Feb 2023

Mizutori M (UNDRR) (2020) Time to say goodbye to "natural" disasters. Preventionweb, https://www.preventionweb.net/experts/oped/view/72768. Last access 01 Sept 2020

Tagliacozzo S, Farabollini P, Morsut C, Casarotti C, Di Bucci D, Polese M, Giuliani F, Hamidova E, Rebora N, Dolce M, Prota A, Viegas DX (2022) Good practices in building back better and leave no one behind. ROADMAP Project—Thematic Paper 3. https://doi.org/10.57580/TP3DOI

Tiernan A, Drennan L, Nalau J, Onyango E, Morrissey L, Mackey B (2019) A review of themes in disaster resilience literature and international practice since 2012. Policy Design Pract 2(1):53–74. https://doi.org/10.1080/25741292.2018.1507240

United Nations (2015) Transforming our world: the 2030 agenda for sustainable development, A/RES/70/1. https://sustainabledevelopment.un.org/post2015/transformingourworld/publication. Accessed 14 Dec 2022

United Nations (2017) The New Urban Agenda, A/RES/71/256, Habitat III and United Nations. https://habitat3.org/wp-content/uploads/NUA-English.pdf

United Nations (2022) Sustainable Development Goal 11: make cities and human settlements inclusive, safe, resilient and sustainable. https://sustainabledevelopment.un.org/sdg11

UNDRR (United Nations Office for Disaster Risk Reduction) (2022) Global assessment report on disaster risk reduction 2022: our world at risk: transforming governance for a resilient future. UNDRR, Geneva

UNDRR (2019) Global assessment report on disaster risk reduction, United Nations Office for Disaster Risk Reduction (UNDRR), Geneva

UN-ESCAP (2017) Disaster resilience for sustainable development: Asia-Pacific disaster report 2017. UN-ESCAP, Bangkok

UNFCCC (United Nations Framework Convention on Climate Change) (2022a) Sharm el-Sheikh implementation plan, draft decision-/CMA.4. https://unfccc.int/documents/621908.

UNFCCC (2022b) Funding arrangements for responding to loss and damage associated with the adverse effects of climate change, including a focus on addressing loss and damage, draft decision-/CP.27-/CMA.4. https://unfccc.int/documents/624440

UNFCCC (2016) The Paris Agreement. UNFCCC, Bonn. https://unfccc.int/documents/184656

UNISDR (United Nations International Strategy for Disaster Risk Reduction) (2015) Sendai framework for disaster risk reduction 2015–2030. United Nations Office for Disaster Risk Reduction, Geneva. https://www.undrr.org/publication/sendai-framework-disaster-risk-reduction-2015-2030

UNISDR (2005) Hyogo framework for action 2005–2015: building the resilience of nations and communities to disasters. UNISDR, Geneva. https://www.unisdr.org/2005/wcdr/intergover/official-doc/L-docs/Hyogo-framework-for-action-english.pdf

UN-HABITAT (2020) World cities report 2020: the value of sustainable urbanization. UN-HABITAT, Nairobi

Williams SW (1997) 'The Brown agenda': urban environmental problems and policies in the developing world. Geography 82(1):17–26. http://www.jstor.org/stable/40572862

Flood Risk in Urban Areas: Disaster Resilience Assessment and Governance

Leaving Nothing to Chance: Reducing Flood Risk by Evaluating Simulation Exercises in Urban Contexts

Eleonora Gioia and Fausto Marincioni

Abstract Disasters triggered by natural hazards pose an increasing threat globally, thus, preparing an effective emergency response is crucial and should include simulation exercises with all the stakeholders involved. Evaluation of such exercises is especially important to collect and capitalize performance feedbacks from a plurality of point of view. The objective of this work is to evaluate the flood field exercises held at urban scale in five Emilia-Romagna, Marche, and Abruzzo municipalities. Specifically, is proposed the application of an existing three-dimensional method to analyze the management (*structures*), the networks (*relations*), and the proactivity (*sensemaking*) of the organizations, necessary to deal with both planned and unplanned situations. Nine face-to-face interviews, with both the organizers and the attendants of the exercises, were carried out and qualitatively analyzed. Results showed overall positive reactions to the events, the response of the participants, the relations built, and the proactivity toward the unexpected scenarios. Nonetheless, the need for flood codified alarms, strengthened procedures, back-up communication tools, more realistic scenarios, and more targeted training was raised mostly by the attendants, thus stressing the importance of an evaluation activity extended to all the actors. Future research could deepen the outcomes through the application of quantitative techniques also to different scales and disasters.

Keywords Disasters triggered by natural hazards · Field exercise · Evaluation · Flood · Italy

1 Introduction

Disasters are a global issue, with increasing number of recorded events causing substantial human and economic losses (CRED and UNDRR 2021). These phenomena are the results of nature and human interactions, specifically when

E. Gioia (✉) · F. Marincioni
Università Politecnica delle Marche, Ancona, Italy
e-mail: e.gioia@staff.univpm.it

B. Dahiya et al. (eds.), *Disaster Resilience and Human Settlements*,
Advances in 21st Century Human Settlements,
https://doi.org/10.1007/978-981-99-2248-2_2

natural or human-induced hazards affect unprepared human settlements and vulnerable population (Antronico and Marincioni 2018). Widening this perspective in the broader context of the Anthropocene, it becomes clear that the increasing frequency and intensity of extreme events is linked to a poor environmental management and to a lack of funding to implement effective sustainable development policies (United Nations 2021). Not surprisingly, one of the 17 United Nations Sustainable Development Goals is to reduce the human and economic losses caused by disasters (United Nations General Assembly 2015). More precisely, within goal 11 (make cities and human settlements inclusive, safe, resilient, and sustainable), the target 11.5 reads: "*by 2030, significantly reduce the number of deaths and the number of people affected and substantially decrease the direct economic losses relative to global gross domestic product caused by disasters, including water-related disasters, with a focus on protecting the poor and people in vulnerable situations.*" From the standpoint of cities, this requires planning a "New Urban Agenda" (United Nations General Assembly 2016), in which material, living, and social systems (Bowen and Gleeson 2019) are harmonized to improve the resilience to disasters and climate change.

Pivotal to such purpose is the development of a culture of disaster risk reduction to consolidate correct procedures and behavior of both emergency organizations and citizens, and to prepare the response to the potential consequences of extreme events (Farabollini et al. 2018; Marincioni et al. 2019; UNDRR 2020). As a matter of fact, part of the objectives of the Sendai Framework for Disaster Risk Reduction 2015–2030 focus on enhancing disaster preparedness and increasing communities' resilience (UNISDR 2015). Among the possible activities toward such goal is the implementation of drills and simulation exercises, such as table-top or field training, to ensure that the emergency plans work efficiently when disaster strikes (Alexander 2002). Exercises have indeed the multiple purposes of testing the disaster plans and procedures, the personnel training and knowledge, and the communication systems and equipment in a safe and realistic environment (Alexander 2002; Beerens and Tehler 2016; Peterson and Perry 1999).

Evaluation of the exercises is especially important in order to have performance feedback from the stakeholders involved. If the debriefings after the operations are conducted by and limited to the professionals involved in the simulations, they might be focused only on the strict application of plans and procedures, while deviations might be evaluated negatively or generally hidden or minimized (Wybo 2008). On the contrary, it is desirable an evaluation that provides a comprehensive overview of the exercise, that reaches the complexity of the stakeholders and of the organizations involved during emergencies (Wybo 2008), possibly performed by experienced staff members, third-party, or external observers. In this regard, the academic research might play a crucial role, though keeping in mind that in the end the outcomes should be capitalized in actual emergency plans (Beerens and Tehler 2016).

The interest in simulation exercises and related evaluation is increasing in various fields, especially in the areas of health and social sciences (Beerens and Tehler 2016). Usually, these studies use social science data collection methods such as questionnaires (Gilmore et al. 2007; Kaji et al. 2008) and interviews (Berlin and

Carlström 2013; Cranmer et al. 2014). A large part of the research is focused on the description of a specific exercise, such as design, scenario, lesson learned, and method employed for the evaluation (Beerens and Tehler 2016; Reddin et al. 2021). Less effort is dedicated to the development of a comprehensive methodological perspective aimed at understanding and assessing disaster exercises in general (Kim 2013; Lonka and Wybo 2005; Sinclair et al. 2012; Wybo 2008) and natural hazards in particular (Beerens et al. 2010; Khorram-Manesh et al. 2014; Kim 2013; Savoia et al. 2009).

The main objective of this work is to evaluate the flood simulations exercises held at urban scale in the Emilia-Romagna, Marche, and Abruzzo pilot areas of the EU Project LIFE PRIMES (Preventing flooding Risks by Making Resilient communities). Such study is specifically intended to apply a broad assessment of the overall structures, relations, and sensemaking of the emergency response emerged from such simulations. This is important for several reasons: (i) to enhance the level of dialogue between the institutions and the voluntary organizations involved in the flood emergency and between the institutions and the citizenship; (ii) to improve the effectiveness of the flood emergency management; (iii) to foster the growth of a community informed of the best practices to adopt in case of flood emergency.

Specifically, the chapter is divided into the following sections: (a) introduction; (b) description of the study areas and the simulations exercises; (c) description of the methodology used to evaluate the exercises; (d) analysis of the results and discussions; (e) conclusions and recommendation for future implementation.

2 The Flood Simulations Exercises

The study was conducted in five Italian municipalities participating in the European Life PRIMES project (Preventing flooding Risks by Making Resilient communities—LIFE14 CCA/IT/001280). The purpose of the project, conducted from October 2015 to December 2018, was to form flood risk resilient communities in specific Italian pilot municipalities located in the Emilia-Romagna, Marche, and Abruzzo regions. Specifically, the project aimed at increasing the citizens perception of risk and their involvement in the definition of adaptation plans (http://www.lifeprimes.eu/). During this period, the municipalities involved in the project, with the support of the related Regional Civil Protection Departments, organized 5 flood field exercises. Such exercises were held in: Sant'Agata sul Santerno and Imola for the Emilia-Romagna region; Senigallia and San Benedetto for the Marche region; Pineto for the Abruzzo region (Fig. 1). The pilot municipalities were chosen among those most subject to river and coastal floods, especially as a consequence of the ongoing climate changes (EU Life PRIMES 2016).

In the Emilia-Romagna region, the nearby municipalities of Sant'Agata sul Santerno (Fig. 2a) and Imola (Fig. 2b) are located along the river Santerno, which historically caused several floods. From 2009, the frequency and intensity of these events increased, so that the floods of 2013 and 2014 exceeded the past records

Fig. 1 Localization of the municipalities which held the flood field exercises: Sant'Agata sul Santerno and Imola (Emilia-Romagna region); Senigallia and San Benedetto del Tronto (Marche region); Pineto (Abruzzo region)

and critically affected the dense urban and intensively agriculture areas (EU Life PRIMES 2016).

In the Marche region, Senigallia (Fig. 2c) is located near the mouth of the Misa river from which experienced several historical floods: in 1940, 1955, 1976, 2011, and lastly in 2014; the latter had the harshest consequences on the territory, causing 3 casualties and extensive damage to residential urban areas, industrial areas, and tourist facilities along the coast (EU Life PRIMES 2016). A similar scenery can be described for San Benedetto del Tronto (Fig. 2d), located on the mouth of the Tronto river and characterized by a dense urban fabric. From a historical point of view, San Benedetto del Tronto suffered significant floods in 1975 and 1992 and other minor events in 1927, 1928, 1935, 1992, and 2011 (EU Life PRIMES 2016).

In the Abruzzo region, the municipality of Pineto (Fig. 2e) is crossed by the Vomano river, which in the past has flooded several times mainly due to closed canals coursing the quite dense urban area. The town is also crossed by the Calvano stream which, despite being dormant for most of the year, on rainy or heavy rain days collects abundant water from the hinterland, in particular from neighboring municipalities, and overflows averagely every two years. As a matter of fact, in the years 1992, 1999, 2009, 2011, 2012, and 2013 Pineto was affected by flood events (EU Life PRIMES 2016).

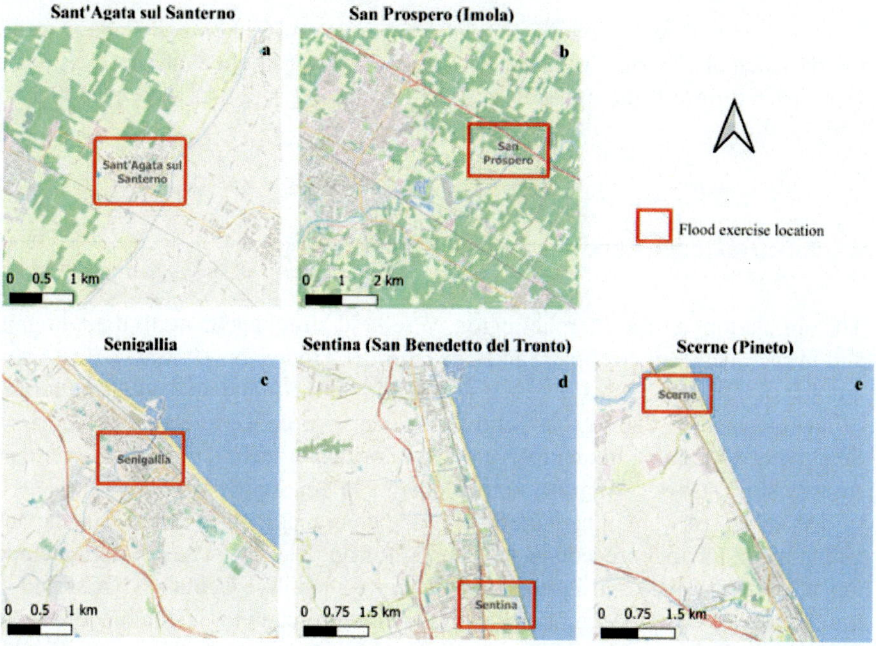

Fig. 2 Detailed localization of the areas which held the flood field exercises: Sant'Agata sul Santerno (**a**), Imola (**b**), Senigallia (**c**), San Benedetto del Tronto (**d**), and Pineto (**e**). Maps show the urban and agricultural fabrics of the sites (© OpenStreetMap base maps)

The following paragraphs describe the flood simulation exercises designed in the above-mentioned communities of the Emilia-Romagna, Marche, and Abruzzo regions.

3 Simulation Exercises in the Emilia-Romagna Region

The simulations in the Emilia-Romagna region were held simultaneously in the district of San Prospero (Imola) and in Sant'Agata sul Santerno on 18 November 2017. A total number of 129 participants were registered: 5 regional/municipal public authorities or technicians and 18 citizens or civil protection volunteers in San Prospero; 6 regional/municipal public authorities or technicians and 100 citizens (mostly students) or civil protection volunteers in Sant'Agata sul Santerno.

The exercises were aimed at testing the regional and local civil protection system and at providing training for the participants, in order to improve the emergency response in case of flood. As reference scenario was considered the historical event of the Santerno river flood in 20 September 2014, which was caused by few hours of high intensity rainfall, with return periods equal to or greater than 200 years, and involved the riverside. Therefore, a flood has been simulated in the neighborhood of

San Prospero and near a middle school in Sant'Agata sul Santerno. In San Prospero 7 families, namely 16 people, were evacuated. In Sant'Agata sul Santerno, the students were headed toward the upper floor. Finally, training activities were performed in both locations.

4 Simulation Exercises in the Marche Region

The simulations in the Marche region were held simultaneously in the Senigallia and in the district of Sentina (San Benedetto del Tronto) on 19 May 2018. A total number of 868 participants were registered: 14 regional/municipal public authorities or technicians and 280 citizens (mostly students) or civil protection volunteers in Senigallia; 14 regional/municipal public authorities or technicians and 560 citizens (mostly students) or civil protection volunteers in Sentina.

The exercises were aimed at applying the emergency plans of schools and activating the municipal procedures envisioned for flood events. The program included the activation of the Civil Protection Municipal Operative Centers (COCs) and the training of the voluntary components of the civil protection system. In addition, the students of a primary school in Senigallia were headed toward the upper floor, while the students of a primary and a middle school in Sentina were evacuated.

5 Simulation Exercise in the Abruzzo Region

The simulation in the Abruzzo region was held simultaneously in the district of Scerne (Pineto) on 16 June 2018. A total number of 86 participants, 8 regional/municipal public authorities or technicians, and 46 citizens or civil protection volunteers were registered.

The exercise was aimed at testing the response capacities of the civil protection system in the municipality of Pineto and at providing training for the participants, in order to improve their response in the event of a flood emergency. As reference scenario was considered the historical flood occurred in November 2017, which was caused by intensity rainfall and caused widespread damages in many coastal municipalities, such as Pineto. The exercise involved both the regional and the local civil protection system and simulated the flooding of a resort and of an underpass connecting the state road and the sea. The scenario considered the evacuation of part of the resort and the rescue and recovery of a motor vehicle stuck in the underpass.

6 Methodology

For data collection, two face-to-face interviews were performed for each flood exercise simulation: one with a municipal public authority or technician, which organized the exercise (hereinafter "organizers"), and one with a citizen, which attended the exercise (hereinafter "attendants"). A total number of 9 interviews (the attendants in Imola declined the interview) were carried out by the authors of this study, which participated to all the simulations as external observers. The aim was to observe the overall organizational behavior during the field exercises, based on a plurality of points of views.

The method, based on the conceptualizations of Jacques and Specht (2006) and Wybo (2008), consists of a three-dimensional evaluation model. The first dimension is the *structures* layer, which is the management layer, namely what is planned by the organization, objective and measurable, such as procedures and tasks. The second dimension is the *relations* layer, which is the network layer, namely the interactions among the actors involved in the structure layer, either verbal or non-verbal, formal or non-formal. The third dimension is *sensemaking* layer, which is the proactive layer, namely the capacity of taking the initiative to invent a form of cooperation that was not defined in the plan. While the first dimension of the model is related to the ability of the organizations to deal with planned and known situations, the last two dimensions are related to the ability to deal with deviations and unforeseen scenarios (Wybo 2008). As such, this model allows to overcome the limits of evaluating exercises in specific case studies and tries to produce knowledge about disaster evaluation exportable to different fields and contexts (Beerens and Tehler 2016). The framework of the interviews follows the three-dimensional model and is reported in the Annex 1.

Data analysis has been achieved through a qualitative text analysis of the transcripts of the conducted interviews.

7 Results and Discussion

The outcomes of the flood simulation exercises are described in the following sections in relation to the three-dimensional evaluation model.

7.1 Structures

The analysis of the structures layer in the context of the flood simulations shows a general positive reaction to the events from both the organizers and attendants interviewed in the three Regions (Table 1).

Overall, the exercises were considered useful, especially for the achievements of the established objectives, the response of the participants, and for the codification

Table 1 Summary of the outcomes of the Emilia-Romagna, Marche, and Abruzzo flood exercise simulations, in relation to the structures dimension. The table shows the highlights of the interviews to both the organizers (O) and the attendants (A). The results are focused on positive outcomes, problems and shortcomings, emerging necessities, and suggestions for the improvements of the exercises

	Emilia-Romagna	Marche	Abruzzo
Positive outcomes			
O	• Very positive and useful • The voice alert messages to the population (Alert System) have worked very well, we have had direct feedback. They arrived, and people understood them • People managed to evacuate in due time • Roles, timing, and procedures have been respected as planned • The Regional web portal and the radio links were effective and efficient	• I think the exercise went very well, especially for the students' response • Through this exercise we have codified the relationships, the contacts, the communications that take place between us and the school, while we had only some messages and the call • For the occasion, they created a WhatsApp group of parents' representatives in which the head of security sends the situation to the parents • Everything went as planned • The students were praiseworthy in their participation and response • Nobody was afraid even if they waited about an hour • Procedures, roles, and timing have been respected • Technologies were effective • Positive exercise • There was a good response from the school, both from the teaching and non-teaching staff and from the students • It was useful because at the end of a training course. What has been said in the previous months has been put into practice • Procedures, roles, and timing have been respected • Technologies were effective, even after simulating a block of telephone communications. We then used radios and WhatsApp	• Very positive and useful • The voice alert messages to the population (Alert System) have worked very well, we have had direct feedback. They arrived, and people understood them • People managed to evacuate in due time • Roles, timing, and procedures have been respected as planned • The Regional web portal and the radio links were effective and efficient

(continued)

Table 1 (continued)

	Emilia-Romagna	Marche	Abruzzo
A	• The exercise was useful because it helped to make everyone aware of the fact that there is another real risk, besides the earthquake and the fire for which we do two exercises a year • It seems to me that the exercise took place correctly • Students remember things that have a strong impact and having seen the civil protection vehicles outside, the sandbags, … was important • Students were mostly interested during the training • The established procedures were respected • The alarm has been heard	• Everything succeeded • Procedures, roles, and timing have been respected • The simulation of an accident, in this case a child who got hurt when climbing stairs, went well • Overall the exercise was good, very useful and the stated objectives have been achieved • Procedures, roles, and timing have been respected • Technologies were effective • The tools adopted for the transmittal/reception of the warnings in this case were certainly very valid, because they were overabundant	• They did not say anything new that we did not already know, but it is always useful to review this phase and we hope it will not happen again • The exercise was useful • The established procedures have been respected • Technologies for the alert are good
Problems and shortcomings			
O	• None	• None	• None
A	• The procedures were a little milder compared to other evacuation drills for earthquake and fire, for which we exit in an orderly manner by an external staircase	• In a real situation it would not be possible to have the same type of tools because people would be involved in several fronts	• We expected more people
Emerging necessities			
O	• None	• None	• None
A	• None	• Perhaps the presence of radios at school could also be useful for communication between us, considering that the school is divided into two wings and two floors	• None

(continued)

Table 1 (continued)

	Emilia-Romagna	Marche	Abruzzo
Suggestions for improvement			
O	• None	• None	• None
A	• None	• None	• None

of behaviors and procedures that must be used in case of flood emergencies. In fact, only the attendants representative of the Abruzzo region affirmed to be already aware of what to do in case of flood, even before the participation to the simulation, and to actually expect more participant people. Nonetheless, the attendants of the other regions, in which the exercises were held also in schools, said that for the students these events have been very important because of their strong emotional impact.

For all the interviewed people the warning and evacuation procedures, the roles, and the timing have been respected as planned. However, in the opinion of the Emilia-Romagna attendant, the procedures in the school should be strengthened because they are apparently milder if compared to those for earthquakes and fires. As a matter of fact, the schools are obligated to perform two exercises for earthquakes and fires per year, while nothing is foreseen or codified for floods, although the schools are located in flood-prone areas. Despite that, people managed to evacuate in due time.

Furthermore, the technologies and tools used for the transmittal and reception of the warnings, from the regional civil protection web portals to the radio links and the smartphone texts (via WhatsApp), were considered effective, efficient, and overabundant. In some cases these were tested for the first time. One of the attendants interviewed in the Marche region argued that in a real situation it would not have been possible to rely on the same type of tools because people would be involved in several fronts. Another attendant suggested to provide the schools of radios for internal communications, especially between different buildings.

7.2 Relations

The analysis of the relations layer in the context of the flood simulations indicates the interactions between stakeholders as the most appreciated aspect from both the organizers and attendants interviewed in the three Regions (Table 2).

For all the interviewed people, the relations were constructive, and the response of the participating groups was coordinated and proper. That was possible because there was synergy between the regional and municipal components, the civil protection and the volunteers, the civil protection and the citizens, the teachers and the children.

The exchange of communications was clear and timely. Telephone, radio, oral communication, and acoustic communication functioned as expected. A school of the Marche region also successfully simulated oral communication from window to window of two different buildings. To further strengthen the communications with the civil protection, emerged from the schools the need to have more than one cell phone number to contact via WhatsApp in case the telephone lines would not be operational. Moreover, the communication between municipality and parents has been deepened. Indeed, one of the objectives set in place for the exercise was precisely to avoid parents that autonomously decide to pick up the children from school when a flood warning is issued, thus putting themselves in danger.

Finally, in all the Regions the organizers claimed that shortly before the exercises the civil protection plan was revised and updated. On the other hand, the participants

Table 2 Summary of the outcomes of the Emilia-Romagna, Marche, and Abruzzo flood exercise simulations, in relation to the relations dimension. The table shows the highlights of the interviews to both the organizers (O) and the attendants (A). The results are focused on positive outcomes, problems and shortcomings, emerging necessities, and suggestions for the improvements of the exercises

	Emilia-Romagna	Marche	Abruzzo
Positive outcomes			
O	• Good and constant contacts with the Civil Protection of Emilia-Romagna, both in terms of people and in terms of logistics and structure • Synergy between the municipal component and the voluntary work of 4 associations, also since the operational nucleus of Civil Protection of the Red Cross has just joined the Civil Protection associations of the Province of Bologna • Communications were clear and precise	• The exchange of communications was clear and timely • The school also simulated verbal communication from window to window of two buildings • The volunteers explained, interacted with the children, they were very helpful and prepared • We have deepened the communication between the Municipality and possibly the parents, because now does not exist. One of the objectives that we have set in place for this exercise was precisely the goal of not having parents come to pick up the children • The response of the participating groups was coordinated and proper • Communications were punctual and clear	• The participating groups, both volunteers and citizens, responded in a coordinated manner • Communications were clear • The Civil Protection plan was recently revised, and the Mayor mentioned it in his introductory greetings

(continued)

Table 2 (continued)

	Emilia-Romagna	Marche	Abruzzo
A	• Students understood what they had to do, certainly it was a particular situation, not a daily context • A previous work has been done for the management of a visually impaired student, with cognitive delay, that is very anxious and when in new situations is shaken	• In this case, the exchange of communications (telephone communication, verbal communication, sound from the trumpet, and all clear) worked • The children and the teachers were calm • The response of the participating groups was coordinated and proper • Communications were precise and clear • We have updated the emergency plan • For two special situations with wheelchair children in primary schools we have created ad hoc procedures. A child in a wheelchair has been carried to arms, simulating the breakdown of the lift. The chair was also brought	• Participants were coordinated • Communications were clear
Problems and shortcomings			
O	• None	• We were ready for a first-grade student who always has difficulties during the exercises but unfortunately the parents did not consider it appropriate to make her come. She will grow and then sooner or later will participate	• None
A	• None	• This year we have a student who, particularly when there is evacuation, goes into a crying crisis. The family had been warned, we thought of a collaboration with the family to talk with the girl, to reassure her, and instead the child did not come at school	• None

(continued)

Table 2 (continued)

	Emilia-Romagna	Marche	Abruzzo
Emerging necessities			
O	• None	• None	• None
A	• None	• It would be useful to have some cell phone number to contact via WhatsApp in case the telephone lines were not working. In this case the number was only one, but in an emergency it could not be reached	• None
Suggestions for improvement			
O	• None	• None	• None
A	• None	• None	• None

declared that in the schools have been created ad hoc procedures for special situations, such as securing children with wheelchairs or visually impaired.

7.3 Sensemaking

The analysis of the sensemaking layer in the context of the flood simulations shows that, according to both the organizers and attendants interviewed in the three Regions, the participant groups were proactive (Table 3).

Overall, there were no accidents or elements of chaos, except for minor setbacks. Even in these last cases, the organizations proved to be able to deal with the unexpected situations and to find alternative solutions. Also, there were no incorrect behaviors or scenes of fright, even among the students, because they knew how to behave during earthquakes and fires drills and they were immediately ready to get in line and go out.

Both organizers and attendants were satisfied with the type of simulations and they suggested to repeat them, in addition to permanent training, even in other contexts (e.g. private sectors). However, numerous necessities and suggestions for improvements emerged, especially from the participants. For example, the attendant in Emilia-Romagna recommended to simulate more realistic scenarios. As a matter of fact, most of the students of the school selected for the exercise are usually placed on the upper floors so that in case of flood emergency they should have been told to stay in their rooms. Moreover, emerged the need to codify the flood emergency with a specific acoustic sound and to understand how to better manage the exits from school. Basically, it is essentially to establish or reexamine in depth the various emergency procedures. This have also arisen from the interviews of the organizers. Finally, most of the participants thought that the information given during the training sessions were overabundant and everlasting.

8 Conclusions

Evaluation of the simulation exercises is crucial to assess its effectiveness and to refine the success of a disaster response.

In this study was proposed the application of a three-dimensional method to examine the general organization of flood field simulations held at urban scale in the Emilia-Romagna, Marche, and Abruzzo pilot areas of the EU Project LIFE PRIMES (Preventing flooding Risks by Making Resilient communities). The analysis was focused on the management (*structures*), the networks (*relations*), and the proactivity (*sensemaking*), in order to assess the ability of the stakeholders to deal with both planned and unplanned situations. The authors, as external observers, carried out two face-to-face interviews for each flood exercise simulation, one with the organizers

Table 3 Summary of the outcomes of the Emilia-Romagna, Marche, and Abruzzo flood exercise simulations, in relation to the sensemaking dimension. The table shows the highlights of the interviews to both the organizers (O) and the attendants (A). The results are focused on positive outcomes, problems and shortcomings, emerging necessities, and suggestions for the improvements of the exercises

	Emilia-Romagna	Marche	Abruzzo
Positive outcomes			
O	• No unexpected events or elements of chaos	• In one wing they forgot to turn off the lights and the third-grade children reported it • There were no accidents or chaos • There were no accidents or chaos	• During the exercise phase there were also small setbacks, such as when a winch had been trapped, but the volunteers proved to be able to find alternative solutions and complete the activity
A	• There was no incorrect behavior • There was no chaos	• There have been no panic scenes because children know how to behave and are ready to get in line and go out. We organize at least two evacuation tests per year for earthquake and fire • There were no accidents or chaos	• There were no accidents
Problems and shortcomings			
O	• None	• None	• None
A	• Today we worked in a situation that is not so realistic for us, because our classrooms are on the first floor and we rarely find ourselves all on the ground floor so that we have to go above. The evacuation on the upper floor was not useful • In my opinion, too much information all at once during the training	• None	• The presentation was a bit too prolonged
Emerging necessities			

(continued)

Table 3 (continued)

	Emilia-Romagna	Marche	Abruzzo
O	• Repeat the exercises for the volunteers and the population • Permanent training, especially for children and teenagers, thinking about elementary and middle school	• We must codify the first call with the operating room on the river level of the Bettolelle hydrometer. This is a criticality that we knew but the exercise was the starting point to deepen it • In schools there should be a person who controls all the rooms to make sure nobody is left (e.g. a child who remains in the bathroom)	• None
A	• I was thinking of asking in my classes if there was any information that had particularly interested them, and in depth it based on their interest • We should establish a sound, a bell, different, especially for floods. Students usually react better when they know what it is • We could better coordinate the care of the visually impaired student, by establishing more than one person, even a classmate, to support him • We need to understand how to manage the exit from school, even at the end of school hours	• Examine in depth the various emergency procedures in the plexuses. Considering the exercise and looking at the plan, you realize that some things could be better specified	• It would be necessary to know in time if the wave arrives so that we can save things. We lost all the reception documentary with a past flood. At the base of everything you should monitor, clean, and do not throw things inside the river
Suggestions for improvement			
O	• It would be important to do more exercises and training in large buildings, both public and private, but starting from the schools	• I would not have added anything else because the drills must be targeted, and they do not have too much stuff. If the topic is limited we can be more punctual in the solicitations and in the answers	• None

(continued)

Table 3 (continued)

	Emilia-Romagna	Marche	Abruzzo
A	• More training. It would be worth pointing out that if a flood happens and the students are in the classroom, from what I understand, they must stay there, they are already safe on the upper floor • It would be nice to be able to do some projects for schools to inform about the behavior to be taken in case of emergency • To better pass information on everyone, you need to say fewer things and in more time, otherwise the students cannot understand it • The flood simulation should be done on a normal day in which they are in the classroom and they know that there is an alarm	• In my opinion there was nothing that could have been tried with another method or another strategy • The primary school could have been tested, but today there are no children because it is Saturday	• None

and one with the attendants, with the aim of building the evaluation on a plurality of points of views.

The qualitative analyses showed, from both the organizers and the attendants interviewed, an overall positive reaction to the events, great satisfaction for the response of the participants and for the relations built, and constructive proactivity toward the unexpected scenarios and the emerged needs for improvements.

Nonetheless, some problems and some suggestions arose from this investigation. The most important are the need for flood codified alarms and strengthened procedures, the necessity of extra communication tools in case the envisioned ones are damaged, the request of simulations based on more realistic scenarios, and the demand for more targeted information during the training sessions. It is worth noting that the majority of the suggestions arose from the attendant citizens, thus stressing the importance of an evaluation activity extended to all the participants.

In conclusion, this study allowed to implement a three-dimensional model to examine flood field exercises, held in different urban contexts, in order to produce exportable knowledge about disaster evaluation. Future research could be directed to deepen the outcomes through the support of quantitative, measurable, methods and to export the proposed method to additional scales and type of disasters.

Acknowledgements This research was funded by the EU Life PRIMES project (Preventing flooding Risks by Making Resilient communities-LIFE14 CCA/IT/001280).

Annex 1

Model interview for the evaluation of the simulations.

A. **ADMINISTRATIVE INFORMATION (to be completed by the interviewer)**

   a. Name of the interviewee: ____________________
   b. Place of interview: ____________________
   c. Date of interview: ____________________

B. **GENERAL INFORMATIONS**

(1) **What is your job?**

   a. What do you normally do?
   b. Is it part of the civil protection system?
   c. Do you have a role of responsibility and/or coordination?

(2) **Briefly describe the exercise.**

   a. What were the general objectives?
   b. What was the reference scenario?
   c. Which groups of people were involved? E.g.: technicians, citizens, students, teachers, tourists, …

d. What response actions were planned?

(3) **What was your role in the exercise?**

a. Did you participate in the organization of the exercise?
b. Where was it during the event?
c. Did he have roles of responsibility and/or coordination? Against whom?

(4) **Have you participated in other exercises?**

a. Are these recent exercises?
b. What kind of exercises? What were the scenarios?
c. What was your role?

C. **STRUCTURES**

(5) **How would you rate the exercise as a whole?**

a. Have the objectives been achieved?
b. Do you think the exercise was useful? Why?
c. How was the response of the participating groups?

(6) **Have the pre-established procedures been respected?**

a. Have the warning procedures set by the civil protection plan been activated?
b. Have the evacuation procedures set by the civil protection plan been activated?
c. Were the roles respected?

(7) **Do you think that the technologies and tools (e.g. IT tools, telecommunication tools, ...) used have been effective?**

a. How do you judge the tools used to branch / receive alerts?
b. Have additional support tools been or would be needed?

D. **RELATIONS**

(8) **How do you judge the response of the participating groups (e.g.: technicians, volunteers, teachers, students, citizens, ...)?**

a. Do you think that the participating groups have responded in a coordinated way?
b. Do you believe that the participating groups have adopted correct behaviors?
c. Have there been any unexpected interactions between the participants? If yes, have they had positive or negative effects?

(9) **How do you judge the exchange of communications?**

a. Have communications been clear?
b. Did the communications arrive on time? Have the timetables been respected?

c. Did the communication exchange comply with the procedures laid down in the civil protection plan?

(10) **What strategies have been adopted to avoid mistakes?**

a. Are organizational meetings planned?
b. Has the civil protection plan been examined or investigated?
c. Has the evacuation plan been examined or investigated?

E. **SENSEMAKING**

(11) **Have there been any unexpected events?**

a. If so, what kind of nature? How were they dealt with?
b. If not, what could they have been? How do you think they would have been addressed by the participants?

(12) **Were there any elements of chaos?**

a. There were moments of tension? If so, how could they have been avoided?
b. Do you think that the participants reacted well to the stress of the events?

(13) **Finally, what are your suggestions?**

a. Are you generally satisfied with the type of exercise?
b. What else could have been tested?
c. What do you think there is the greatest need to face an emergency? E.g.: training, effectiveness of communications, exercises.

References

Alexander D (2002) Principles of emergency planning and management. Oxford University Press on Demand

Antronico L, Marincioni F (2018) Introduction. In: Antronico L, Marincioni F (eds) Natural hazards and disaster risk reduction policies. Geographies of the Anthropocene, Il Sileno Edizioni, pp 11–14. https://www.ilsileno.it/geographiesoftheanthropocene/wp-content/uploads/2019/05/Chapter-1-1.pdf

Beerens RJJ, Tehler H (2016) Scoping the field of disaster exercise evaluation—a literature overview and analysis. Int J Disaster Risk Reduct 19(September):413–446. https://doi.org/10.1016/j.ijdrr.2016.09.001

Beerens RJJ, Kolen B, Helsloot I (2010) EU FloodEx 2009: an analysis of testing international assistance during a worst credible flood scenario in the North Sea area. WIT Trans Ecol Environ 133:241–255. https://doi.org/10.2495/FRIAR100211

Berlin J, Carlström E (2013) The dominance of mechanistic behaviour: a critical study of emergency exercises. Int J Emerg Manag 9(4):327–350. https://doi.org/10.1504/IJEM.2013.059878

Bowen WM, Gleeson RE (eds) (2019) Models for thinking about human settlements BT—the evolution of human settlements: from pleistocene origins to anthropocene prospects. Springer International Publishing, pp 3–20. https://doi.org/10.1007/978-3-319-95034-1_1

Cranmer H, Chan JL, Kayden S, Musani A, Gasquet PE, Walker P, Burkle FM, Johnson K (2014) Development of an evaluation framework suitable for assessing humanitarian workforce competencies during crisis simulation exercises. Prehosp Disaster Med 29(1):69–74. https://doi.org/10.1017/S1049023X13009217

CRED & UNDRR (2021) 2020: The non-COVID year in disasters. http://hdl.handle.net/2078.1/245181

EU Life PRIMES (2016) Report B1: summary of data and state of the art about hydraulic risk observed in the three Regions. http://www.lifeprimes.eu/index.php/scenari-climatici-report/?lang=en

Farabollini P, Lugeri FR, Lugeri N (2018) Humankind and risk: a difficult history. In: Antronico L, Marincioni F (eds) Natural hazards and disaster risk reduction policies. Geographies of the Anthropocene, Il Sileno Edizioni, pp 88–103. https://www.ilsileno.it/geographiesoftheanthropocene/wp-content/uploads/2019/05/Chapter-5-1.pdf

Gilmore GD, Schwan WR, McLaughlin MK (2007) An assessment of emergency preparedness in western Wisconsin. WMJ Off Publ State Med Soc Wis 106(2):71–77

Jacques J-M, Specht M (2006) Cognition towards crisis: the blind man held a handful of snow … and concluded that white was cold. Int J Emerg Manag (IJEM) 3(1):21–39. https://doi.org/10.1504/IJEM.2006.010278

Kaji AH, Langford V, Lewis RJ (2008) Assessing hospital disaster preparedness: a comparison of an on-site survey, directly observed drill performance, and video analysis of teamwork. Ann Emerg Med 52(3):195–201.e12. https://doi.org/10.1016/j.annemergmed.2007.10.026

Khorram-Manesh A, Angthong C, Pangma A, Sulannakarn S, Burivong R, Jarayabhand R, Örtenwall P (2014) Hospital evacuation; learning from the past? Flooding of Bangkok 2011. Br J Med Med Res 4(1):395–415. https://doi.org/10.9734/bjmmr/2014/5059

Kim H (2013) Improving simulation exercises in Korea for disaster preparedness. Disaster Prev Manag Int J 22(1):38–47. https://doi.org/10.1108/09653561311301961

Lonka H, Wybo J-L (2005) Sharing of experiences: a method to improve usefulness of emergency exercises. Int J Emerg Manag 2(3):189–202. https://doi.org/10.1504/IJEM.2005.007359

Marincioni F, Gioia E, Zoppi M, Vittadini E (2019) Food management in disasters: the case study of the earthquakes of 24 august 2016 in Central Italy. In: Farabollini P, Lugeri FR, Mugnano S (eds) Earthquake risk perception, communication and mitigation strategies across Europe, vol 2, issue 2. Geographies of the Anthropocene, Il Sileno Edizioni, pp 172–192. https://www.ilsileno.it/geographiesoftheanthropocene/Earthquake

Peterson DM, Perry RW (1999) The impacts of disaster exercises on participants. Disaster Prev Manag Int J 8(4):241–254. https://doi.org/10.1108/09653569910283879

Reddin K, Bang H, Miles L (2021) Evaluating simulations as preparation for health crises like CoVID-19: insights on incorporating simulation exercises for effective response. Int J Disaster Risk Reduct 59:102245. https://doi.org/10.1016/j.ijdrr.2021.102245

Savoia E, Testa MA, Biddinger PD, Cadigan RO, Koh H, Campbell P, Stoto MA (2009) Assessing public health capabilities during emergency preparedness tabletop exercises: reliability and validity of a measurement tool. Public Health Rep 124(1):138–148. https://doi.org/10.1177/003335490912400117

Sinclair H, Doyle EE, Johnston DM, Paton D (2012) Assessing emergency management training and exercises. Disaster Prev Manag Int J 21(4):507–521. https://doi.org/10.1108/09653561211256198

UNDRR (2020) Design and conduct of simulation exercises—SIMEX. Words Action 8:43

UNISDR (2015) Sendai framework for disaster risk reduction 2015–2030. In: United Nations Office for Disaster Risk Reduction. https://www.undrr.org/publication/sendai-framework-disaster-risk-reduction-2015-2030

United Nations (2021) Our common agenda: report of the secretary-general. In: United Nations. https://www.un.org/en/content/common-agenda-report/assets/pdf/Common_Agenda_Report_English.pdf

United Nations General Assembly (2015) Transforming our world: the 2030 agenda for sustainable development web.pdf. https://sdgs.un.org/sites/default/files/publications/21252030

United Nations General Assembly (2016) New urban agenda. http://uploads.habitat3.org/hb3/NUA-English.pdf

Wybo J-L (2008) The role of simulation exercises in the assessment of robustness and resilience of private or public organizations, pp 491–507. https://doi.org/10.1007/978-1-4020-8489-8_23

Urban Growth and Increasing Flood Impact in the City of Palma: A Loss of Resilience Capacity

Miquel Grimalt-Gelabert, Joan Rosselló-Geli, and Joan Bauzà-Llinàs

Abstract The flood resilience capacity has been decreasing in Palma since the nineteenth century. Among the causes, the growth of the city toward coastal areas and former agricultural lands, both flood-prone locations once avoided. The increase of small flood events since the start of the twenty-first century highlights the inability of city authorities and inhabitants to cope with the existing risk and to find solutions to solve the problem. Results show that the urban expansion, largely without control during the final decades of the twentieth century, is the main cause of flood-related impacts and damages, while the lack of a public and private answer only increases the resilience failure.

Keywords Floods · Resilience · Urbanization · Impact · Mallorca

1 Introduction

Resilience determines the persistence of relationships within a system and is a measure of the ability of these systems to absorb changes of state variables, driving variables, and parameters, and still persist (Holling 1973). Regarding floods, resilience is the ability to resist a flood and its consequences, trying to minimize their impact and to live with floods instead to fight against them.

M. Grimalt-Gelabert · J. Bauzà-Llinàs
Climaris Research Group, Geography Department, Universitat de Les Illes Balears, 07122 Palma, Spain
e-mail: miquel.grimalt@uib.es

J. Bauzà-Llinàs
e-mail: joan.bauza@uib.eu

J. Rosselló-Geli (✉)
Climaris Research Group, Universitat de Les Illes Balears, 07122 Palma de Mallorca, Spain
e-mail: joan.rossellogeli@uib.es

B. Dahiya et al. (eds.), *Disaster Resilience and Human Settlements*,
Advances in 21st Century Human Settlements,
https://doi.org/10.1007/978-981-99-2248-2_3

Such approach is not new, but flood management often neglected it, preferring a technical answer in terms of the construction of flood defenses like dykes, embankments, and channelization of streams (Zevenbergen et al. 2020). The result in case of extreme floods was the failure of large-scale infrastructures due to the poor design or a lack of proper maintenance (Zevenbergen et al. 2020).

Currently, floods are largely affecting urban areas because of the increasing pressure of urbanization throughout the world. In Europe, floods could affect half a million people in 2050 and damages can climb up to 40 billion EUR (Sofia et al. 2017).

The expected increase of weather-related disasters coupled with the expansion of large human settlements, like cities, in hazard prone areas will add pressure on the resilience capacity of the affected societies, both on individual and societal terms, unless proper measures are taken (Rangari et al. 2021). Around the world, urban resilience against flooding is becoming an important part of the planning policies, to help inhabitants cope with this risk (Zuñiga Igarza, 2018).

In Spain, the increase of urbanized land since the past century led to a loss of the resilience capacity of the population, as urban societies do not know the risk thus taking no actions to protect themselves. In fact, a survey in southeastern Spain shows how a large number of people thinks that flood management should be organized by the government or by city councils (Olcina Cantos et al. 2010). Even the future is expected to be worse as the societal capacity to cope with increasing risks will be smaller as the number of people living in urban areas increases (Serrano Rodríguez 2017) and issues like water scarcity arise (Depraetere et al. 2020).

Flood resilience includes mitigation and adaptation measures, which change with time (Zhu et al. 2021). A combination of methods is nowadays usual, aiming to provide the capabilities to adapt and overcome the flooding risk affecting urban areas. Worldwide proposals attempting to address major risks and manage disasters through governances are developed at different levels, such as national, regional, or local (UNISDR 2015; United Nations 2021).

The aim of this chapter is to show how the city of Palma, through adaptive measures, faced the flood risk in the past and how such measures are ineffective today due to the urban sprawl into flood-prone areas, reducing the resilience capability of their residents.

The chapter is structured as follows: First, an explanation of the research area is included, and then historical flood events and the provided answers are presented, followed by the urban development of the twentieth century and the resulting flooding problems, with some examples of the reduction of resilience. Finally, some conclusions are included.

2 Research Area

The city of Palma, historically known as Ciutat de Mallorca, is the largest of the Balearic Islands. It is located in the South-West of Mallorca, lying on a bay with the

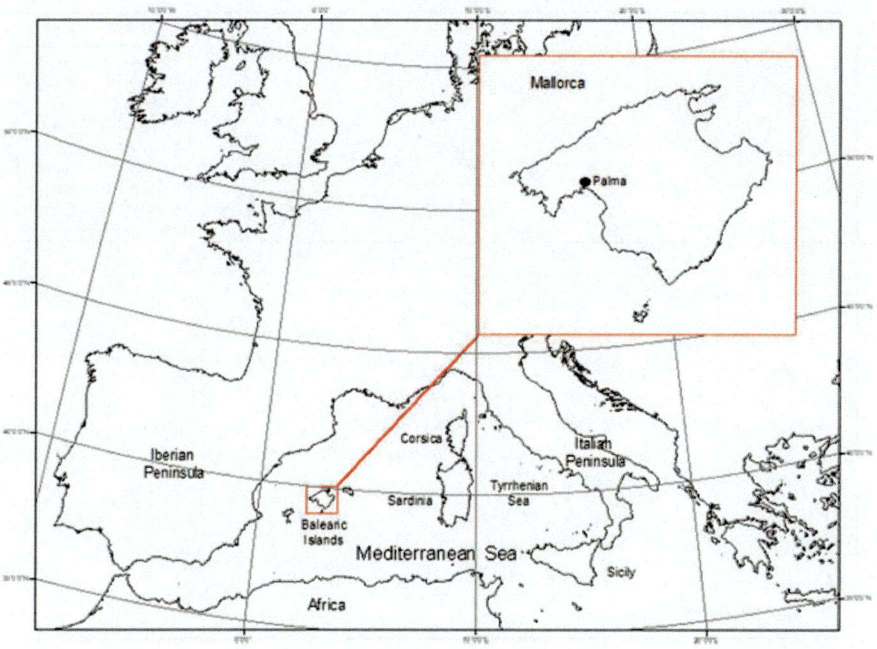

Fig. 1 Location of Balearic Islands within de Mediterranean basin and location of Palma within Mallorca

same name (Fig. 1). The original town was built on a cliff that overlooks the mouth of a stream, *sa Riera*, which ended on Palma's bay.

Once a prehistoric settlement, the oldest remains of human occupation are from Roman times (second century BC). Since the nineth century, with the Muslim domination, the city is the capital of the island and became its main human settlement, both in political and economic terms.

From a geographical point of view, Palma is located on the westerly limit of an alluvial plain, delimited by the *Tramuntana* range and the *marina de Llucmajor*. The climate is Mediterranean, with mild winters, hot summers, and a seasonal distribution of rainfall mainly in autumn and driest months in summer.

The hydrological system is based on ephemeral streams, known locally as *torrents* that run on catchments shaped by the geology and the climate of the area.

The city is located in a subsidence depression, closed by the *Tramuntana* mountain range in the Northwestern area and by *es Pla* relief eastwards. The southernmost part of the plain is where the main streams of the basin flow into the sea. The three primary watercourses are *sa Riera* (with a catchment area of 58.23 km^2), *torrent de Bàrbara* (40.62 km^2), and *torrent Gros* (208.46 km^2).

Sa Riera flows toward the sea directly from the *Tramuntana* mountain range, enclosed by steep banks, which are the result of a tectonic fault. Both the *Gros* and *Bàrbara* streams run through an alluvial plain, which opens to the Palma Bay.

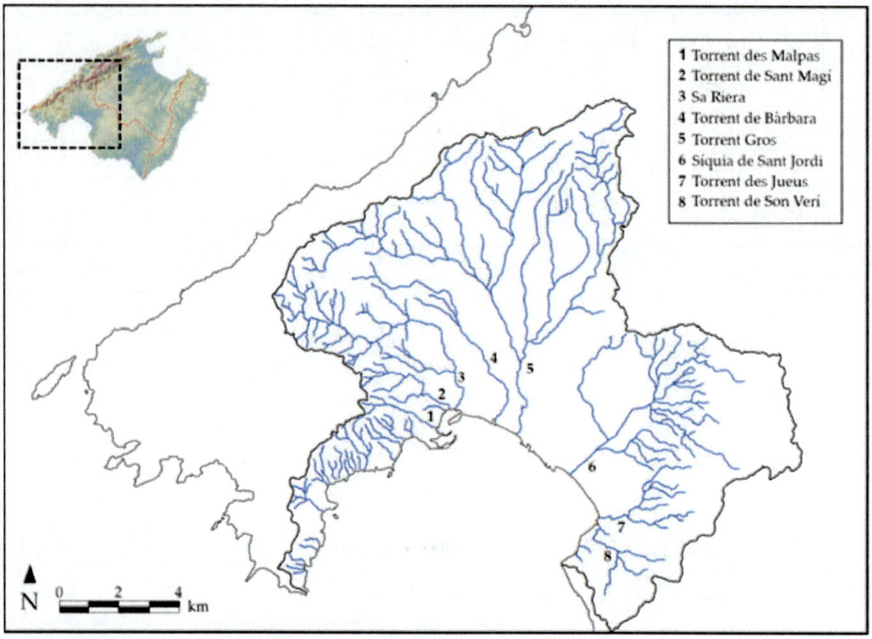

Fig. 2 Watercourses in Palma´s basin

Addtionally, there are several smaller watercourses in Palma's basin, including *torrent des Malpas, torrent de Sant Magí, Síquia de Sant Jordi, torrent des Jueus, and torrent de Son Verí* (Fig. 2).

3 Historical Flood Impacts and Adaptative Measures

3.1 Urban Area

The original settlement was located on a hill at 17 MASL, where currently seats the *Almudaina* castle and *La Seu*, the city's cathedral. The location had the purpose to control a sea inlet where *sa Riera* stream mouth was located. Such high location offered protection as well as access to water resources. During the Roman domination, the city grew easterly, occupying again high grounds, away from flooding events (Fig. 3). Under the Roman rule, the city never expanded westwards of *sa Riera* banks (Vallori, 2019).

After 902, under the Muslim domination, there was an urban growth which lead to the construction of a suburb west of *sa Riera*. Such configuration incorporated the stream into the city limits and originated a town divided into two areas: the lower city, westwards of *sa Riera* and on lower grounds; and the high city, the original urban

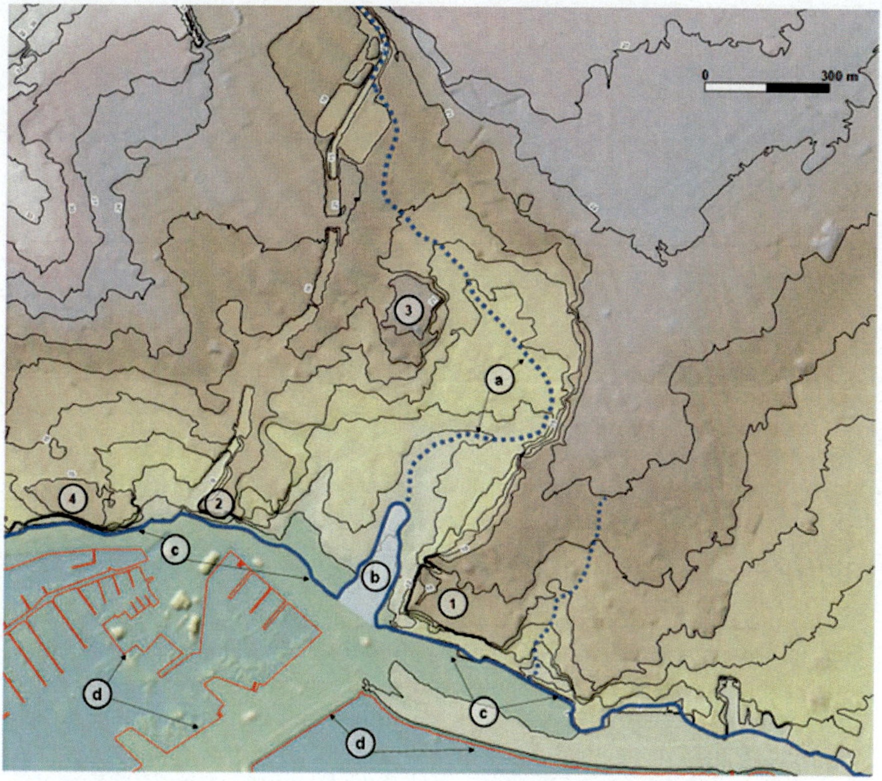

Fig. 3 Main natural landforms in the area of the original nucleus of the City of Palma

nucleus, expanded northwards and eastwards but located on high grounds, sitting at the top of a hilly zone.

After the Catalan conquest of 1229, the city structure did not change, even after becoming one of the largest cities of the Western Mediterranean because of an economic and demographic growth between the thirteenth and fourteenth centuries.

The location of a torrential stream across the town area caused great problems, ranging from a defensive point of view but also regarding the communication between the two main zones, linked through bridges and also in terms of healthiness. Furthermore, the natural harbor located at the mouth was progressively modified by alluvial sedimentation, leading to the research of a new port location.

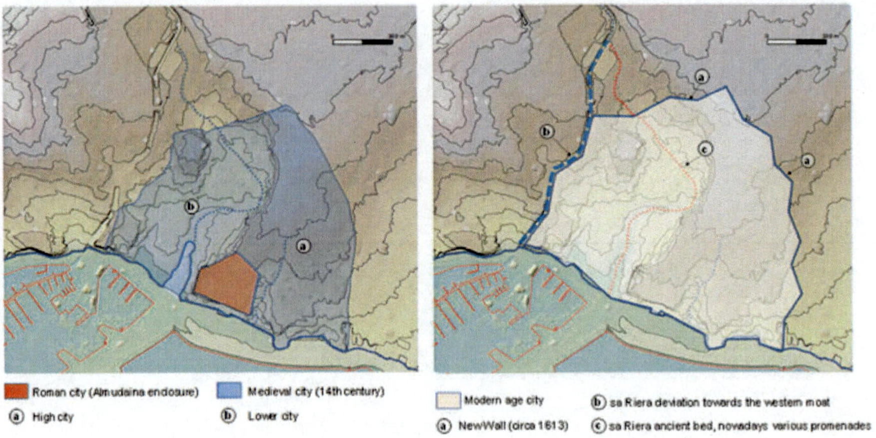

Fig. 4 Limits of the City of Palma during Roman, Medieval, and Modern times

Historical impact of flooding.

During the Middle Ages, *sa Riera* caused two important flood events. The 14th of October 1403 one is the largest, occurring after two days of rainfall. The floodwaters broke the rampart and flooded almost completely the lower city (Fig. 4) (Grimalt, 1989). In some zones, the water levels reached 7 m high above street level and when the rampart facing the sea collapsed, it produced a wave that destroyed a large amount of buildings. According to chronicles, the fatalities reached 5000 people and 500 buildings were destroyed, as well as all the bridges. Another flood happened the 30th of September 1444, affecting the areas closest to the stream banks and destroying some bridges.

During 1613, the streambed was deviated outside the city ramparts, using the western wall as a limit. Despite such change, floods kept the original watercourse during events in the seventeenth century (September 28, 1618, September 20, 1620, and November 7, 1635); in the eighteenth century (November 14, 1734, September 3, 1750, and October 17, 1750); and finally in the nineteenth century, when the October 4, 1850 flood was the last one that followed the old stream course. The widening of the new channel and the creation of obstacles to stop the floodwaters highlight an adaptation model based on a trial and error process.

Because of the deviation of *sa Riera* bed, since 1850 a large flood has not affected the historical city center, even if the old watercourse is still a thalweg across the urban area. Nevertheless, another result of the deviation was the limitation of the working capacity of the downstream area, where the streets act as channels and cause small flooding after intense rainfall events. Furthermore, on the western side of the stream, effluents that end into the new channel had direction changes thus leading to located flooding in some areas.

3.2 Rural Area

The rural area surrounding Palma was not heavily populated, as defensive requirements did not allow the building of suburbs close to the city ramparts. The only exception was *Santa Catalina* suburb, of medieval origin, located on a glacis and not affected by flood events, and small farming areas eastwards of the ramparts (*la Soledat* and *es Hostalets*), both far away from the ramparts and also away from floodwaters (Fig. 5).

The rural area closest to the city was occupied by orchards, watered by a complex irrigation system using spring water (*Horta de Dalt*). Such area could be affected by floods resulting from high waters from two streams, *el torrent Gros* and *el torrent de Bàrbara*, thus implying that the number of buildings of these zones was small, to avoid damages and danger.

East of the city was the *pla de Sant Jordi*, a large wetland where the main lagoon was the *prat de Sant Jordi*. It was used for agricultural activities but population was

Fig. 5 Location of Palma's suburbs and neighborhoods

scarce and lived on the highest areas, hills of alluvial deposits and sand dunes, thus avoiding possible flooding risks.

Outside the city limits, inhabitant's populated zones without risk, selecting carefully where were located to build their lodgments or used traditional prevention and mitigation features (Grimalt and Rosselló 2018b).

4 Twentieth-Century Growth: New Areas, New Risks

During the first years of the twentieth century, the city ramparts were demolished. Such action changed the shape of the city, allowing the development of neighborhoods following an enlargement plan, being *Santa Catalina* or *es Camp den Serralta* examples of such development (Fig. 5).

The bourgeoisie desire to find new locations, away from the old city, for vacation purposes led to the occupation of areas close to the mountains, such as *Sant Agustí, El Terreno, Son Armadans, sa Vileta,* and *Son Roca*, and seaside locations like *es Molinar* or *Ciudad Jardín*, both located in the alluvial plain of Palma.

Moreover, during the first decades of the past century, former farmlands were developed into residential areas in *sa Indioteria* (also in the alluvial plain) and *s'Aranjassa* and *sa Casa Blanca*, both in a drained marshland.

After the 1950s, and related to the new tourism industry, of paramount importance in Mallorca, the enlargement grew exponentially, occupying flood-prone areas, formerly avoided by the population. Neighborhoods along the watercourses appeared in *Verge de Lluc, Son Ametler, Son Ximelis,* and *sa Gruta*. Moreover, tourism infrastructures like hotels and apartments were built close to the seaside, occupying the mouths of the streams (*s'Arenal* and *la Ribera*). The coastline was also changed, with the building of the Maritime Promenade of Palma and later the *Molinar* Promenade. In both cases, roads covered the mouths of the streams and the beds were narrowed. In the 1960s the gentrification of the center of Palma provoked the movement of the industrial activities toward the suburbs. The result was the creation of industrial states, away from residential zones, but located again in flood-prone locations (*Son Castelló* and *Can Valero*).

Finally, the increase in mobility of local population and tourists led to the construction of large road networks, which also affected flood plains and the surface drainage system of the basin. Another example of a large infrastructure located in a flood-prone area is the airport of *Son Sant Joan*, built over drained marshes.

Because of all that spatial changes, nowadays a large area of the city occupies flood-risk zones (Fig. 6). The loss of historical memory and the arrival of inhabitants who do not know the landscape where they live increases the danger and the inability to cope with flooding, thus explaining the loss of resilience. In that sense, Palma is an example of the failure of public authorities to create a governance to manage the existing risk, both in terms of disaster reduction policies and investing in disaster risk resilience (UNISDR 2015). Furthermore, there is no preparedness to answer a possible disaster in terms of recovery and rehabilitation.

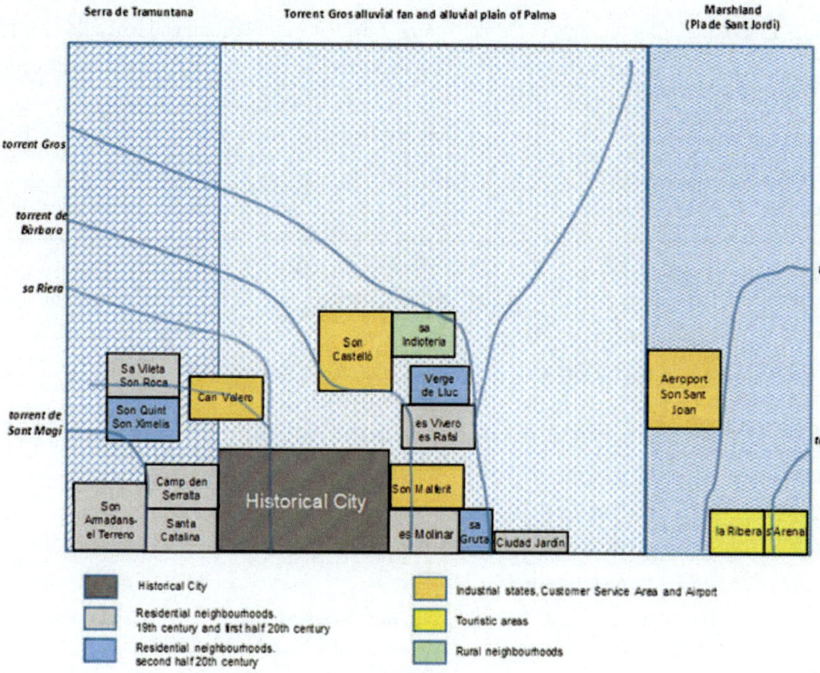

Fig. 6 Cartogram of the main flood areas of Palma linked to the stream network and landforms

In Palma, the evolution from a rural to an urbanized settlement was the result of an economic and social dynamism, which did not take into account factors as risk or vulnerability (Bowen and Gleeson 2019). Even today, such concepts, alongside sustainability, are often forgotten by private entrepreneurs and public authorities, thus leading to the need of changes in terms of planning, management, and governance, if the challenges related to climate change are to be solved in the near future (United Nations 2017).

Results. The measures against flooding

A large number of flood events have affected the municipality of Palma since the Middle Ages. Even if chronicles explain the affectation of such events, it is in the twentieth century when it is possible to gather more information about events and the related damages. The arrival of daily press, common since the nineteenth century, allows researchers to gain a valuable knowledge of the dates and locations as well as damages related to flooding.

Table 1 includes all the main events that have taken place in Palma's basin since 1933 for every affected catchment. It includes a damage category, divided according to three levels:

Catastrophic: Destruction of bridges and infrastructures, heavy damages in buildings, large number of cars carried away by water.

Table 1 List of main flood events in Palma. *Source* Adapted from Grimalt and Rosselló (2011)

Date	Catchment	Affected areas	Damage category
November 7, 1933	torrent Gros	sa Indioteria, es Molinar, Ciudad Jardín	Serious
November 7, 1933	torrent de Bàrbara	es Molinar	Serious
September 29, 1934	sa Riera	Historical City, Santa Catalina, Camp den Serralta, sa Vileta, Son Roca	Catastrophic
September 29, 1934	torrent de Sant Magí	Santa Catalina	Catastrophic
September 29, 1934	torrent des Malpas	el Terreno	Catastrophic
April 16–17, 1942	torrent Gros	sa Indioteria, es Molinar, Ciudad Jardín	Serious
April 16–17, 1942	torrent de Bàrbara	es Molinar	Serious
October 4, 1944	East of Pla de Palma	sa Casa Blanca	Moderate
September 9, 1955	East of Pla de Palma	sa Casa Blanca	Moderate
September 15, 1962	torrent des Jueus	s'Arenal	Catastrophic
September 25, 1962	torrent des Jueus	s'Arenal	Catastrophic
September 25, 1962	torrent Gros	sa Indioteria, es Molinar, Ciudad Jardín, Son Ametler, es Rafal	Catastrophic
September 25, 1962	torrent de Bàrbara	es Molinar, es Rafal	Catastrophic
September 25, 1962	sa Riera	Historical City, Santa Catalina, sa Vileta, son Roca	Catastrophic
September 25, 1962	torrent de Sant Magí	Santa Catalina	Catastrophic
October 12, 1962	sa Riera	Historical City	Moderate
October 18, 1974	torrent des Jueus	s'Arenal	Serious
January 14, 1978	torrent de Bàrbara	es Molinar, Polígon Son Castelló	Moderate
January 14, 1978	torrent Gros	es Molinar, Ciudad Jardín	Moderate
October 17, 2007	sa Riera	el Hoyo	Moderate
December 15, 2008	torrent Gros	es Molinar, sa Gruta, Ciudad Jardín, rural areas	Moderate
December 15, 2009	torrent de Bàrbara	Son Malferit	Moderate
December 19, 2016	East of Pla de Palma	sa Casa Blanca, la Ribera	Moderate

Serious: Exceeded infrastructures, buildings damaged, and some cars carried away.

Moderate: Flooded buildings and cars, traffic interruptions.

There has been a large period without events from the 1980s until the twenty-first century. The listed cases since 2000 are all within the moderate category but with more floods affecting Palma. They can be classified as urban floods, non-related to catchments but with high rainfall intensities and creating small impacts—basically the flooding of streets and underpasses, which cause daily life disturbances ranging from minutes to a couple of hours (Table 2).

Because of the flooding events affecting the municipality, the societal answer currently used in Palma does not follow a simple model. What is clear is that the traditional solution to avoid flood-prone areas has been abandoned, as the urban planning regulations did not exclude from development those spaces that are flood-prone. In fact, some of the recent built areas (like *sa Gruta* or *Son Malferit*) have a high risk of flooding.

The public administration actions against the risk of floods may vary, from an absolute ignorance of the danger to the use of technical solutions. A classification

Table 2 Urban floods affecting Palma. Source: Adapted from Grimalt and Rosselló (2018a)

Date	AFFECTED AREAS
September 6, 2001	prat Sant Jordi, s'Arenal
July 13, 2002	Historical city
August 20, 2007	Historical city, Son Castelló, es Vivero
September 24, 2007	Son Castelló
October 4, 2007	Historical city
October 12, 2007	Son Ferriol
May 9, 2008	es Vivero
May 18, 2008	Son Sardina, Can Valero
June 11, 2008	Son Armadans, Historical city
September 1, 2008	Son Armadans
September 22, 2008	es Rafal, es Vivero
October 12, 2008	Son Castelló
September 28, 2009	sa Casa Blanca
January 8, 2010	Historical city
May 3, 2010	Can Valero, la Soledat, s'Arenal, son Malferit
May 31, 2011	es Vivero, Son Castelló
November 4, 2011	sa Vileta, Can Valero
August 26, 2013	–
October 29, 2013	Son Castelló,
June 17, 2014	Historical city, es Molinar, es Rafal, Son Castelló
September 4, 2015	Son Malferit, Historical city, Son Armadans, Ciudad Jardin, es Molinar, Aeroport Son Sant Joan
Septemer 30, 2015	es Rafal, Verge de Lluc
September 24, 2016	sa Vileta, son Quint, Can Valero
September 15, 2017	Historical city, es Rafal, es Vivero
August 17, 2018	S'Arenal, Sant Jordi, Historical city

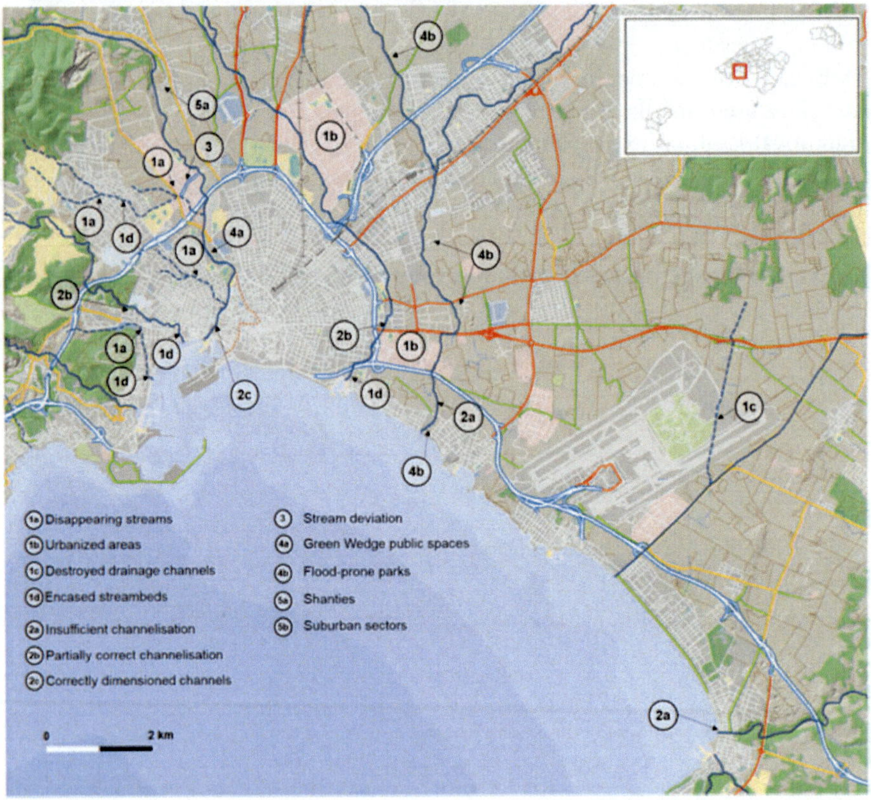

Fig. 7 Spatial location of impact areas according to type of action

of actions can be listed, from the most dangerous to the ones that show the best adaptation level. Such actions are both public and private initiatives (Fig. 7).

1. Areas where the undertaken actions worsen the risk or do not improve it:

1a Zones where the runoff channels have disappeared when integrated on the sewage water system or into the street's system. It is typical on the Western side of Palma, where the streams run in steep valleys. In that case, the streambed is urbanized and becomes a street. Some examples are located in historical suburbs like *Camp den Serralta, sa Vileta, Son Roca*, and neighborhoods like *el Terreno* and *Son Armadans*, both placed on the foot of *Bellver* hill. Another example is the *Can Valero* industrial state.

1b Plain landscapes where urbanization or road network disturb the runoff of flooding waters. Such effects can be identified in places where buildings were built over alluvial plains (*Son Castelló* and *Son Malferit* industrial states) causing the interference with the historical flood channels.

1c Former marshlands that were dried up and current urbanization has destroyed or disabled the drainage channels. Examples of such behavior are located close to the *Son Sant Joan* airport and the road network surrounding this area.

1d Areas where the torrential streambeds are covered, either being converted on a drain pipe, either being encased on a narrow concrete bed, which is unable to contain the water in case of a flood. This category can be located at the end of the *Sant Magí* and *Aigua Dolça* streams in *el Terreno*, and in *sa Vileta,* and, finally, at the end of *Barbara* stream in *es Molinar*.

2. Areas where the channels have been encased on concrete beds, allowing a free water flow. Usually the beds of a same stream show a high variability of capacity, from cross-sections that allow the passing of ordinary floods and will only overflow when facing extreme events, to cross-sections, which are clearly insufficient to contain a flood.

2a Insufficient channels which are common in the largest catchment of the *torrent Gros* basin. The torrent's final stretch has a cross-section of 28 m^2, which is too small considering historical floods affecting the catchment, like the one of September 1962, when the flood peak was calculated to be 500 m^3/s, an estimated value from witnesses' reports and comparison with values of other flood events in Mallorca (Grimalt-Rosselló, 2020). The same streambed has sections upriver reaching 36 m^2 but mostly are smaller, around 20 m^2.

2b Partially correct channelization. In that case, the correlation between flood peaks of historical events, the bed cross-sections, and the catchment area are not as negative as in case 2a but the channels are unable to hold large flood values. The best example is the *Barbara* stream, also located on the eastern side of Palma's bay. The recent urbanized stream banks have a cross-section of barely 24 m^2 (*Son Malferit*) surrounded by roads but, upriver, the sections are located between buildings and have only 15 m^2 in old built areas. The variability between areas with acceptable channel widths and those that are insufficient can also be found in the *Sant Magí* stream, where the bed upriver has a 12 m^2 section, which increases to over 30 m^2 between roads but ends close to its mouth with cross-sections of 12 m^2 between buildings and only 8 m^2 while running under roads.

2c Correctly dimensioned channels. Beds able to resist large floods. Only one case can be found in Palma, this being the historical deviation of *sa Riera* which, after the demolition of the city ramparts in 1902, runs within a depth and wide bed, with a 100 m^2 cross-section that allows the runoff on large peaks, higher than 500 m3/s. This solution avoided flooding during large events affecting Palma in 1934, 1942, and 1962, even if sometimes the banks were almost overcome.

3. Large deviation works. In recent times, there have been public works to deviate a channel, the *sa Vileta* stream, using a bypass to avoid the intersection with the cemetery and the ring road around Palma. Such works contrast with the

headwaters area of the stream, which has been heavily occupied by buildings and streets.

4. Streambeds close areas, not built upon. Such measure allows avoiding the flood risk and can be related with the historical answer to avoid flood-prone zones to build human settlements.

4a A main example is the *Falca Verda*, built at the end of the twentieth century and nowadays continued toward the city limits. Old buildings close to the bed were demolished thus creating open spaces that do not interfere with a possible water spill. Such action was combined with a large concrete bed and the geomorphic characteristics of the stream (within a depth valley) thus helping to create a building-free zone.

4b In newest built upon areas, a similar solution was provided, building public parks on the stream banks. In that case, the plain landscape and the insufficient channel cross-section contribute to allow the flooding of such areas.

5. Finally, a new problem has arisen recently. It is related with the prohibition to build close to the stream banks, mandatory since the start of the twenty-first century, and has resulted in the conversion of such spaces in marginal areas and places with an unlawful urban development.

5a Shacks areas are located within high flood-risk areas. An example is the suburb known as *el Hoyo*, on the middle course of *sa Riera* over quaternary fluvial terraces. Its location on farmlands, abandoned after the 1962 flood, and the large number of shanties convert it on a critical dangerous place.

5b Along the banks of the *torrent Gros* there are suburban sectors, where the farmlands barns have become housing units.

5 Conclusions

In this chapter, a historical evolution of the flood resilience in Palma has been presented. A review of the urbanization of the city revealed how the inhabitants coped with the risk in the past, basically avoiding flood-prone locations, while public works of deviation were undertaken to solve the problem caused by the stream, *sa Riera*, which crossed the center of the old city and flooded occasionally.

Once the city outgrew its ramparts, entering the twentieth century, the urbanization pressure led to the occupation of risk zones, even if such areas were mostly vacation houses, thus occupied only temporally. When the tourism industry arrived to Mallorca, the increase of population and the lack of public policies led to an intense urban pressure on all the available land, even if it was prone to flooding. In the past 60 years, Palma has witnessed a large increase of urbanized land while losing the resilience against floods. Nowadays, public expects that authorities provide solutions while authorities implement some policies which cannot alleviate the existing problems. The mixture of technical solutions (like deviation works or widening of beds) and policies like forbidding constructions close to the stream banks do not solve a recurring problem around the city, which is expected to be worse in the future, related to the impact of climate change in the Mediterranean region.

References

Anònim (1859): Memoria de las inundaciones de la Riera. Sacada de los noticiarios contemporáneos (1403–1850). Imprenta de Esteban Trías. Palma. 24 pp

Bowen WM, Gleeson RE (2019) Prospects for human settlements in the anthropocene: Challenges to equity, sustainability and agency. In: The evolution of human settlements. Palgrave Macmillan, Cham. https://doi.org/10.1007/978-3-319-95034-1_12

Campaner y Fuertes, Álvaro (1881) Cronicón Mayoricense. J Colomer y Salas. Palma. XVI. 661 pp

Depraetere C, Soulis KX, Tsesmelis DE, Avgoustidis G, Spilanis I (2020) Impacts of climate change on the evolution of water resources in the context of the Mediterranean Islands using as example two Aegean Sea islands: consequences for touristic activities in the future. In: Grimalt-Gelabert M, Micallef A, Rosselló-Geli J (eds): The Anthropocene and Islands: vulnerability, adaptation and resilience to natural hazards and climate change. Geographies of the Anthropocene, 3(2), 143–182

Ginard, A., Morata, J., Picornell, C., Seguí, J.M. (1989): Un plànol de Palma de l'any 1613 circa. Treballs de Geografia, 42, 27–34. http://ibdigital.uib.es/greenstone/collect/treballsGeografia/index/assoc/Treballs/_de_Geog/rafia_19/89v42p02.dir/Treballs_de_Geografia_1989v42p027.pdf

Grimalt Gelabert M (1989) Les inundacions històriques de sa Riera. Treballs de Geografia. 42, 19–26. ISSN 1133–181X http://ibdigital.uib.es/greenstone/collect/treballsGeografia/index/assoc/Treballs/_de_Geog/rafia_19/89v42p01.dir/Treballs_de_Geografia_1989v42p019.pdf

Grimalt M, Rosselló J (2011) Anàlisi històrica de les inundacions a les Illes Balears segons la guia metodològica del Ministerio de Agricultura, Alimentación y Medio Ambiente. Conselleria d'Agricultura, Medi Ambient I Territori. Govern de les Illes Balears. Palma, 442 pp+atlas

Grimalt M, Rosselló J (2018a) Inundaciones en la ciudad de Palma de Mallorca: distribución de los episodios y tipos de tiempo asociados. El clima: Aire, Agua, Tierra y Fuego. Asociación Española de Climatología. Murcia, 245–258

Grimalt M, Rosselló J (2018b) Traditional flood mitigation measures in Mallorca. In: Antronico L, Marincioni F (eds). Natural Hazards and Disaster Risk Reduction Policies. Geogr Anthropocene, 1(2), 243–260

Grimalt M, Rosselló J (2020) InunIB: Analysis of a flood database for the Balearic Islands. Eur J Geogr 11(3):6–21. https://doi.org/10.48088/ejg.m.gri.11.3.6.21

Holling CS (1973) Resilience and stability of ecological systems. Annu Rev Ecol Syst 4:1–23

Olcina Cantos J, Hernández Hernández M, Rico Amorós AM, Martínez Ibarra F (2010) Increased risk of flooding on the coast of Alicante (Region of Valencia, Spain). Nat Hazards Earth Syst Sci. 10, 2229-2234, doi: https://doi.org/10.5194/nhess-10-2229-2010

Peña, Pere d'Alcàntara (1891) Inundaciones de la Ciudad de Mallorca (1403). Boletín de la Sociedad Arqueológica Luliana, IV, 151–154 https://prensahistorica.mcu.es/es/catalogo_imagenes/grupo.do?path=1002712036

Rangari VA, Umamahesh NV, Kumar Patel A (2021) Flood-hazard risk classification and mapping for urban catchment under different climate change scenarios: A case study of Hyderabad city. Urban Climate 36:10079. https://doi.org/10.1016/j.uclim.2021.100793

Serrano Rodríguez A (2017) España, un territorio diverso poco resiliente ante los riesgos futuros. Revista de Obras Públicas, 3587, 22–32 http://ropdigital.ciccp.es/pdf/publico/2017/2017_mayo_3587_03.pdf

Sofia G, Roder G, Dalla Fontana G, Tarolli P (2017) Flood dynamics in urbanised landscapes: 100 years of climate and humans' interaction. Nat Sci Rep, 7:40527. https://doi.org/10.1038/srep40527

UNISDR (2015) Sendai Framework for Disaster Risk Reduction 2015–2030, http://preventionweb.net/files/43291_sendaiframeworkfordrren.pdf

United Nations (2017): New Urban Agenda. Habitat III Conference Report. Online at: http://uploads.habitat3.org/hb3/NUA-English.pdf

United Nations (2021) Our common agenda-Report of the Secretary-General. U N Publ, New York, p 76

Vallori Márquez, Bartomeu (2019) La ciutat romana de Palma. Una topografía arqueològica. Ajuntament de Palma. Col·lecció Regne de Mallorca núm 3. ISBN 789–84–95267–09–2. 374 pp

Zevenbergen C, Gersonius B, Radhakrishan M (2020) Flood Resilience. Philos Trans R Soc 378:20190212. https://doi.org/10.1098/rsta.2019.0212

Zhu S, Li D, Huang G, Chhipi-Shrestha G, Nahiduzzaman KM, Hewage K, Sadiq R (2021) Enhancing urban flood resilience: a holistic framework incorporating historic worst flood to Yangtze river delta, China. Int J Disaster Risk Reduct, 61, 102355. https://doi.org/10.1016/j.ijdrr.2021.102355

Zuñiga Igarza LM (2018) Resiliencia urbana ante inundaciones por intensas lluvias en contribución al desarrollo urbano equilibrado. Arquitectura y Urbanismo, XXXIX, 1, 39–50, https://www.redalyc.org/articulo.oa?id=376858935004

Addressing the Impacts of Inland Floods on Informal Housing in Honiara, Solomon Islands

Iftekhar Ahmed

Abstract In this chapter, building resilience of informal housing is explored based on the case of Honiara, Solomon Islands. Informal housing is severely impacted by natural hazards throughout the world. Institutional approaches towards informal housing vary widely, but there is now growing consensus that instead of eviction and relocation, in-situ upgrading offers positive outcomes. Given the limited work on informal settlements in the Pacific region compared to other parts of the Global South, a research agenda is identified focusing on this region. Much of the discourse on the Pacific region deals with coastal climate change impacts relating to global activities within the current Anthropocene Epoch, however, inland floods with damaging impacts, as evident for example in Honiara, is less covered, however, they are also related to the Anthropocene. This chapter thus focuses on the case study of Honiara, a rapidly urbanising city with a significant proportion of its population residing in informal housing. This is an issue that is strongly linked to global frameworks; there is of course a direct connection to the Sendai Framework for Disaster Risk Reduction, and also to the Sustainable Development Goals (SDGs), especially SDG 13 on Climate Action, and also SDG 10 on Reduced Inequalities and SDG 11 on Sustainable Cities and Communities. Flood impacts on informal housing built without legal tenure on exposed sites are pronounced because of the use of vulnerable building materials without following resilient construction guidelines. Local land-use planning and building codes largely do not recognise informal housing. There are nonetheless opportunities for building resilience of informal housing by drawing on the initiatives, networks and skills of informal housing residents. Thus, a potential framework for informal housing improvement is proposed based on such contextual factors. There is a need for further research to develop a comprehensive suite of design and construction guidelines specifically applicable for informal settlements in Honiara, which could also have relevance for the wider Pacific region.

Keywords Floods · Honiara · Informal housing · Resilience · SDGs · Upgrading

---

I. Ahmed (✉)
School of Architecture and Built Environment, University of Newcastle, Callaghan, Australia
e-mail: ifte.ahmed@newcastle.edu.au

B. Dahiya et al. (eds.), *Disaster Resilience and Human Settlements*,
Advances in 21st Century Human Settlements,
https://doi.org/10.1007/978-981-99-2248-2_4

1 Introduction

The widespread and rapid urbanisation process in the Global South over the last few decades has been accompanied by the growth of informal settlements characterised by impoverished and difficult housing conditions, and vulnerability to natural and human-induced hazards. In response to the widespread absence of affordable urban land, and housing and services provided through the formal sector, informal settlements are typically established without governmental sanction and beyond formal planning regulations and building codes (Abunyewah et al. 2017; Bah et al. 2018; Jones 2017; Keen and Jones 2020). Housing in informal settlements presents a critical challenge in the urban settlements field, an issue widely explored for a long time with diverse approaches implemented around the world, yet the challenge persists, thus pointing to the need for further in-depth and intensive research.

More than a billion people around the world live in informal settlements (UN-Habitat, 2016a), with about 80% concentrated in Asia and Africa (United Nations Statistics Division 2019). While the vast agglomerations of population in these continents draw institutional and media attention, it is less recognised that the smaller and scattered island nations in the Pacific region are also urbanising rapidly, accompanied by the growth of informal settlements (Connel and Keen, 2020); in many of the Pacific Island Countries (PICs), urbanisation is occurring at more than three times the global rate (Kuruppu 2016). The Solomon Islands is a case in point—although still predominantly rural, a high urbanisation rate of close to 5% has resulted in 25% urbanisation in 2020, jumping from 20% a decade ago (Plecher 2020; UN-Habitat 2012); 35–40% of the growing population in the capital city, Honiara, live in informal housing (Kiddle and Hay 2017; Tonkin & Taylor 2020a; UN-Habitat 2012) (see Fig. 1). Within the Pacific region, the most extensive informal settlements are in Melanesian cities including Honiara (Jones & Sanderson 2017), thus, the selection of Honiara as a case study for future research offers potential for gaining valuable lessons. Much of the existing body of work on informal settlements, both research and practice, has been conducted in contexts other than the PICs, and the case study in this chapter addresses this gap by focusing on the Solomon Islands where housing in informal settlements faces a significant challenge. This case study is linked to key global frameworks—it clearly relates to the Sendai Framework for Disaster Risk Reduction (UNISDR 2015), with the proposed approach in Sect. 7 on flood risk reduction. It also has implications for the Sustainable Development Goals (SDGs) (UNDESA, 2022)—addressing climate change related floods is linked to SDG 13 on Climate Action; reducing the vulnerability of the urban poor dwelling in informal settlements is linked to SDG 10 on Reduced Inequalities; and dealing with this critical urban challenge relates to SDG 11 on Sustainable Cities and Communities.

In this chapter, issues relating to building resilience to floods of housing in informal settlements is explored, drawing on the case study of Honiara, Solomon Islands. Firstly, an overview of the flood risk context of greater Honiara and the impacts of floods on housing is provided, drawing from the recent most severe flash flood of 2014. A gap analysis is then presented, particularly of the policy framework regarding

Fig. 1 Informal housing in Honiara along the Mataniko River (*Source* Leeanne Marshall)

land-use planning and building codes, allowing the identification of potential opportunities for supporting flood resilience. The chapter then concludes with a proposed framework for housing improvement in informal settlements of Honiara, which has relevance for the wider Pacific region.

2 Disasters and Informal Housing

The growing trend of disasters worldwide have significant impacts on informal settlements. Such impacts are widely documented (see for example, Abunyewah et al. 2017; Asian Development Bank 2016; Carrasco and Dangol 2019; Jones 2017; McCallin & Scherer 2015; Sanderson et al. 2020) and most clearly evident in rapid-onset disasters such as earthquakes. In large-scale earthquakes, such as in Haiti in 2010, around 220,000 people were killed and nearly 300,000 houses were devastated in a situation where in the capital city, Port au Prince, almost 90% of the population were living in informally constructed housing (Blaranova and Christiaens 2012; United Nations 2012). In the 2015 earthquakes in Nepal, more than 600,000 buildings were destroyed where 80% of settlements were characterised as informal (von Meding et al. 2017).

Other than inadequate construction that is particularly vulnerable to earthquakes, the location of informal housing is also often a risk factor. Such housing is often established on marginal land on coastal locations, riverbanks or water bodies, exposing them to floods and water logging (Dahiya 2012); in addition to the locational risk factor, housing in such settlements is often of vulnerable construction and is also impacted by floods (Ahmed 2014; Carrasco and Dangol 2019; Dangol and Carrasco 2019). While the loss of lives and damage to property and assets is usually less in floods than rapid-onset disasters such as earthquakes, they tend to occur on a regular basis annually and over time have a cumulative impact and economic cost, which the urban poor can ill afford. Furthermore, spurred by human-induced global climate change within the current Anthropocene Epoch, floods are increasing in magnitude and becoming more widespread (Braganza 2012; Power et al. 2017), creating an amplifying risk scenario for informal housing. Indeed, the Anthropocene has witnessed the increase, frequency and magnitude of floods worldwide (Razavi et al. 2020), and small island countries such as the Solomon Islands face a double threat from both coastal and inland floods due to respectively sea level rise and erratic rainfall patterns. This chapter therefore focuses on floods and informal housing in the Pacific region where this relationship has been comparatively less studied, drawing from the case of Honiara, Solomon Islands, which has relevance for the wider Pacific region.

3 Institutional Approaches to Addressing Informal Housing Challenges

The diversity of institutional approaches towards informal housing has been succinctly captured by Bah et al. (2018, p. 222–223) as ranging from "… benign neglect, laissez-faire, forced eviction and demolition, resettlement or relocation, slum upgrading programs, and the adoption of enabling strategies" (also see Arimah 2010; Jones 2017; Weksea et al. 2011). A common approach followed by many governments in the Global South is to demolish informal housing and evict their residents. Even approaches termed as "slum rehabilitation" or "slum improvement" (Jones 2017, p. 2) that attempt to convey well-intended policies can in reality be eviction programs. For example, the Tamil Nadu Slum Clearance Board, India, essentially an eviction program, employs such terms (see website http://www.tnscb.org/). There have been advocates against such an approach since the 1970s (for example Turner, 1972) and there are many organisations around the world that seek to prevent forced evictions and advocate the rights of informal settlement residents (for example, the Asian Coalition for Housing Rights). Some key approaches are discussed below.

3.1 Upgrading

The concept of upgrading should be highlighted in relevance to this chapter, that is, improvement of informal housing instead of eradication, now promoted by prominent international agencies such as the World Bank (2011) and United Nations (UN-Habitat 2014). Typically slum upgrading projects and activities include "provision of basic services such as housing, streets, footpaths, drainage, clean water, sanitation, and sewerage" (Cities Alliance 2014). Insecure tenure is a key aspect of informal housing, and it is essential that upgrading initiatives are accompanied by regularisation of tenure, otherwise physical investments can be lost in eviction drives; even upgrading self-initiatives of informal settlement residents are threatened by eviction (see for example Ahmed 2007; Dongol and Day, 2017). Increasingly, prominent agencies are supporting the improvement of tenure security in upgrading programs (see for example World Bank 2020). Most upgrading projects focus on infrastructure and services (see for example Ahmed 2016; Bah et al. 2018), and very few deal with housing improvement, although housing is such an important and valuable asset, a reason for focussing on it in this chapter.

3.2 Relocation Versus In-Situ Upgrading

A consequence of the widely practiced clearance of informal settlements and eviction approach in the Global South is involuntary relocation, which may sometimes be accompanied by resettlement on alternative sites. There is extensive documentation of such resettlement projects that indicate a wide range of negative social, economic and psychological outcomes for the displaced and resettled communities ranging from disruption of social and community networks, to difficulty of accessing livelihood opportunities, to trauma, isolation and marginalisation (Carrasco and Dangol 2019; Jones 2017; McCallin & Scherer 2015; Partridge 1989). In many cases, the new settlements are untenable for habitation by the displaced urban poor communities and are abandoned or pre-empted by groups with higher incomes (Ahmed and Shaw, 2010; Bah et al. 2018).

Relocation and resettlement are widely practiced in a post-disaster context with the primary aim of reducing future disaster risk; again, the mixed outcomes of such projects have been widely discussed (see for example, Ahmed et al. 2020; Ahmed and Shaw, 2010; Carrasco and Dangol 2019; Jha et al. 2010; McCallin & Scherer 2015; Oliver, 1991). The importance of relocation from areas with high disaster risk and exposure, and also anticipated future climate change impacts, is acknowledged in various sources (see for example Bah et al. 2018; McEvoy et al. 2020; Weerasinghe et al. 2014; World Bank 2013). Nonetheless, as pointed out by McCallin & Scherer (2015), forced eviction should not be the basis for relocation and in-situ upgrading for disaster risk reduction should be the first option; only in situations of extreme risk should relocation be considered. In the case of Honiara in this chapter, the significance of such an in-situ upgrading approach for flood resilience is underscored.

4 Methodology

This chapter is principally based on a literature review including an overview of the wider background and the specific case study on Honiara, Solomon Islands. Drawing from the above overview of the broader context, the key points below indicate the research gap, thereby identifying a research agenda and the methodological approach.

4.1 Identifying a Research Agenda

The significance of informal settlements in the Global South and the bulk of related research and action is reflected less in the smaller urban centres in the Pacific region compared to elsewhere, indicating an important area for research. While it is recognised that disasters cause severe impacts to housing in informal settlements, much of that recognition focuses on rapid-onset disasters such as earthquakes, while floods which have persistent, regular and incrementally cumulative impacts, particularly in the Pacific region, are less featured in the discourse. Much of the discussion focuses on coastal inundation in the Pacific region due to sea level rise, but there is very limited discussion on inland floods. There is a need to advance a wider understanding of this hitherto less covered hazard in the Pacific context within the wider global framework discussed in the previous sections.

Among the diverse institutional approaches regarding informal settlements, in-situ upgrading with a view towards building resilience to persistent and widespread disasters such as floods, offers a potential route for addressing the long-term challenges in informal settlements. The nature and form of this approach can possibly be developed through participatory research engagement with informal settlement communities. The literature and work on upgrading of informal settlements focuses largely on physical improvement of infrastructure and services, and comparatively less on housing. There is thus an important need for in-depth exploration of housing improvement options.

These gaps allow identifying a research agenda encapsulated in a key research question: What would be the nature of a framework for in-situ upgrading of housing in informal settlements in the Pacific region?

4.2 *Methodological Approach*

The seminal work of Yin (2017) on case study research has informed the development of this chapter. The case study on Honiara, Solomon Islands, presented here is drawn from work done in partnership with Tonkin & Taylor, a New Zealand based environmental and engineering firm, as part of a larger project on 'Honiara Flood Risk Management Study and Plan' undertaken for the World Bank (Tonkin & Taylor 2019). It also draws from research and community level investigations in Honiara informal settlements supported by the Australian Council for International Development (ACFID) (Bruce and Marshall 2018, 2020). The case study was also informed by consultations with experts involved in developing the 'Honiara Urban Resilience and Climate Action Plan' (Trundle and McEvoy 2016).

The case study approach as followed here allows a contextual understanding and exploring a research problem in detail, the findings of which can have relevance in similar contexts and can potentially be interpreted in and adapted for other contexts and situations. The research question above delineates a research agenda consisting of three methodological components in successive stages:

- Development of an informal settlements housing improvement framework by drawing on recent work in Honiara. This is presented below in the form of a case study drawn from a literature review and other research engagements.
- The framework can be the basis of future field-based research where it would be tested and validated through participatory engagement with communities and organisations in Honiara. It could thereby directly inform policy and practice.
- At the next stage, the outcomes of the above two stages can be assessed for wider relevance in other PICs through a larger multi-country research project. This would also indicate opportunities for adaptation elsewhere in the Global South.

5 Floods and Informal Housing in Honiara, Solomon Islands

The Solomon Islands is located in the Melanesian group of islands in the South Pacific Ocean, consisting of an archipelago of six major islands and many smaller islands (Encyclopaedia Britannica 2022) (see Fig. 2). As in most of the other PICs, the Solomon Islands faces a set of natural hazards, among them most prominently tropical cyclones. Floods too are increasingly more frequent and severe, a pressing hazard that is not recognised widely; much of the discussion is on climate change impacts leading to coastal flooding, but the problem of inland flooding due to heavy rainfall and inadequate drainage, and its impacts due to the exposure of growing density of urban settlements is less discussed. Serious and persistent impacts by inland floods occur, particularly on informal settlements.

Fig. 2 Map of Solomon Islands showing location of Honiara (*Source* Mapcruzin.com; copyright-free retrieved on 23 April 2022 from https://mapcruzin.com/free-maps-solomon-islands/solomon_islands_sm_2008.gif)

5.1 Honiara Urban Profile

Honiara is the capital and largest city of the Solomon Islands with an area of 22 square kilometres and a growing population of more than 92,000 in 2021, that is, more than an eighth of the population of the whole country (National Statistics Office 2022; Tonkin & Taylor 2019). It is located on the largest island of Guadalcanal, which is also the name of the province (see Fig. 2). With less than 25% of the people living in urban areas, it is still a largely rural nation, however, the urban population is growing rapidly at about 4.5% per year (Macrotrends 2022). Livelihoods are reliant on subsistence agriculture and marine resources (Encyclopaedia Britannica 2022), and timber is the main export item (PHAMA, n.d.), which has led to deforestation in some areas and consequent floods and landslides in urban areas. Access to some of the islands is difficult and limited, and roads are generally unpaved and few in number within most of the islands, presenting transportation challenges. The Solomon Islands is categorised as a 'least developed country' by the United Nations (2021).

Estimates suggest that there are about 4,000 households living in informal settlements in Honiara and a substantial proportion of the housing, about 35–40%, is informal (Kiddle and Hay 2017; Tonkin & Taylor 2020a; UN-Habitat 2012), that is, built on land without legal tenure and building permits. There are also informal settlements in the wider province of Guadalcanal. These settlements are growing rapidly at a rate higher than that of urban growth (UN-Habitat 2016b). Many of the settlements have been established for a long time over 30–40 years (Kiddle and Hay 2017), and the government seems to have accepted their existence as a reality of the urban condition.

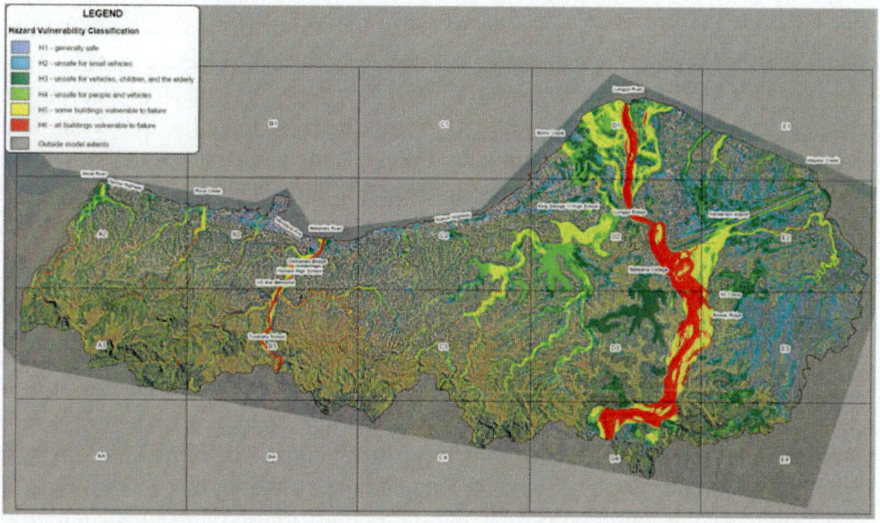

Fig. 3 Flood risk assessment map showing the levels of vulnerability along the three main river catchments (*Source* Tonkin & Taylor)

5.2 Honiara Flood Risk Profile

Flash floods triggered by heavy rainfall frequently impact Honiara, the capital of Solomon Islands, and its surrounding urban areas. One of the most devastating flood events was in 2014, and there have also been subsequent floods; climate change and human actions on the environment within this Anthropocene Epoch might potentially magnify the future flood risks. Drawing from a flood risk assessment by Tonkin & Taylor (2020a, b), it is evident that although the flooding is concentrated around the three catchments of the Lungga, Mataniko and White rivers (see Fig. 3), inundation can be widespread in many areas, particularly those closer to Lungga River with an estimated 1.5–3.0 m flood depth in an 1% Annual Exceedance Probability (AEP) flood event (Tonkin & Taylor 2020a). The extent of flooding in the Lungga catchment is the maximum, however, but housing is much less at risk there (around 2.5%) compared to the Mataniko catchment (more than 43%) because of the higher density of the built environment. The White River catchment also has a high number of houses at risk (around 21%), but only a small number of houses have been categorised at the highest risk level (Tonkin & Taylor 2020b).

5.3 Flood Impacts on Housing

Nearly 250 houses in Honiara were damaged or destroyed in the 2014 floods (Government of Solomon Islands, 2014). Although this appears to be a small proportion of the city's housing stock, it comprised a substantial part of the almost US$30 million damage and loss in the housing sector in the Guadalcanal province (GoSL, 2014). Housing on the banks of the Mataniko River, particularly the Koa Hill area, was most severely impacted (GoSL, 2014). Within the Lungaa, Mataniko and White River catchments, about 4,700 buildings including housing, are expected to be inundated by more than 1.0 m inundation with a quarter of those at higher risk levels in a 1% AEP flood (Taylor & Tonkin, 2020a and b). Flood impacts are the most pronounced on informal housing that is often built with vulnerable materials and without following resilient construction guidelines. Such informal housing is expected to persist over the long term (Keen and McNeil 2016).

Floods impact housing due to depth, duration and, particularly in flash floods, horizontal force created by flood currents. In Honiara, floods often occur concurrently with landslides, hence slope stability is an important factor to consider in housing improvement initiatives, especially because many houses are built on steep slopes. Floodwater causes various degrees of damage from staining of walls to structural collapse depending on the type of building. Foundation settlement causing cracks and failures in different parts of the building is common and shallow foundations can become unstable due to scouring of soil-cover. Documentation from Tonkin & Taylor indicates such impacts including house piers detaching from floors and toppling shallow foundation footings and piers. Estimates from the 2009 census indicate that only 21% of housing in Honiara have a concrete or brick floor (GoSL, 2014), while the rest is built of corrugated iron sheet, timber or other less durable materials. Even concrete structures if not built properly can be damaged, as evident from the Tonkin & Taylor photographs mentioned above.

6 Challenges to and Opportunities for Building Flood Resilience of Informal Housing in Honiara

As elsewhere around the world in Global South countries, informal housing in Honiara faces various challenges posed by natural hazards such as floods. Nonetheless, there are also opportunities for building resilience. These challenges and opportunities are discussed below.

6.1 Key Challenges

There is hardly any zoning or land-use planning in Honiara; building standard controls, particularly in the extensive informal settlements is generally lacking (McEvoy 2019; UN-Habitat 2012). There are some initiatives by the government for relocation away from flood-prone areas, such as the April Ridge project initiated after the 2014 flood to relocate nearly 250 households from the Mataniko riverbank to an area safer from floods; the new location presented similar hurdles as discussed above in Sect. 3.2, including procedural delays, lack of adequate infrastructure and services, and lack of livelihoods and financial support (Ha'apio and Gonzalez 2018; Keen and McNeil 2016). There is no plan of the government for a long-term, large-scale resettlement program at the urban or national level (McEvoy 2019). In the meantime, the informal growth and densification of existing settlements continue, leading to settlement on steep slopes (McEvoy 2019; Trundle et al. 2016), which in addition to the impact from strong flow of floodwater are vulnerable to landslides, creating a multi-hazard situation. The lack of tenure for the bulk of residents, inhibiting investment for housing improvement in fear of eviction or demolition, has resulted in poor quality and vulnerable housing (Bruce and Marshall 2018; Trundle et al. 2016). Upgrading housing for flood resilience in such a context can clearly present powerful challenges.

The existing national policy framework relating to housing has major limitations. The Building Ordinance expected to be followed for a building permit was legislated in 1960 and amended in 1995 (Tonkin & Taylor 2020a), but it is somewhat dated in terms of dealing with the present situation; in any case, it is generally not followed in informal housing. The National Building Code of Solomon Islands (NBCSI) and its companion document, the Home Building Manual of Solomon Islands (HBMSI), were published in 1990, but not being legislated (Gwilliam 2019), do not have much bearing on the bulk of housing in Honiara. Although these documents are used by some built environment professionals, hard copies or online copies are not easily available (Gwilliam 2019) and limited internet access is also a constraint. At the city council, there are only a few building inspectors and with the rapid urban growth, they do not have the capacity to monitor the building process; although there is a provision to issue notices for breaches of the Building Ordinance, that is seldom applied (such urban planning limitations have also been noted by UN-Habitat 2012). The lack of river catchment-wide monitoring and uncoordinated building activity means that ad hoc construction of flood defences by upstream households affects the flow downstream (McEvoy 2019) and solid waste dumped into the river washes downstream and affects the communities there (Bruce and Marshall 2018).

The housing in most informal settlements is built with mixed materials - corrugated iron sheet, timber, leaf thatching, etc. - and even cardboard and plastic in poorer housing (Bruce and Marshall 2018; McEvoy 2019) (see Fig. 4). Almost 50% of the housing is of a single floor, sometimes elevated on stilts, with timber frame construction, though concrete posts are used by those who can afford them (Bruce and Marshall 2018; McEvoy 2019; Trundle and McEvoy 2016). However, the NBCSI and HBMSI do not recognise the reality of such housing—the mixed use of materials and traditional construction (Gwilliam 2019). Additionally, provisions for floods and other hazards are not included (Gwilliam 2019) and the Building Ordinance does not have a requirement for a minimum Finished Floor Level (FFL) elevated above the flood level together with a necessary freeboard (Tonkin & Taylor 2020a).

Fig. 4 A mix of materials is used in informal housing in Honiara (*Source* Tonkin & Taylor)

6.2 *Potential Opportunities*

At the institutional level, an updated NBCSI in conjunction with the HBMSI is reportedly being developed and with the expected enactment of the Infrastructure Management Bill, it may have the power to be enforced (Gwilliam 2019; Tonkin & Taylor 2020a). Acknowledging the limitations in monitoring capacity and the challenges in informal settlements discussed above, the new NBCSI has the potential for flood risk reduction of new housing if adequate provisions are included.

At the community level, since the massive 2014 flood people have become aware and there is evidence that many houses have been built by raising on timber or concrete columns (Bruce and Marshall, 2018; McEvoy 2019) (see Fig. 5); in settlements with high social cohesion and a relatively stronger sense of tenure security, as observed in the Ontong Java neighbourhood, such adaptive capacity results in relatively durable housing raised above flood level on concrete posts (Trundle et al. 2016). Such initiatives are generally undertaken without formal institutional guidelines and support, which point to the opportunity for community-based engagement for flood resilience. There are also widespread community skills in construction, especially carpentry, which could be the basis for community-based skills development for flood-resilient construction. Community Development Committees exist in most communities (McEvoy 2019), which could be focal points for community-based programs for housing improvement. People in Honiara, even from informal settlements, can access loans from banks for building materials (McEvoy 2019), so there is the potential for developing a microcredit program for housing improvements.

Fig. 5 An informal house raised on concrete posts being built (*Source* Leeanne Marshall)

7 Potential Framework for Informal Housing Improvement

Drawing on the above contextual observations, a potential framework for a housing improvement program is proposed here, in line with the Sendai Framework for Disaster Risk Reduction (UNISDR, 2015) and addressing the Sustainable Development Goals (specifically SDGs 10, 11, 13). Because of the preponderance of informal housing in Honiara and its vulnerability to floods, housing improvement should be prioritised for such housing. It is being suggested that the Mataniko River catchment should be prioritised because it is at the highest level of flood risk as mentioned above. It will require developing detailed community profiles and conducting localised risk assessments, and then replicating the process in the other catchments. The application of a minimum 300-mm freeboard above the 1% AEP level should be specified in the forthcoming NBCSI, and the necessary FFL will need to be applied according to such localised assessments.

Although tenure security is often posited as a prerequisite for housing improvements in Honiara (see for example Trundle et al. 2016), in the context of informal settlements there it is a complex process with diverse arrangements (Bruce and Marshall 2018) and with such settlements expected to remain (Keen and McNeill

2016) and arguably continue growing, approaches for housing improvement and property modification will need to work with this reality. Also, with relocation to safer locations being a limited option, in-situ upgrading and retrofitting will need to be considered, even though this might prove challenging.

A participatory community-based approach will need to be followed for such in-situ initiatives. There are many examples of projects in other developing countries that follow such an approach to housing improvement and upgrading for resilience in informal settlements, which offer lessons that can be translated to Honiara (for example, the author was involved in such a project in Bangladesh, see Ahmed 2016). The existing Community Development Committees in informal settlements should be the focal points to initiate and implement community-based projects, where capacity building for resilient construction by building on existing construction skills (e.g. carpentry) through a training-of-trainers (ToT) approach might lead to positive long-term outcomes.

While the forthcoming NBCSI offers an opportunity to incorporate guidelines for flood-resilient and affordable construction, as in the past it is likely to be based on Australian and New Zealand standards where the comprehensive array of requirements might prove onerous and difficult to implement in the low-income socio-economic context of Honiara informal settlements. Standards based on manufactured materials might not be completely relevant where traditional materials such as thatching and non-standard timber is used. At a minimum, if only the disaster resilient aspects of the code can be implemented, it would save lives and property. Such an approach was followed in a project led by the author, where a 'grey building handbook' was developed for informal sector builders, where a suite of safe building options is presented to match different incomes and site conditions, and extensively used visual material for ease of comprehension and communication (Ahmed et al. 2018). The author also produced such an illustrated handbook for flood-prone areas of Bangladesh (Ahmed 2005). Such a simple manual, handbook or design guide targeted specifically for the Solomon Islands should be produced, providing "step-by-step guidelines, and training to construct resilient houses (safer construction) and to strengthen existing building stock" (McEvoy 2019, p 19). Capacity building of informal sector builders and the suggested manual/handbook should provide clear and illustrated instructions for new housing and retrofitting.

As a basic guideline for new housing, flood-resilient construction should pay specific attention to the lower part of the house. Codes should provide guidelines for construction of foundations, footings and posts, and it should be ensured that they are embedded into the ground at adequate depth to avoid toppling due to scouring by floods. To establish a secure base for footings, compacted sand filling is necessary, and if possible, soling with brick or concrete block. The posts should be connected securely with the flooring structure and other building components so that the piers do not become dislocated in floods.

Retrofitting of existing housing in the higher risk zones, particularly where flood depth could be above 3 m in a 1% AEP (H6) (Tonkin & Taylor 2020b), may not be feasible (Tonkin & Taylor 2020a), but in the other risk zones with lower depth, it could be an option. There is evidence that many houses are already raised on stilts (Bruce and Marshall 2018; McEvoy 2019) and houses that are not should be encouraged to do so by providing incentive. A key element in a retrofitting initiative should include adding diagonal bracings to provide sturdiness to the posts of elevated houses against strong horizontal floodwater currents. Again, this is a practice that some households follow spontaneously (see Fig. 6), but usually not in a technically adequate way and thus technical support and encouragement for wider replication can be provided. Retaining walls on steep slopes (see for example Nilsalam et al. 2019) and a stormwater drainage system are ways to reduce floods impacts and will require technical guidance and catchment level planning to assess the impact on hydrological flow. Households should be encouraged to build an overhead shelf, loft or mezzanine for safely storing valuable belongings and food during floods; in houses that are not elevated, a raised space or platform within the house should be built. Such practices are followed in other flood-prone countries (Ahmed 2017) and can be replicated in Honiara.

The above suggestions are not exhaustive, they will require detailed community-based investigations to develop a comprehensive suite of design and construction guidelines specifically applicable for informal settlements, which can then inform the forthcoming NBCSI and HBMSI. It would be necessary to conduct a Building Audit Program (Honiara City Council and Solomon Islands Government, n.d.) for detailed documentation and risk assessments.

To back such initiatives for housing improvement and property modification, there is a need for incentives such as grants and loans. Microcredit programs are widespread throughout the developing world and offer possibilities for Honiara based

Fig. 6 Bracings used for sturdiness in a house in Honiara (*Source* Tonkin &Taylor)

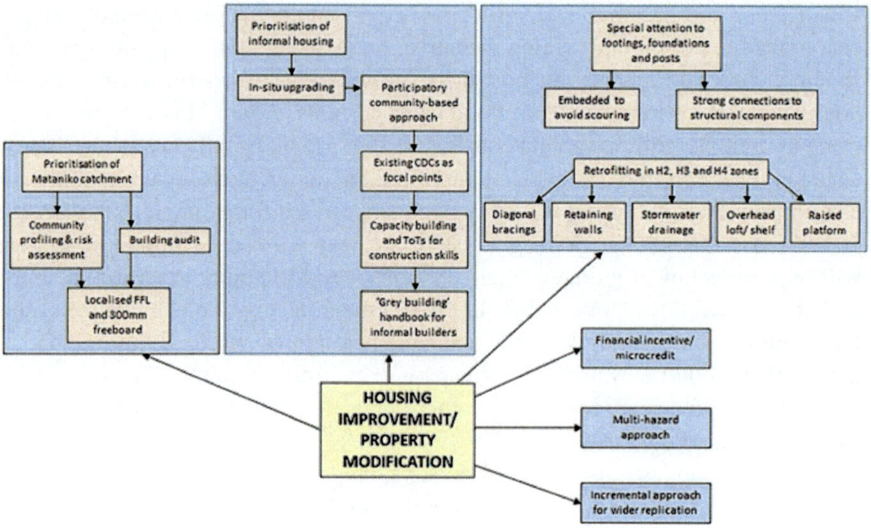

Fig. 7 Key elements of the framework for informal housing improvement in Honiara (*Source* Author)

on the existing system of loans from banks for building materials. Such loans can be combined with capacity building and provision of technical support.

Although floods present a key risk in Honiara, there are other hazards such as cyclones and landslides that can happen in conjunction with floods. Therefore, a multi-hazard approach will need to be followed, again, informed by localised risk assessments. Guidelines for construction to enable resilience to a wider range of hazards will need to be incorporated into the forthcoming NBCSI, as suggested in a review of the existing code (Gwilliam 2019).

Figure 7 shows the key elements of the framework proposed here. This framework might prove ambitious; it would require compromises to achieve some of its key goals. It would also require an incremental approach, building gradually on small successes and replicating across the three catchments in Honiara that all face varying levels of flood risk. The effectiveness of the framework would be underpinned by building capacity both at the institutional and community levels.

8 Conclusion

The intractable challenges of informal housing in the Global South vis-à-vis vulnerability to hazards such as floods is reflected as a microcosm in PICs such as the Solomon Islands where the additional layer of extreme sensitivity and exposure to anthropogenic climate change impacts necessitate a specific suite of contextualised measures to address these challenges. This is all the more important in a situation where the bulk of institutional and international aid efforts focuses on coastal hazards linked to climate change while the highly damaging impacts of inland floods tends to be overlooked. The lack of formal status of informal housing serves as a deterrent to governmental and non-governmental agencies to engage in addressing the risk communities residing in such housing face. Existing institutional instruments for land-use planning and building codes hence do not cover informal housing, although given current trends, such housing can be expected to proliferate.

In response to this situation, this chapter posits a framework for building resilience of informal housing to inland floods based on the case study of Honiara, which may have relevance in the wider Pacific region. The suggestions comprising the framework are not exhaustive, it would require detailed community-based research to develop a comprehensive suite of design and construction guidelines specifically applicable for informal settlements in Honiara, as already proposed by the author and co-authors (Ahmed et al. 2021), which can inform the forthcoming NBCSI and HBMSI. It would be necessary to conduct a Building Audit Program as suggested by the Honiara City Council and Solomon Islands Government (n.d.) for detailed documentation and risk assessments. The flood risk assessment work undertaken by Tonkin & Taylor (2019) is an important initiative that can serve as a basis for implementation of the framework. Compared to Asian megacities, the scale of the problem is smaller in Honiara even with the recent rapid urban growth, and it would prove timely and effective to address the problem at this juncture before it becomes overwhelming and unmanageable. Importantly, reducing the flood risk of informal settlements would contribute to the performance of the Solomon Islands towards addressing the targets of the Sendai Framework for Disaster Risk Reduction and the Sustainable Development Goals (SDG 10, SDG 11 and SDG 13), and help in its government's national reporting on these global frameworks with consequent potential for access to future development funding.

Acknowledgements This chapter draws on technical advisory work undertaken by the author for a flood risk assessment project conducted by Tonkin & Taylor, New Zealand, supported by the World Bank. Advice and resources that informed this chapter and are utilised here were kindly provided by Darryn McEvoy, RMIT University; Alexei Trundle, University of Melbourne; and Leeanne Marshall, Australian Red Cross.

References

Abunyewah M, Gajendran T, Maund K (2017) Profiling informal settlements for disaster risks. Procedia—Engineering 212:238–245

Ahmed I (2005) Design and construction of housing in flood-prone rural areas of Bangladesh. Asian Disaster Preparedness Center, Bangkok

Ahmed I (2007) Urban poor housing in Bangladesh and potential role of ACHR. Retrieved on 18 August 2021 from http://www.achr.net/upload/downloads/file_20112019162941.pdf

Ahmed I (2014) Factors in building resilience in urban slums of Dhaka, Bangladesh. Procedia—Econ Financ, 18, 745–753

Ahmed I (2016) Building resilience of urban slums in Dhaka, Bangladesh. Procedia—Social Behav Sci 218:202–213

Ahmed I (2017) Flood adaptations in the Asian vernacular. In: Piesik S (ed) Habitat: Vernacular architecture for a changing planet. Thames & Hudson, London, pp 508–513

Ahmed I Marshall L, Alenge H (2021) Addressing disaster risk in informal settlements. In J. Twigg and Babister E (Eds). Roadmap for research: A collaborative framework for humanitarian shelter and settlements assistance. Washington D.C.: InterAction

Ahmed I, Gajendran T, Brewer G, Maund K, von Meding J, Kabir H, Faruk M, Shrestha HD, Sitoula NR (2018) Understanding the opportunities and challenges of compliance to safe building codes for disaster resilience in South Asia: The cases of Bangladesh and Nepal (report). Kobe, Japan: Asia Pacific Network for Global Change Research

Ahmed I, Maund K, Gajendran T (2020) Disaster resilience in South Asia: Tackling the odds in the sub-continental fringes. Routledge, Abingdon, England

Arimah BC (2010) Slums as expressions of social exclusion: Explaining the prevalence of slums in African countries. Nairobi: U N Hum SettlS Program

Asian Development Bank (2016) The emergence of Pacific urban villages: Urbanization trends in the Pacific islands. Asian Dev Bank, Manila

Bah EM, Faye I, Geh ZF (2018) Slum upgrading and housing alternatives for the poor. In: Bah EM, Faye I, Geh ZF (eds) Housing market dynamics in Africa. Palgrave Macmillan, London, pp 215–253

Blaranova L, Christiaens B (2012) Project proposal: Community development Delmas 30. Port-au-Prince, Haiti: Int Fed Red Cross Red Crescent Soc

Braganza K (2012) A land of (more extreme) droughts and flooding rains? Retrieved on 30 September 2020 from https://theconversation.com/a-land-of-more-extreme-droughts-and-flooding-rains-5184

Bruce L, Marshall L (2018) What does neighbourhood level urban resilience look like in Honiara? A case study (report). Deakin (ACT), Aust: Aust Counc Int Dev

Bruce L, Marshall L (2020) What Does neighbourhood-level urban resilience look like in Honiara. In: Sanderson D, Bruce L (eds) Urbanisation at risk in the Pacific and Asia: Disasters, climate change and resilience in the built environment. Routledge, New York, pp 74–88

Carrasco S, Dangol N (2019) Citizen-government negotiation: Cases of in riverside informal settlements at flood risk. Int J Disaster Risk Reduct 38:1–10

Cities Alliance (2014). Slum upgrading. Retrieved on 30 September 2020 from http://www.citiesalliance.org/About-slum-upgrading

Connell J, Keen M (2020) Urbanisation at risk: Urban resilience in Pacific island countries. In: Sanderson D, Bruce L (eds) Urbanisation at risk in the Pacific and Asia: Disasters, climate change and resilience in the built environment. Routledge, New York, pp 3–21

Dahiya B (2012) Cities in Asia, 2012: Demographics, economics, poverty, environment and governance. Cities 29(Supplement No. 2): S44–S61. https://doi.org/10.1016/j.cities.2012.06.013

Dangol N, Carrasco S (2019) Residents' self-initiatives for flood adaptation in informal riverbank settlements of Kathmandu. Int J Disaster Risk Reduct, 40, https://doi.org/10.1016/j.ijdrr.2019.101156

Dangol N, Day J (2017) Flood adaptation by informal settlers in Kathmandu and their fear of eviction. Int J Saf Secur 7(2):147–156

Encyclopaedia Britannica (2022). Solomon Islands summary. Retrieved on 23 April 2022 from https://www.britannica.com/summary/Solomon-Islands

Government of Solomon Islands (GoSL) (2014) Rapid assessment of the macro and sectoral impacts of flash floods in the Solomon Islands. World Bank, Washington DC

Gwilliam R (2019) Regional diagnostic study of constraints in the application of building codes in the Pacific: Solomon Islands (report). Asian Dev Bank, Manila

Ha'apio MO, Gonzalez R (2018) Limits and barriers to transformation: A case study of April Ridge relocation initiative, East Honiara, Solomon Islands. In Filho WL (Ed). Clim Chang Impacts Adapt Strat Coast Communities (pp 455–470). Cham, Switzerland: Springer

Jha AK, Barenstein JD, Phelps PM, Pittet D, Sena S (2010) Safer homes, stronger communities: A handbook for reconstructing after natural disasters. World Bank, Washington DC

Jones P (2017) Formalizing the informal: Understanding the position of informal settlements and slums in sustainable urbanization policies and strategies in Bandung, Indonesia. Sustainability, 9(8), https://doi.org/10.3390/su9081436

Jones P, Sanderson D (2017) Urban resilience: Informal and squatter settlements in the Pacific region. Dev Bull 78:11–15

Keen M, McNeil A (2016) After the floods: Urban displacement, lessons from Solomon Islands. Aust Natl Univ, Canberra

Keen M, Jones P (2020) Resilience in Pacific towns and cities: The social dimensions of change. In: Sanderson D, Bruce L (eds) Urbanisation at risk in the Pacific and Asia: Disasters, climate change and resilience in the built environment. Routledge, New York, pp 121–138

Kiddle GL, Hay I (2017) Informal settlement upgrading: Lessons from Suva and Honiara. Dev Bull 78:25–29

Kuruppu N (2016, September 28) Turning the tide on urbanisation policy in the Pacific islands. Retrieved on 28 September 2020 from https://unu.edu/publications/articles/urbanisation-in-pacific-islands.html#:~:text=But%20despite%20their%20comparatively%20lower,three%20times%20the%20global%20average

Macrotrends (2022) Solomon Islands urban population 1960–2022. Retrieved on 23 April 2022 from https://www.macrotrends.net/countries/SLB/solomon-islands/urban-population#:~:text=Solomon%20Islands%20urban%20population%20for,a%204.59%25%20increase%20from%202018

McCallin B, Scherer I (2015) Urban informal settlers disaplaced by disasters: Challenges to housing responses (report). Norwegian Refugee Council, Geneva

McEvoy D (2019) Climate resilient Honiara: Technical scoping of engineering actions and community profiles (report). Fukuoka, Japan: UN-Habitat and Melbourne: RMIT University

McEvoy D, Barth B, Trundle A, Mitchell D (2020) Reflecting on a journey from climate change vulnerability assessments to the implementation of climate resilience actions: Honiara, Solomon Islands. In: Sanderson D, Bruce L (eds) Urbanisation at risk in the Pacific and Asia: Disasters, climate change and resilience in the built environment. Routledge, New York, pp 53–73

National Statistics Office (2022). Projected population by province 2010–2025. Retrieved on 23 April 2022 from https://www.statistics.gov.sb/statistics/social-statistics/population

Nilsalam Y, Tara A Vahanvati M (2019) Climate resilient Honiara: Climate resilient spaces—nature-based solutions. Fukuoka: UN-Habitat and Melbourne: RMIT University

Oliver-Smith A (1991) Successes and failures in post-disaster resettlement. Disasters 15(1):12–23

Partridge WL (1989) Involuntary resettlement in development projects. J Refug Stud 2(3):373–384

PHAMA (n.d.). Fact sheet: Solomon Islands sawn timber exports. Honiara: PHAMA (Pacific Horticultultural & Agricultural Market Access Program)

Plecher H (2020) Urbanization in the Solomon Islands 2009–2019. Retrieved on 13 July 2020 from https://www.statista.com/statistics/728723/urbanization-in-the-solomon-islands/

Power SB, Murphy B, Chung C, Delage F, Ye H (2017) Droughts and flooding rains already more likely as climate change plays havoc with Pacific weather. Retrieved on 1 October 2020 from https://theconversation.com/droughts-and-flooding-rains-already-more-likely-as-climate-change-plays-havoc-with-pacific-weather-71614

Razavi S, Gober P, Maier HR, Brouwer R, Wheater H (2020) Anthropocene flooding: Challenges for science and society. Hydrol Process 34:1996–2000

Sanderson D, Bruce L, Sitko P (2020) Climate and disaster risks, challenges and opportunities for resilient Pacific towns and cities. In: Sanderson D, Bruce L (eds) Urbanisation at risk in the Pacific and Asia: Disasters, climate change and resilience in the built environment. Routledge, New York, pp 22–33

Shaw J, Ahmed I (2010) Design and delivery of post-disaster housing resettlement programs: Case studies from Sri Lanka and India (report). Clayton (VIC), Australia: Monash Asia Institute

Tonkin, Taylor (2019) Honiara flood risk management study and plan: Inception report. Auckland: Tonkin & Taylor

Tonkin, Taylor (2020a) Honiara flood risk management study: Working paper 4—property modification options (report). Auckland: Tonkin & Taylor

Tonkin, Taylor (2020b) Honiara flood risk management study and plan: flood hazard maps. Auckland: Tonkin & Taylor

Trundle A, McEvoy D (2016) Honiara urban resilience & climate action plan. Honiara: Honiara City Council and Solomon Islands Government

Trundle A, McEvoy D, Mitchell D (2016) Land tenure, climate vulnerability, & adaptive capacity project: Honiara, Solomon Islands—case study report. Fukuoka, Japan: UN-Habitat

Turner JFC, Fichter R (1972) Freedom to build: Dweller control of the housing process. Macmillan, New York

UNDESA (United Nations Depart of Economic and Social Affairs) (2022). Sustainable development. Retrieved on 1 September 2022 from https://sdgs.un.org/goals

UN-Habitat (2012). Solomon Islands: Honiara urban profile. Nairobi: UN-Habitat

UN-Habitat (2014). Participatory slum upgrading programme (PSUP): Halving the number of slum dwellers by 2020. Nairobi: UN-Habitat

UN-Habitat (2016a). Slum almanac 2015/2016a: Tracking improvement in the lives of slum dwellers. Nairobi: UN-Habitat

UN-Habitat (2016b). Draft Honiara city-wide informal settlement analysis. Honiara: Ministry of Lands, Housing and Survey

UNISDR (United Nations International Strategy for Disaster Reduction) (2015) Sendai framework for disaster risk reduction. UNISDR, Geneva

United Nations (2012) Key statistics: Haiti relief. UN Off Secr-General's Spec Advis, New York

United Nations (2021) List of least developed countries. Retrieved on 23 April 2022 from https://www.un.org/development/desa/dpad/wp-content/uploads/sites/45/publication/ldc_list.pdf

United Nations Statistics Division (2019) Make cities and human settlements inclusive, safe, resilient and sustainable. Retrieved on 1 October 2020 from https://unstats.un.org/sdgs/report/2019/goal-11/

von Meding J, Shrestha HD, Kabir H, Ahmed I (2017) Nepal earthquake reconstruction won't succeed until the vulnerability of survivors is addressed. The Conversation. Retrieved on 1 October 2020 from https://theconversation.com/nepal-earthquake-reconstruction-wont-succeed-until-the-vulnerability-of-survivors-is-addressed-87335

Weerasinghe S, Martin S, Turk V, Riera J, Franck M, McAdam J, Ferris E (2014) Planned relocation, disasters and climate change: Consolidating good practices and preparing for the future. U N High Comm Refug, Geneva

Weksea B, Steyn G, Otieno F (2011) A review of physical and socioeconomic characteristics and intervention approaches of informal settlements. Habitat Int 35:238–245

World Bank (2011) Urban poverty and slum upgrading. Retrieved on 24 July 2015 from http://web.worldbank.org/WBSITE/EXTERNAL/TOPICS/EXTURBANDEVELOPMENT/EXTURBANPOVERTY/0,,menuPK:341331~pagePK:149018~piPK:149093~theSitePK:341325,00.html

World Bank (2013) Building resilience: Integrating climate and disaster risk into development. World Bank, Washington DC

World Bank (2020) Kenya receives $150 million to improve living conditions for 1.7 million residents in urban informal settlements. Retrieved on 28 September 2020 from https://www.worldbank.org/en/news/press-release/2020/08/07/kenya-receives-150-million-to-improve-living-conditions-for-17-million-residents-in-urban-informal-settlements

Yin RK (2017) Case study research and applications: Design and methods. Sage Publications, Thousand Oaks (USA)

Adaptive Capacity Analysis of Flood Prone Regions in Bihar, India

Vinita Yadav and Zeeshan Ibrar

Abstract Disasters are calamitous events bringing destruction and devastation to life and property. It influences the psychological, socio-economic, political, and cultural state of the populace of the affected area. Anthropocene changes the frequency, magnitude, and spatial distribution of disasters with time. These disasters undermine the progress toward Sustainable Development Goals (SDGs). Disaster risk reduction and sustainable development are closely linked and interdependent. The 2030 Agenda for Sustainable Development acknowledges the need to reduce the risk of disasters. SDG 11 and target 11.5 specifically focus on protecting the people with low income and residing invulnerable situations due to water-related disasters. Seventy-Six percent population of Bihar state in India lives under the constant threat of flood. In 2008, Bihar experienced severe floods affecting more than 2.3 million people in the northern region. From 2000 to 2019, floods affected more than 137 million population and caused 5,800 deaths in Bihar state. The local government plays a vital role in building resilience at varied spatial scales through strengthening adaptive capacity. This study aims at weighing the adaptive capacity of a district using disaggregated data at the community development block level for the flood-affected Kishanganj district of Bihar. The linkage of adaptive characteristics and vulnerabilities has been assessed using the frameworks of the World Bank and the International Union for Conservation of Nature. Adaptive Capacity Index determines critical infrastructure for adaptation at the community development block level in a district. It is calculated using the composite index approach for quantifying adaptive strategies. Analytical Hierarchy Process is used to rank adaptive strategies to suggest the institutional framework for coping with the disaster. The paper concludes that the Terhagachh community development block is the most vulnerable as it has the least adaptive measures in place whereas the Pothia community development block is more resilient to floods. The study ranked the institutional assistance and setup aspect as the topmost factors for determining the adaptive capacity of any region. The paper suggests using the Adaptive Capacity Index approach to monitor adaptive capacity and resource utilisation in flood-affected regions.

V. Yadav (✉) · Z. Ibrar
School of Planning and Architecture, Department of Regional Planning, New Delhi, India
e-mail: V.YADAV@SPA.AC.IN

B. Dahiya et al. (eds.), *Disaster Resilience and Human Settlements*,
Advances in 21st Century Human Settlements,
https://doi.org/10.1007/978-981-99-2248-2_5

Keywords Flood adaptive capacity · Coping strategy · Natural hazards · Resilient region · SDGs

1 Introduction

Disasters are categorised as geophysical, meteorological, hydrological, climatological, biological, and extra-terrestrial. In this paper, the first four categories are used to analyse the macro data. Disasters are a product of social vulnerability and pose a phenomenal threat to human life, leading to social and economic setbacks. During the last decade (2010–2020), disasters' frequency and magnitude of losses have increased worldwide. EM-DAT (2021) has registered more than 15,750 different types of disasters accounting for 32.5 million deaths between 1900 and 2020 in the world. Seventy-one percent (11,262) of these disasters had occurred in only the last three decades which resulted in 1.87 million deaths and affected more than 5.9 billion populations. The hydrological disasters constitute of maximum share i.e., forty-two percent (4,795) of the total natural disasters between 1991 and 2020. These hydrological disasters affected the lives of three billion populations and caused damage to six hundred billion US dollars.

The Intergovernmental Panel on Climate Change (IPCC) predicted that there would be an increase in the frequency of flood hazards due to climate change, and floods would become the most frequent disaster in various parts of the world (Cannon and Schipper 2014). India has recorded 462 disasters, out of which forty-seven percent (218) is due to hydrological disasters i.e., flooding, from 1990 to 2020. The floods also accounted for 32% of the global death count. In India, about 40 million hectares of land, which is one-eighth of the country's geographical area, is prone to flooding. From 1953 to 2018, a total of 471 million hectares of area and 2167 million population got affected, 0.1 million human lives were lost, and properties were damaged worth 60.48 billion US dollars[1] due to floods in India. Between 1990 and 2018, annual flood damage was approximately 56.11 billion U.S dollars which is 14 times more (4.36 billion) than the damage faced from 1953 to 1989 (CWC 2019). The impact of the flood on human life is very complicated. The flood creates negative repercussions on both health and livelihood. Adaptation to reduce disaster-related risk helps to cope with hazards (Yates 2010). Social and economic aspects are important and should be considered for effective mitigation of risk due to floods. The Sustainable Development Goal (SDG) number 11 focuses on making cities and human settlements safe, resilient,and sustainable. It aims to reduce the number of deaths and direct economic loss caused by water related disasters with a focus on protecting the poor and people in vulnerable situations (NITI Ayog 2021, p. 5–8). Sendai Framework for Disaster Risk Reduction (2015–2030) focuses on integrated and inclusive policies to address social, cultural, economic, environmental, and institutional factors to prevent and reduce hazard exposure and disaster vulnerability,

[1] US Dollar = (469,596.71 crore Indian rupees*77.64 1 USD to INR).

improve preparedness for response and recovery, and further strengthen resilience (UNISDR 2015). The framework has emphasised risk-sensitive adaptive capacity development as a means of preventing and mitigating new emergent disaster risks to develop a resilient region (Saja et al. 2020).

In India, for risk-sensitive assessment, data is collated at state, district and Community Development Block (CDB) levels. CDB is an administrative unit within a district which is important for planning, and creating a resilient region. Hence, indicators are aggregated at the CDB level to analyze the district's adaptive capacity. The paper is structured into five parts. Part one introduces the issue followed by discussions related to the concept and approaches to comprehend adaptive capacity in part two. The methods used for the meta-analysis of the selected case are explained in part three. Part four describes the key drivers for assessing the adaptive capacity identified in the case area and the last (fifth) part concludes and recommends the ways to improve the quantitative indices of adaptive capacity to cope with floods.

Society has responded to floods in a variety of ways. It resists, recovers, or adapts through structural and non-structural measures. Anthropocene affects the interaction between society and floods that lead to changes in the human-flood dynamic. In the dynamics resulting from the society and flood events, there is a need to understand anthropocene-related empirical and theoretical research exploring the impact of floods (Di Baldassarre et al. 2016). This research aims to assess the drivers of adaptive capacities and analyse the institutional setup in the flood-affected region. To achieve the aim, the objectives are (1) to identify the aspects and indicators for analysing adaptive capacity, and (2) to assess adaptive capacity using a framework to improve the livelihood in the flood-affected region. The paper's uniqueness is in its usage of secondary data to construct an adaptive capacity index which was earlier used by Maldonado and Moreno-Sánchez (2014) and Thathsarani and Gunaratne (2018) using primary data only.

2 Approaching Adaptive Capacity—Literature Review

Smit et al. (2001) defined 'adaptation to environmental change' as the change in the social, ecological, and economic system in the response to unexpected changes due to external impact. Nelson et al. (2007, p 398) inculcated the following emergent research areas on adapting to environmental changes:

(i) Identification of system thresholds, their limitations, and barriers
(ii) Experimentation on technological options
(iii) Rational process of risk assessment, and
(iv) Public and private sector role, in implementation and governance of adaptation strategies.

Guillén Bolaños et al. (2016) and Few et al. (2017) showed that the availability and accessibility of resources lead to an adjustment in society which results in ecological and social resilience at varied scales. The local communities have found resourceful ways to withstand the adverse conditions (floods) affecting the farming and livelihood of the people. The communities tend to adapt to natural phenomena and learn to live in a flooding situation. For example, the practice of floating agriculture in Bangladesh allows farmers to increase their agricultural productivity during seasonal flooding (Daszkiewicz 2020). Adaptation of crop-based activities i.e., mixed sowing of two varieties of paddy crops, one early and one normal variety, makes it a unique and useful practice to mitigate the loss of agriculture due to flood in eastern Uttar Pradesh in India (Mishra et al. 2008). Decision-making is the foremost task of adaptation concerning climate change. It involves the process of knowledge, experience, and institutional setup to explore the options and determine the actions. The ability of people to use skills and available resources varies within the community. The adoption of strategies to manage the adverse conditions during natural disasters, such as floods, is what makes certain groups more vulnerable. The climate change adaptation framework, approaches and their descriptions are summarised in Table 1.

Two frameworks i.e., 'Framework for Social Adaptation to Climate Change' (SACC-IUCN) (Marshall et al. 2010) and 'Building Resilience through Adaptive Social Protection' (World Bank, 2010) appear most suitable for developing adaptive capacity and building a climate-resilient region. The IUCN and World Bank frameworks address the vulnerability, sensitivity, and adaptive capacity of vulnerable communities at various scales. The IUCN framework is suitable for regional level analysis as it has a separate set of indicators dealing at both individual and community levels and helps in finalizing the indicators and sub-indicators for understanding adaptive capacity. The World Bank (2017) framework is also used to assess adaptive social protection through the analysis of programs, data and information systems, finance, and institutional arrangements and partnerships. It is in the continuum of the IUCN framework and promotes government-led investment in the resilience capacities of households to constitute the social adaptation system.

The Adaptive Capacity Index (ACI) is a multi-dimensional analytical framework that assesses the degree of institutional capacity at various levels. Based on Hahn et al. (2009)'s proposed framework of the World Bank (2017), part four of the paper assesses and calculates the ACI to estimate the differential impact of the flood on the affected population in community development blocks of a district. The multiple indicators are used to assess the exposure and adaptability to resist the flood's impact on the social and economic characteristics of the flood-affected region.

2.1 Conceptualizing Adaptive Capacity

United Nations Office for Disaster Risk Reduction (UNDRR) defined coping capacity as 'the ability of people, organizations and systems, using available skills and resources, to face and manage adverse conditions, emergencies or disasters' (UNISDR 2009, p. 08). The adaptive capacity variable is the function of vulnerability and is inversely proportional to each other (Marshall et al. 2010; Yates 2010; Morgan 2011; Nguyen et al. 2016). As the adaptive capacity develops, the susceptibility of a

Table 1 Descriptions of climate change adaptation frameworks

Frameworks	Implementing organizations	Descriptions	Source
Participatory Vulnerability Assessment (PVA)	Practical Action, WWF, IUCN Nepal, CECI Nepal, NAVIN)	• Vulnerability and its component i.e. exposure, sensitivity, and adaptive capacity are defined • Develops a mathematical model for vulnerability with community-based data collection	Yates (2010)
A Framework for Social Adaptation to Climate Change (SACC—IUCN)	International Union for Conservation of Nature (IUCN)	• Comprehensive practical framework for vulnerability assessment in the context of coastal communities • IPCC identified vulnerability-related components i.e., exposure, sensitivity, and adaptive capacity • Sets out a series of indicators for adaptive capacity at two levels i.e. individual and community	Marshall et al. (2010)
Climate Vulnerability and Capacity Analysis (CVCA - CARE)	Care International	• Framework adopts the IPCC definitions and does not set out any indicator for exposure and sensitivity • Elaborates on the livelihood assets for determining adaptive capacity • Draws a useful distinction between coping strategies and adaptation	CARE (2019)
Mainstreaming Adaptation to Climate Change in Agriculture and Natural Resources Management Projects	World Bank	• Focuses on carrying out the economic analysis and developing monitoring and evaluation (M&E) mechanism Framework does not address vulnerability to climate change in terms of exposure, sensitivity, and adaptive capacity	World Bank (2010)

(continued)

Table 1 (continued)

Frameworks	Implementing organizations	Descriptions	Source
Building Resilience through Adaptive Social Protection	World Bank	• Framework is composed of four building blocks i.e., programs, data and information systems, finance, and institutional arrangements and partnerships • Explores the series of statistical measures of vulnerability at the individual and community scale	World Bank (2017)
Adaptation Toolkit: Integrating Adaptation to Climate Change into Secure Livelihoods	Christian Aid	• Originates from the DRR perspective and integrates the livelihood approach towards resilience building for climate change • Explains the need to move beyond coping strategies and includes climate risk cycle management	Christian Aid (2010)
Community-based Adaptation to Climate Change	International Institute for Environment and Development (IIED)	• Community-based adaptation framework for climate change incorporating information related to scientific and local climate change • In the M&E mechanism, the focus is on participatory processes and community management instead of centralized analysis	Reid (2016)
Climate change and Environmental Degradation Risk and Adaptation assessment (CEDRA)	Tearfund	• Focuses on the effect of climate change on developmental projects rather than communities • Tackles the identification of climate risk and hazards using scientific data	CEDRA (2009)

Note CECI = Centre for International studies and Cooperation; IUCN = International Union for Conservation of Nature; WWF = World Wildlife Fund; NAVIN = National Association of Village Development Committees in Nepal. *Source* Compiled by Authors community toward disaster risks declines. The social, economic, and cultural factors influence the exposure and vulnerability of the population. The factors such as wealth and its distribution across society, demographics, migration, access to information technology, employment pattern, education, societal values, and governance structures play an important role in formulating adaptation strategies during the disaster (IPCC 2014). The adaptation capacity empowers the community to cope with disasters, thus making it an important aspect of regional development. Lwasa (2010) pointed out the importance of infrastructure adaptation, community-based adaptation, and institutional adaptation for reducing the impacts of climate change. The flexibility and readiness to change is the key factor for adaptive capacity strategies leading to resilience to climate change.

Swanson et al. (2007) calculated the adaptive capacity for the agricultural sector in the Prairie region using census data. The research identified the sectors in rural agroecosystem experiencing high exposure to climate change and ranked vulnerable districts as high or low in adaptive capacity based on eight determinants. The research incorporated economic infrastructure, interconnectivity, technology, information, institutions, skills, wealth, and equity as the main determinants in the study. The adaptive capacity framework should include resources such as natural, financial, social, and institutional capital (Metzger et al. 2005).

Maldonado and Moreno-Sánchez (2014) studied the adaptive capacity of marine protected areas in Latin America. The study developed a composite index framework using three aspects i.e., socio-economic, socio-ecological, and socio-political, to calculate the adaptive capacity of the fishing community at a local level.

McLaughlin et al. (2010) and Kenney et al. (2011) pointed out that a similar methodology should not be applied at various geographical scales. The macro analysis at the national level needs a different approach than the microanalysis at the household level, as the resources are not equally distributed at both scales. The climatic variables are more useful in certain administrative units. The study based on higher administrative units, such as national or state level, is not applicable to lower units i.e., district, community development block, or village levels. The literature addresses the variations of adaptive capacity across scales by focusing on differences between collective and individual/household adaptive capacity. There was no unanimity on the variation of scale affecting the assessment of vulnerability.

Marzi et al. (2018) developed the comprehensive climatic risk index using ACI as a key component. They performed the adaptive capacity analysis at regional and provincial levels and estimated the variability of ACI scores intra-regionally. The study uses four aspects i.e. economic resources, knowledge and technology, institutions, and infrastructure to analyze the ACI but does not include the social aspect as one of its main determinants. Thanvisitthpon et al. (2020) conducted a primary survey in ten flood-prone municipalities in Thailand and proposed an adaptive capacity framework using six components i.e., economic resources, social capital, awareness and training, technology, infrastructure, and institutions and policies for flood-affected urban areas.

ACI can be estimated for assessing the differential impact of floods in the different administrative units of a district. The paper analyses the adaptive capacity index using

IUCN and the World Bank framework by aggregating indicators at the block level in making a region resilient.

3 Methodology

The systematic analysis of flood events requires a comprehensive database. The paper specifically addresses the adaptive capacity measures at the block level for the flood-affected district. The World Bank (2017) framework's dimensions i.e., program, finance, infrastructure, and institutional adaptation, are assessed using data from the Census. These dimensions are used to formulate the aspects to assess adaptive capacity. IUCN framework has been used to identify the key indicators for calculating the composite ACI of social adaptation dimensions at the block and district levels. Adopting AHP, a ranking system of varied adaptive strategies has been used to quantify the flood-affected region. The regional-level data provides an opportunity to validate and contextualize the conclusions based on individual-level data. The selection of the study area is based on the percentage of population and area affected at each settlement level due to flood using the latest available data.

The ACI is calculated in two steps using secondary data at the community development block level. The first step is a selection of indicators and sub-indicators to find out the corresponding values and the next step is indexing through a balanced weighted average approach (Sullivan 2002). In the second step, the community development block level assessment is carried out to determine the capacity of larger social systems to cope with external stimuli.

3.1 Indicators and Data Sources for Calculating Adaptive Capacity

The procedure of selection of indicators for adaptive capacity is based on the theoretical understanding of relationships based on the deductive research approach (Adger et al. 2004). The study of phenomena of adaptive capacity and identifying its relation with the main process of study leads to the identification of indicators for corresponding aspects and variables (Brooks et al. 2005). Under six aspects i.e. educational commitment, economic autonomy, infrastructure development, technological advancement, livelihood diversity, and institutional setups, indicators to measure the adaptive capacity are described as cited in the IUCN framework (Marshall et al. 2010) (Table 2).

To measure adaptive capacity, values for all the listed indicators explained in Table 2, have been collected at the block level. The data related to demographic and socio-economic characteristics of the population has been collected from Part A Village

Table 2 Indicators to measure adaptive capacity at block level

Aspects	Objective	Indicators
Educational commitment	To assess the level of skills within a region that enables learning and adapting to manage hazard uncertainty and analyze their ability to prepare and respond to disasters (Marshall et al. 2010; Mukuna 2015)	Literacy rate
		Number of schools (Primary, Secondary and Senior Secondary)
		Percentage of Villages having educational facilities of any kind
		Pupil-Teacher Ratio
		Student Classroom Ratio
Economic autonomy	To measure the financial security of the individual to absorb the cost of effect (Howden et al. 2007)	Percentage of villages having Agricultural Credit Society
		Percentage of the population served by credit society
		Percentage of Villages having banking facilities
		Percentage of rural population served by banks
		Percentage of the cultivable area to the total area
Infrastructure Development	To quantify the assets that a region or community can rely on during disasters (Marshall et al. 2010)	Number of villages having recreational facilities (Cinema hall, Public Library, and reading rooms)
		Percentage of the population served by medical facilities
		Percentage of Villages having power supply
		Percentage of Villages having paved road
		Percentage of Households having closed or open drainage connectivity for wastewater outlet
		Percentage of Households living in Permanent structures
Access to Technological Advancement	To measure the capacity to anticipate the future through experimentation, learning, and accessibility of information (Yoon 2012)	Percentage of households with Radio and TV
		Percentage of households with a computer
		Percentage of households with mobile phone

(continued)

Table 2 (continued)

Aspects	Objective	Indicators
Livelihood Diversity	To assess the alternative income options (agriculture) other than a dependency on climate-sensitive resource (Ellis and Allison 2001)	Percentage of main workers population in the households industry and others
		Percentage of main Female workers in the household industry and others
		Percentage of marginalized workers population in the household industry and others
		Percentage of female marginalized workers population in household and others
Institutional setups	To assess the social networks and associations as social institutions help in governing and equal distribution of resources within a region or community (Zacarias, 2019)	Number of SHG
		Number of Nutritional Centre -ICDS
		Number of Anganwadi

Source Compiled by Authors based on (Ellis and Allison 2001; Howden et al. 2007; Marshall et al. 2010; Yoon 2012; Mukuna 2015; Zacarias 2019)

and Town directory and Part B Primary Census Abstract from the District Census Handbook of the year 2011, respectively (Census of India 2011a, 2011b).

3.2 *Calculating Adaptive Capacity Index (ACI)*

To calculate adaptive capacity, an overall score is calculated using the indicators for all aspects. All the indicators are measured in different units, and hence, standardization is carried out before proceeding to prepare an index. Hahn et al. (2009) equation has been used for the conversion in standardized form for the study (Eq. 1).

$$\text{Index } S_b = (S_{ib} - S_{min})/S_{max} - S_{min} \quad (1)$$

where,

Index Sb is the index value i.e., 0 to 1 of the indicator for block b,

S_{ib} represents the value of the ith indicator for block,

S_{max} and S_{min} manifest the maximum and minimum value of the ith indicator among all the blocks.

After the standardization of all indicators having different units, the index is prepared using the Composite Index Approach (CIA). The indicators were averaged using Eq. 2 to calculate the value of each major component:

$$M_{bi} = \left[\sum_{i=1}^{n} \text{index } s_b i\right] \Big/ n \quad (2)$$

where,

M_{bi} is equal to one of the six major aspects for 'block b' [Educational Commitment (EC), Economic Autonomy(EA), Infrastructure Development(ID), Access to Technological Advancement(ATA), Livelihood Diversity (LD), Institutional Setups(IS)], index $s_b i$ represents the sub-components, indexed by i, that make up each major component, and n is the number of indicators in each major aspect.

After calculating the value of all major aspects of a block, the average is then obtained by using Eq. 3 to get the block-level aggregate adaptive capacity index.

$$ACI_b = \left[\sum_{i=1}^{6} W_{Mi} * M_{bi}\right] \Big/ \sum_{i=1}^{6} W_{Mi} \quad (3)$$

Equation 3 can also be expressed as:

$$ACI_b = [w_{EC}EC_b + w_{EA}EA_b + w_{ID}ID_b + w_{ATA}ATA_b + w_{LD}LD_b + w_{IS}IS_b]$$
$$w_{EC} + w_{EA} + w_{ID} + w_{ATA} + w_{LD} + w_{IS}$$

where ACI_b is the Adaptive Capacity Index for block b, equals the weighted average of the six major aspects. The weights of each major aspect, w_{Mi}, are determined by the number of indicators that make up each major aspect and are included to ensure that all indicators contribute equally to the overall ACI (Hahn et al. 2009). In this study, there is a total of six aspects with twenty-one indicators in total. the ACI is scaled from 0 to 1 where 0 is the least adaptive capacity and 1 represents the highest adaptive capacity. The least adaptive capacity represents higher vulnerability whereas the highest adaptive capacity reflects lesser vulnerability.

3.3 Ranking of Adaptive Capacity Using Analytical Hierarchy Process (AHP)

By developing the pair-wise comparison matrix for aspects using the scale, a ranking of adaptive strategies is carried out (Saaty 2008). The scale has a sequence of absolute numbers from 1 to 9, for each pair, to represent the preferences. The element weight for each decision factor is calculated and the matrix is then further converted into normalised pair of the matrix. In this step, all the elements of the column are divided by the sum of the columns and the criteria weights (CW) for each row are calculated by taking the average value of the row. The consistency is calculated to check whether the intended values are in order and the weighted sum value and criteria ratio (CR) is calculated. The average of the criteria ratio gives the value of λ max. Based on this, the consistency index (CI) is calculated using the formula:

$$C.I = (\lambda max - N)/N - 1 \quad (4)$$

where N is the number of criteria.

The consistency ratio is calculated by dividing Eq. 4 by the random index followed by calculating the ranking for different aspects of adaptive capacity.

3.4 Case Area Selection

Out of 28 states in India, Bihar is the most vulnerable state, as 16.5% of the entire flood-prone area and 22% of the total flood-affected population within the country resides in the flood plains of the state (BSDMA 2018). In 2019, 7257 villages in 27 districts were flooded affecting around 15 million population in the state resulting in the death of 480 human lives (CWC 2019). Amongst the states, the number of persons affected by flood per unit area is highest in Bihar as compared to other flood-prone states within the country (NRSC 2020). Hence, Bihar state has been selected for in-depth analysis.

Out of the total 38 districts in Bihar, 28 districts are flood-prone (Yadav and Ibrar 2022). Geographically, the River Ganga divides the state into two parts i.e., North Bihar and South Bihar. Out of 28 flood-affected districts, 21 districts lie in North Bihar. Five river basins i.e., River Burhi Gandak basin, Bagmati-Adhwar river basin, Kamla-Balan river basin, River Kosi basin, and River Mahananda basin drain the region. The river basins' results in flooding of 74% of geographical area in north Bihar. Out of these five river basins, the River Kosi basin has 11,410 square kilometers of catchment area within Bihar, of which 10,150 square kilometers are flood-prone. The flood-prone area is roughly 90% of the largest flood-prone river basin in the state (WRD 2020). River Kosi is known for its changing course. In the last 200 years, approximately 150 km of westward movement has been documented (Mishra 1997). The frequent channel movements and high sediment flux has generated a typical landscape in the River Kosi alluvial plains located in North Bihar termed as 'megafan'. The megafan constitutes eight districts that have been considered for the next level of analysis due to the highest flood-impacted area (Mishra and Sinha 2020). Two factors i.e., percentage of blocks and villages affected due to flood and the percentage of the population affected, have been taken into consideration for the selection of a district. Kishanganj and Araria are the districts that experienced the maximum effect of flood in their CDBs and villages in the year 2017 (Table 3).

Kishanganj district covers an area of 1,884 square kilometers and ranks 28 in terms of geographical area in the Bihar state. It shares its boundary with Nepal on the northern side and the state of West Bengal on the northern and eastern sides. The Araria district lies in the West, and the Purnea district in the South-west. In both districts, floods have affected 100% of blocks, but the percentage of the population affected varied from 99% (highest) in Kishanganj district to 53%

Table 3 Percentage of population and settlements affected due to floods

Districts	Total population (in Millions)	Total number of CDBs	Total number of villages	2017			
				Population affected		Settlements Affected	
				Number	Percentage	Percentage of Block	Percentage of Village
Araria	1.6	9	742	0.90	53	100	96
Purnea	1.9	14	1273	1.23	66	93	62
Kishanganj	1.0	7	771	0.97	99	100	86
Katihar	1.8	16	1540	1.0	55	94	82
Supaul	1.3	12	551	0.41	31	67	41
Khagaria	1.0	7	301	0.20	21	86	32
Saharsa	1.1	12	468	0.4	33	83	47
Madhepura	1.2	13	439	0.04	3	85	57

Source Disaster Management Department (2017, p 1) Available at: https://state.bihar.gov.in/main/CitizenHome.html

in Araria district. This is the reason for choosing the Kishanganj district as a case for the in-depth analysis. Kishanganj district is located in the north-eastern part of Bihar. It lies between 25°20’ and 26°30’ N and 87°7’ and 88°7’ E (Fig. 1). The district is comprised of 7 community development blocks viz. Terhagachh, Dighalbank, Thakurganj, Pothia, Bahadurganj, Kochadhamin, and Kishanganj; 3 Statutory towns also the Nagar Parishad i.e. Thakurganj, Bahadurganj and Kishanganj, and 771 villages. The district’s literacy rate is 55.46%, which is 12% less than the national average, making it one of the most backward districts in the state.

4 Assessing and Calculating Adaptive Capacity Index (ACI)

The adaptive capacity for the flood-affected region of the Kishanganj district is assessed using CDB level data. The approaches are inductive and use state-driven measures to assess the capacity. The indicators to assess the CDB level condition considered for the development of adaptation strategies in the Kishanganj district are discussed below.

4.1 Educational Commitment

Educational commitment (EC) connotes the existence of an effective education system within a region that enables learning and adapting to manage flood uncertainty (Swanson et al. 2007). The four indicators i.e. pupil-teacher ratio, student-classroom

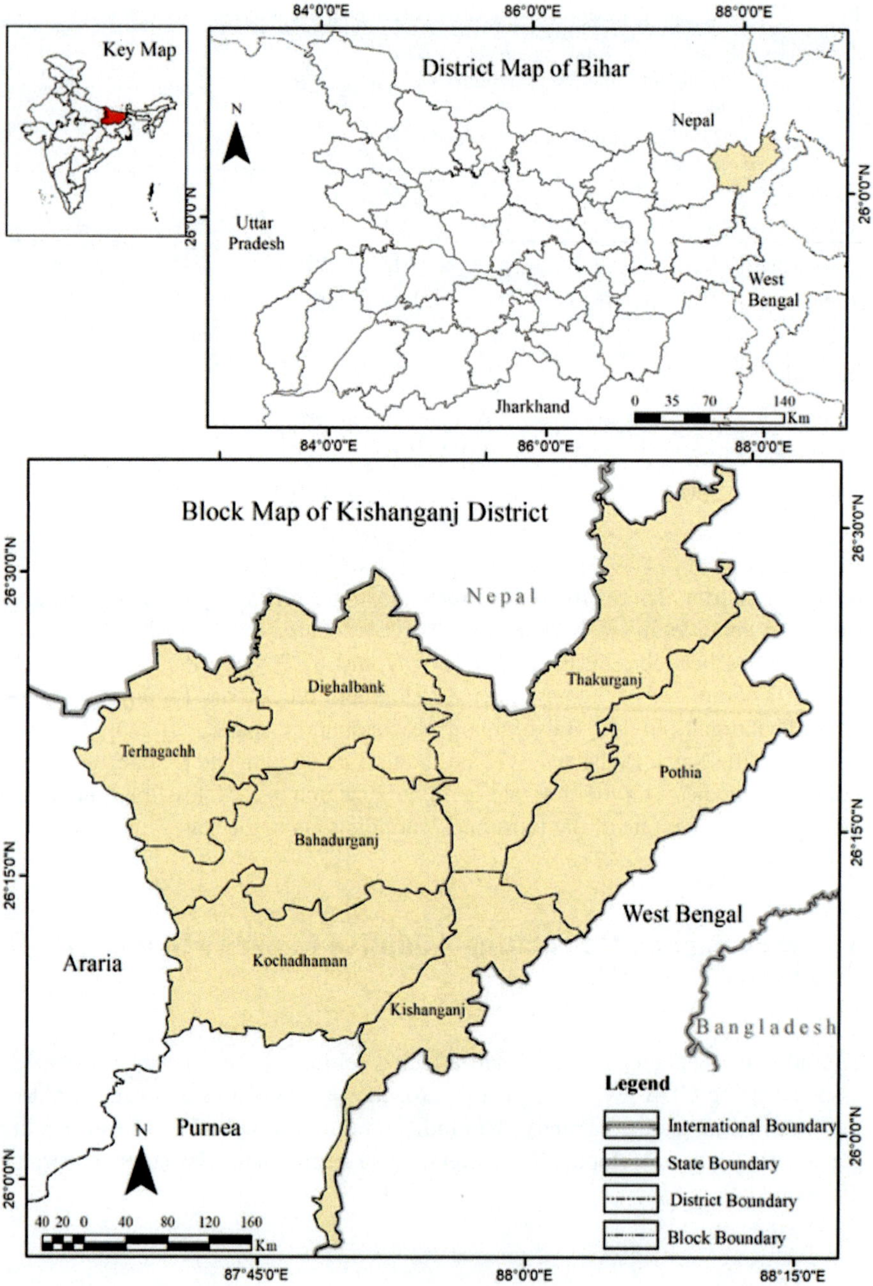

Fig. 1 Community development blocks in the Kishanganj district *Source* Prepared by authors based on the administrative boundary adopted from Census 2011, p 3

ratio, literacy rate, and percentage of villages having educational facilities, have been used to analyse the EC. The risks' perception is a fundamental factor in determining an individual or a community's ability to cope and adapt to change. It is assessed by the educational commitment of the region. Pupil-teacher ratio is highest in Thakurganj and Pothia CDB (57:1). The high ratio indicates that each teacher has to deal with the large number of pupil and conversely, pupil receives less attention from teacher. Kishanganj CDB has a minimum pupil-teacher ratio (39:1) that signifies smaller classes that enable the teacher to pay more attention towards each student. Student-classroom ratio denotes the number of students used one classroom in a school. Higher number suggests large gathering in a classroom, which decreases the time spent by teacher on each student and viceversa (UNICEF, 2016). Pothia CDB has the highest value (46:1), where as Bahadurganj has the lowest (34:1).The CDB wise literacy rate shows that Kochadhamin, with 59.37% literate, is at the top, while Dighalbank CDB with 49.36% is at the bottom (Table 4). Literacy rate is defined as the percentage of literates in the age-group seven years and above. It is calculated by dividing the total number of literate persons aged 7 or above by the total population aged 7 and above and multiplying it by 100. The difference between the highest and lowest literate rate in study area is almost 10%. The gap in the literacy rates of male and female population is 17.5%. The number of schools distributed in each block varies significantly from 97 schools in Kishanganj, the district headquarter, to 306 schools in the Pothia block, which has three non-formal training centres. The percentage of villages having any educational institutions among the CDB, Bahadurganj has the largest proportion of villages with 96.12% of education facility whereas the smallest proportion of villages with the educational facility is observed in Thakurganj CDB with 76.47%. The weighted value of the educational commitment is highest for Pothia CDB and lowest for Terhagach CDB (Fig. 2).

4.2 Economic Autonomy

Economic autonomy is the financial independence and security that helps in coping with damage caused by floods. The aspect is assessed through five indicators i.e., percentage of villages having agricultural society, credit society, banking facilities, and cultivable area to total area (Table 5). The government assistance schemes are an important aspect to analyze financial stability. However, data for government assistance schemes under various grants, such as Indira Gandhi old-age pension, widow pension, Laxmibai Social security pension, and Bihar state disability pension, etc. are not available for all blocks. Therefore, it could not be used as an indicator for this study. The government assistance through different schemes is not robust, as just 0.2% of the population has benefited in the district. A total of 1812 persons in Bahadurganj and 1323 persons in Dighalbank block, respectively are benefited from various pension schemes. The agricultural credit society are the ground-level cooperative credit institutions that provide agricultural loans to the farmers for the various agricultural and farming activities.The agricultural credit society serves only 30.6% of the

Table 4 CDB wise educational commitment in Kishanganj district

Educational commitment	Name of CDBs						
	Terhagachh	Dighalbank	Thakurganj	Pothia	Bahadurganj	Kochadhamin	Kishanganj
Pupil Teacher Ratio	41	49	57	57	42	41	39
Student Classroom Ratio	36	39	44	46	34	36	42
Literacy Rate	54.48	49.36	51.59	51.26	58.13	59.37	53.78
Percentage of Villages having educational facilities of any kind	88.46	90	76.47	89.36	96.12	88.11	88.24
Weighted value of major component, $M_{bi} = [\sum^{n}_{i=1} \text{index } s_b i] / n$	**0.35**	**0.42**	**0.51**	**0.71**	**0.51**	**0.47**	**0.43**

Source Census of India (2011a) p 57,73; NIEPA, (2021)

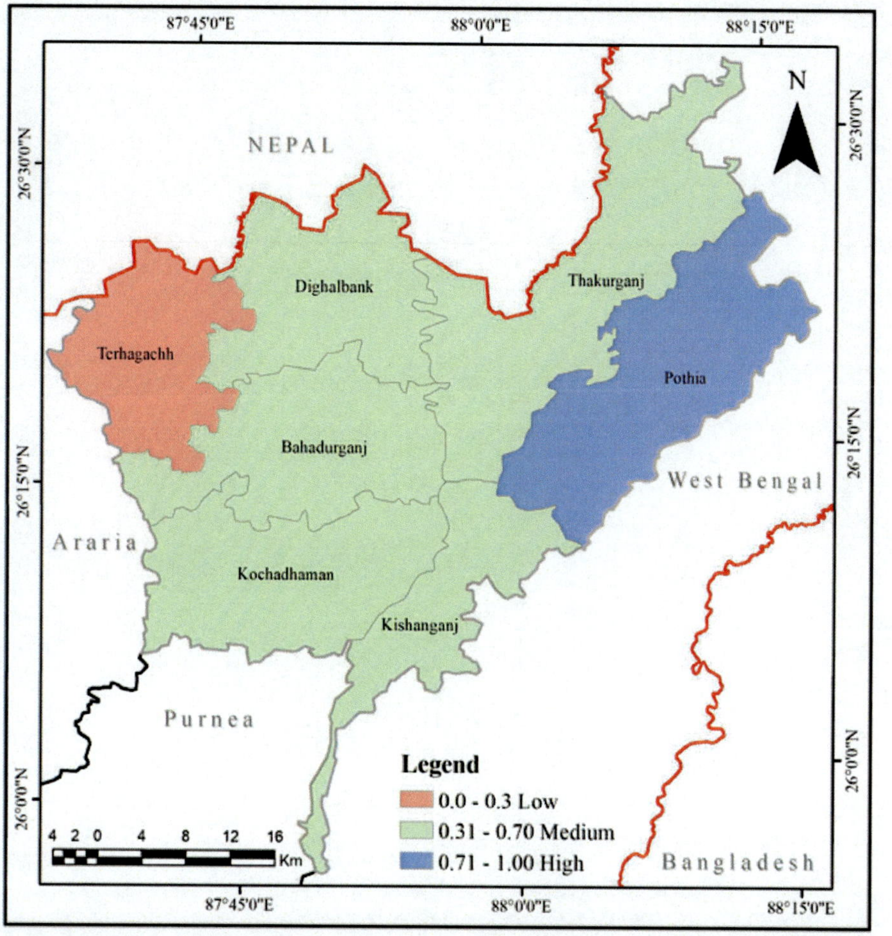

Fig. 2 Spatial variation of educational commitment in CDB of Kishanganj district *Source* Prepared by Authors

district's rural population, with the proportion varying from 25% in the Kishanganj block to 40.5% in the Bahadurganj block. Out of 732 inhabited villages in the district, only 77 villages (10.5%) have banking facilities, and 154 villages (21%) have agricultural credit societies. The banking system of the Kishanganj district caters to only 15.6% of the rural population. The proportion of the rural population served by the banking system varies from a minimum of 3.6% in the Kishanganj block to a maximum of 23.4% in the Pothia block (Table 5). The population of the district is not financially secure which hampers the ability to cope with the effect of flooding, making it more vulnerable to future floods. The weighted value of the economic autonomy is highest for Bahadurganj and Kochadhamin CDB (0.88) and lowest for Kishanganj CDB (0.40) (Fig. 3).

Table 5 CDB wise economic autonomy, Kishanganj district

Economic autonomy	Name of CDBs						
	Terhagachh	Dighalbank	Thakurganj	Pothia	Bahadurganj	Kochadhamin	Kishanganj
Percentage of villages having Agricultural Credit Society	19.23	12.5	24.37	23.4	29.13	18.88	14.71
Percentage of population served by credit society	31.83	26.79	30.27	28.81	40.54	30.05	25.14
Percentage of Villages having banking facilities	6.41	8.75	6.72	19.86	5.83	14.69	2.94
Percentage of Rural population served by Banking facilities	12.61	16.05	17.26	23.38	9.55	18.3	3.59
Percentage of cultivable area to total area	65.99	76.39	68.38	56.93	72.2	78.59	69.05
Weighted value of major component, $M_{bi} = [\sum^{n}_{i=1} \text{index } s_b i] / n$	0.64	0.66	0.75	0.84	0.88	0.88	0.40

Source Census of India (2011a), p 73–76

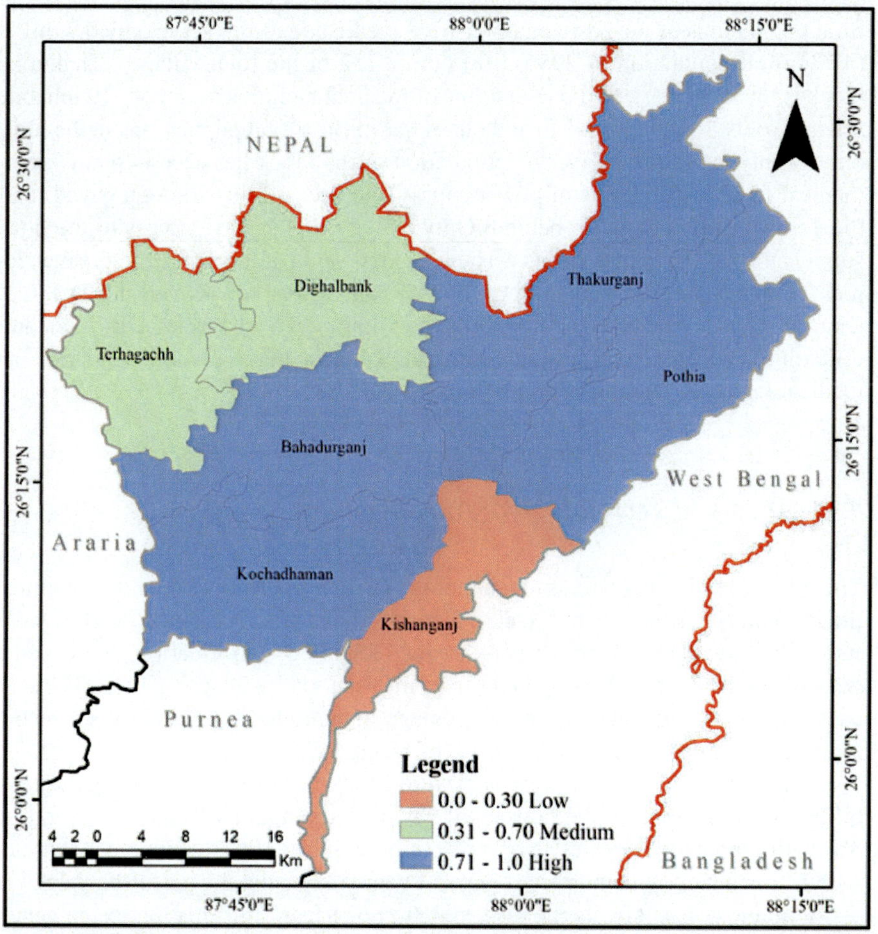

Fig. 3 Spatial variation of economic autonomy in CDB of Kishanganj district *Source* Prepared by authors

4.3 Infrastructure Development

Infrastructure development entails the presence of physical infrastructure which aids in coping capability of community at the time of flood. The infrastructure development is assessed in terms of availability of facilities and termed resource pressure. If the number of resources is higher, then the pressure on a resource will be less, and it will serve the maximum population (Morgan 2011). The absence of resources minimises the capacity to effectively respond to climate change and tends people to migrate to cities. The resource pressure for the CDB is assessed using six indicators i.e., the number of recreational facilities, percentage of the villages served by medical facilities, power supply, and paved road (Table 6). Kishanganj and Pothia blocks do not have any means of entertainment for the population. Out of 1.5 million rural populations of the Kishanganj district, medical facilities serve only 0.7 million (47%) rural population in 295 (40%) out of 732 of the total villages. Thakurganj block is the most neglected one in terms of medical facilities as out of 119 inhabited villages, only 8 villages (6.7%) and out of 0.3 million total populations in the block, only twenty-six thousand (9.8%) population of the block has access to any kind of medical facilities. In terms of power supply and accessibility through paved roads, the Terhagachh block is far behind. Only 5% of villages have access to the power supply and 42% have paved roads. The weighted value of the infrastructure development is highest for Kochadhamin CDB (0.59) and lowest for Terhagach CDB (0.18) (Fig. 4). Thus, every time the flood hits the villages of Terhagach CDB, it becomes difficult for villagers to access safe drinking water, sanitation facilities, market, and medical facilities, leading to higher fatality.

4.4 Access to Technological Advancement

The individuals having access to technology and a propensity to the information are likely to be better equipped to tackle the hazard and can reorganize themselves for future scenarios. The objective of assessing technology, information, and expertise is to assess the approachability and use of information by the population. The study uses block-wise information i.e., percentage of households of each block owning radio, TV, computer, and mobile phone of blocks (Table 7).

Terhagachh block is the most backward block in terms of technology accessibility with only 18% of the household having radio and TV connections and just 3% with computer facilities. None of the blocks has 100% connectivity through mobile, computer (with internet), or even TV and radio, making it a further backward district region. This affects the dissemination of hazard information to the general public which, in turn, affects the implementation of an evacuation plan and other coping strategies. The weighted value of the technological advancement is highest for Kishanganj CDB (0.27) and lowest for Pothia CDB (0.18) (Fig. 5). It is important to recognize the flood prediction methods devised by the communities as they overall affect the adaptive capacity of the region. Some of the indigenous measures for flood prediction adopted by the communities of the Kishanganj district are the bird's nesting pattern, weather intensity level, and movement of stray mammals and reptiles.

4.5 Livelihood Diversity

The possibility to switch between alternative sources of income makes the community resistant to the effect of climate hazards. Individuals tend to diversify their income sources to minimize risk, manage the effect of seasonal events, achieve stability post-disaster, and cope with shocks. The population of the Kishanganj district is primarily

Table 6 Block wise infrastructural development, kishanganj district

Infrastructure development	Name of blocks						
	Terhagachh	Dighalbank	Thakurganj	Pothia	Bahadurganj	Kochadhamin	Kishanganj
Villages having Recreational facilities (Cinema hall, Public Library and reading rooms)	1	2	3	0	1	4	0
Percentage of Villages having medical facilities	61.54	32.5	6.72	56.74	33.01	32.87	76.47
Percentage Villages having power supply	5.13	91.25	31.93	91.49	97.09	76.92	36.76
Percentage of Villages having paved road	42.31	71.25	84.03	98.58	100	82.82	61.76
Percentage of Households having closed or open drainage connectivity for wastewater outlet	8.03	15.04	14.88	10.36	7.75	12.99	25.68
Percentage of Households living in Permanent structures	17.59	24.69	30.25	29.54	31.45	30.14	47.74
Weighted value of major component, $M_{bi} = [\sum^{n}_{i=1}$ index $s_b i] / n$	**0.18**	**0.49**	**0.43**	**0.53**	**0.51**	**0.59**	**0.61**

Source Census of India (2011a), p 152–183

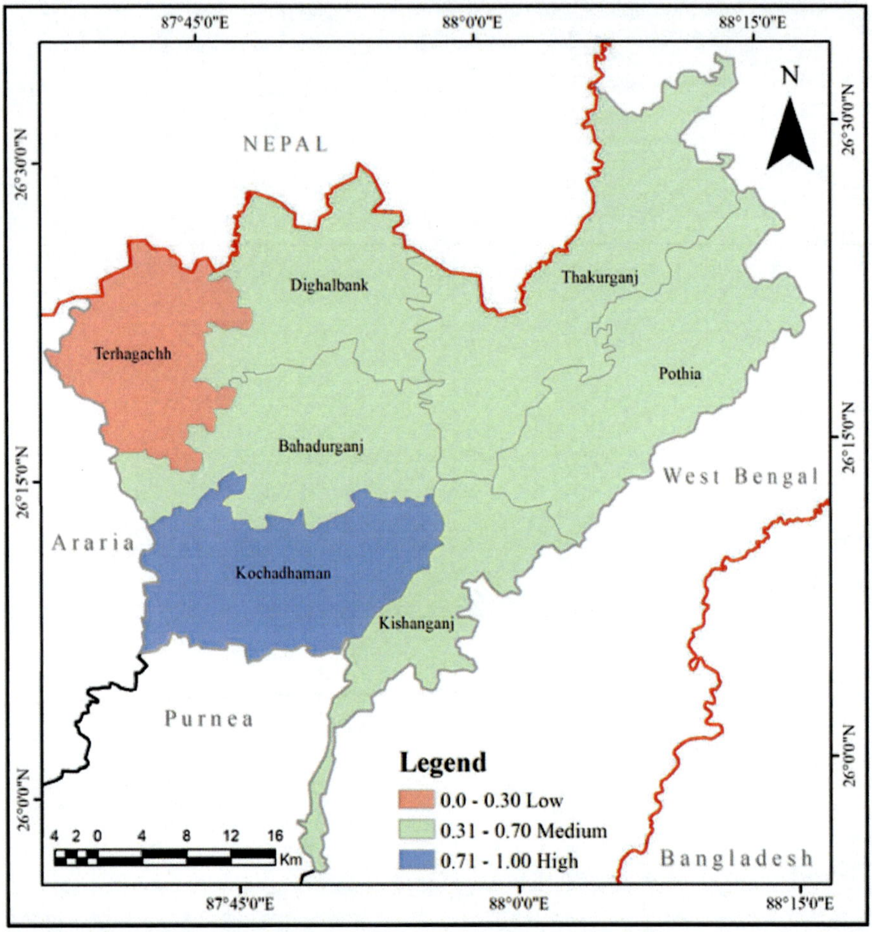

Fig. 4 Spatial variation of infrastructural development in CDB of Kishanganj district *Source* Prepared by authors

dependent upon agriculture-related practices for earning their livelihood. Out of 1.6 million population, around 0.5 million (31%) are total workers, and 1.1 million (69%) are non-workers. The total workers' population of the district is constituted of main workers and marginal workers i.e. 0.4 million and 0.1 million respectively. The Workforce Participation Rate (WFPR) in the district is 22% for main workers and 8% for marginal workers. The average percent age of main workers engaged in household industries and other workers is just 12.6% (Table 8). The highest share of marginalized workers engaged in household industries and other workers is 19% in the Bahadurganj block followed by 17% in both Pothia and Kochadhamin blocks. The maximum share of main female workers in household industries and others is 27% in the Pothia block. In the Terhagachh block, female working as marginal workers is the highest, with a share of about 91%. The weighted value of the infrastructure

Table 7 Block wise technological advancement, Kishanganj district

Technological advancement	Name of blocks						
	Terhagachh	Dighalbank	Thakurganj	Pothia	Bahadurganj	Kochadhamin	Kishanganj
Percentage of households with radio and TV	17.95	23	25.85	22.13	20.11	19.86	37.87
Percentage of household with computer	3.14	3.56	3.82	4.13	3.63	3.96	4.49
Percentage of household with mobile phone	42.78	33.52	31.08	28.6	42.43	41.51	37.79
Weighted value of major component, $M_{bi} = [\sum^{n}_{i=1} \text{index } s_{b}i] / n$	**0.21**	**0.20**	**0.20**	**0.18**	**0.22**	**0.22**	**0.27**

Source Census of India (2011b), p 190–191

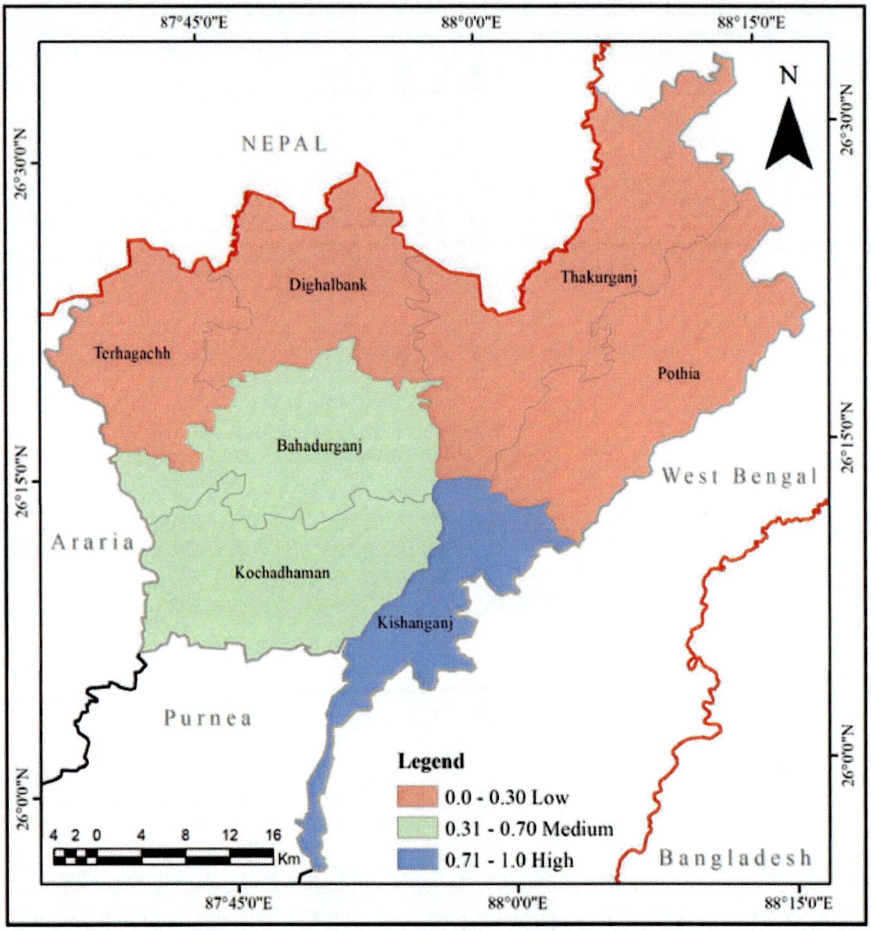

Fig. 5 Spatial variation of technological advancement in CDB of Kishanganj district *Source* Prepared by authors

development is highest for Bahadurganj CDB (0.78) and lowest for Kishanganj CDB (0.31) (Fig. 6).

4.6 Institutional Setups

The institutional setups are an important characteristic of any society and community to effectively respond to the disaster and plan to handle the disturbances. The communities believing and engaging in social networking with associations within

Table 8 Block wise data of livelihood diversity, Kishanganj district

Livelihood diversity	Name of blocks						
	Terhagachh	Dighalbank	Thakurganj	Pothia	Bahadurganj	Kochadhamin	Kishanganj
Percentage of main workers population in hcusehold and others	11	9	10	14	17	12	15
Percentage of main Female workers in household and others	20	25	23	27	24	21	15
Percentage of marginalized workers population in household and others	7	9	14	17	19	17	12
Percentage of female marginalized workers population in household and others	91	66	60	42	60	59	46
Weighted value of major component, $M_{bi} = [\sum^{n}_{i=1} \text{index } s_{b}i] / n$	**0.42**	**0.37**	**0.44**	**0.61**	**0.78**	**0.51**	**0.31**

Source Census of India (2011b), p 26–29

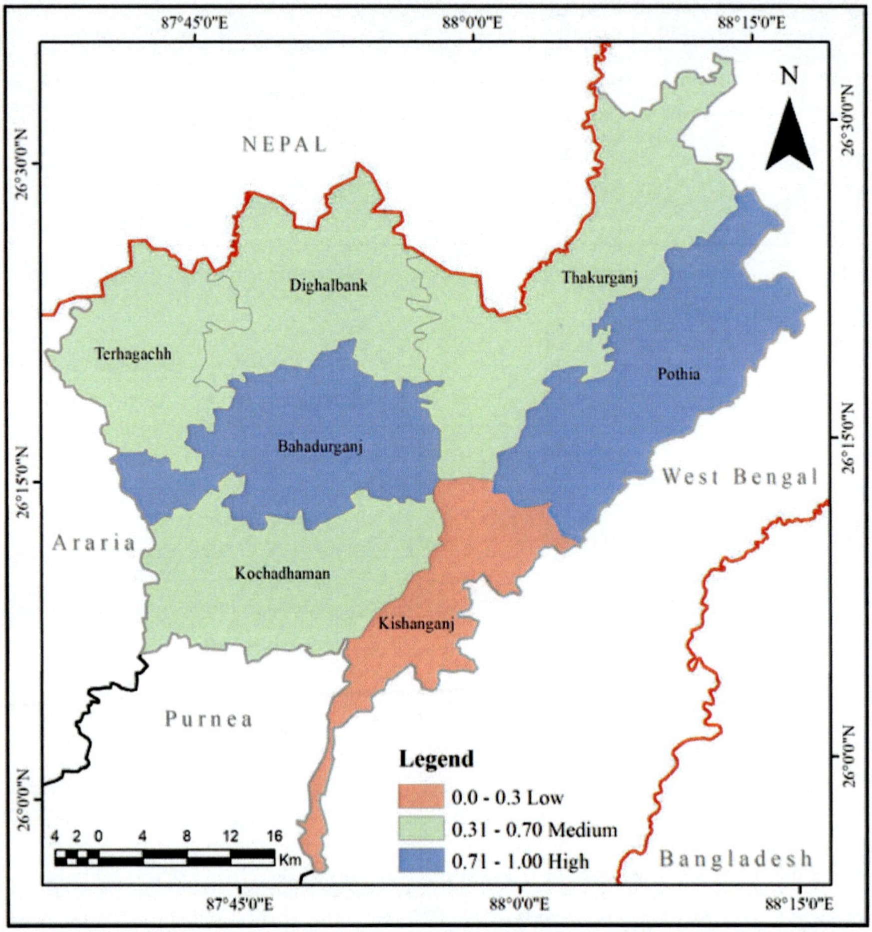

Fig. 6 Spatial variation of livelihood diversity in CDB of Kishanganj district *Source* Prepared by Authors

or outside the community tend to develop a higher capacity to re-organize. Self-help groups are small voluntary associations of people belonging to a similar socio-economic category (Devadasan and Rai 2019). Thus, such social institutions play an important role in creating a resilient region. Kishanganj district has a total of 177 self-help groups catering to just 0.7% of disadvantaged sections of people. The other institutions, such as Integrated Child Development Services (ICDS)[2] and *Anganwadi*[3] plays an important role in providing services to children, pregnant women, lactating mother, etc. Kishanganj district has four ICDS centres, one each located in Terhagachh, Dighalbank, Pothia, and Kochadhamin blocks whereas Kishanganj, Bahadurganj, and Thakurganj blocks do not have such facilities. The government of India has proposed one *Anganwadi* centre for every 400–800 rural population.[4] In the Kishanganj district, there are just 722 centres for the rural population of around 1.1 million, which are approximately 50% of the required number of *Anganwadi* center. Pothia block has the highest number of 91 SHG and 141 *Anganwadi* center. The weighted value of the infrastructural setups is highest for Pothia CDB (1.0) and lowest for Kishanganj CDB (0.06) (Fig. 7) (Table 9).

The Adaptive Capacity Index (ACI) indicators values for each block with minimum and maximum values are computed using Eq. 1 (Sullivan 2002; Hahn et al. 2009; and Bahinipati 2014). The average weighted values for each aspect are calculated for each block using Eq. 2 followed by the calculation of normalized values as shown for the indicators of livelihood diversity. The normalized index value of each block for all aspects is calculated. The aggregated adaptive capacity index values of all the six aspects are represented using composite index approach (Table 10). The spatial variation of all components in the CD blocks of Kishanganj district are shown in Fig. 2. The value for the institutional setups is maximum for Pothia block (1.0) and on the other end of the spectrum, Kishanganj block has the minimum value (0.06). Pothia block also has the maximum value for educational commitment (0.71) where as Terhagachh block has the minimum value for educational commitment indicator (0.35). Access to technology and information is a crucial indicator for calculating overall adaptive capacity as it helps in receiving warning messages for the forthcoming flood and to know government schemes and policies. The government also makes people aware of their schemes and policies through the usage of ICT. Surprisingly, Pothia block has the above-average value of all the indicators except access to technology, which has the minimum value (0.31). Pothia (0.66) and Kochadhamin (0.63) tops the chart of adaptive capacity index, making it the least vulnerable blocks of the district. On the other hand, Terhagachh block is most vulnerable to flood as the adaptive capacity index is minimum i.e. 0.39. This is also reflected in the spider diagram (Fig. 8). The spatial distribution of the Adaptive Capacity Index is shown in Fig. 9.

---

[2] Integrated Child Development Services (ICDS) scheme is initiated by the Government of India to provide package of services such as immunization, health check-up, referral services, pre-school non-formal education, nutrition, health and education to the children below 6 year's age, pregnant and lactating mothers in villages.

[3] Anganwadi is a rural child care centre in India which provides pre-school non-formal education and food to the children.

[4] Mrs. Maneka Gandhi, Minister of State for Women and Child Development gave the information as a response to a starred question in the Lok Sabha on 3 August 2018.

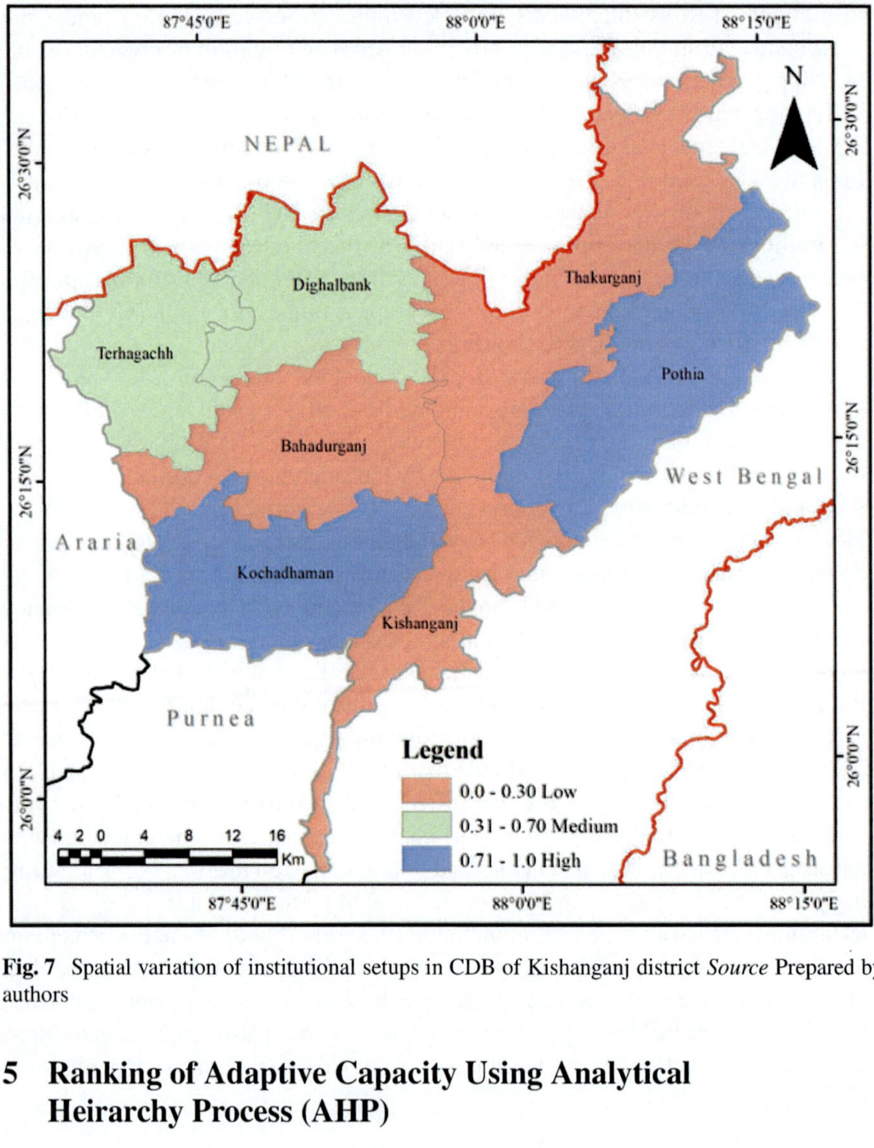

Fig. 7 Spatial variation of institutional setups in CDB of Kishanganj district *Source* Prepared by authors

5 Ranking of Adaptive Capacity Using Analytical Heirarchy Process (AHP)

Decision-making is a complex problem. Hence, the need and purpose of the decision and its criteria and sub-criteria have to be taken into consideration. The stakeholders and affected groups need to be evaluated to decide on the best alternative (Saaty 2008). The AHP is a multi-objective, multi-criteria decision-making process that uses a pairwise comparison process to frame a scale of choices among the set of alternatives (Chakraborty and Mukhopadhyay 2019). It utilizes a basic scale of absolute numbers to represent individual preferences, also known as Saaty's scale of preferences. The pairwise matrix is converted into a normalized pair-wise matrix by dividing all the elements of the column by the sum of the column (Table 11). The next step in this

Table 9 Block wise institutional setup in Kishanganj district

Institutional setups	Name of blocks						
	Terhagachh	Dighalbank	Thakurganj	Pothia	Bahadurganj	Kochadhamin	Kishanganj
Number of Self Help Groups	23	5	11	91	8	18	21
Number of Nutritional Centre -ICDS	1	1	0	1	0	1	0
Number of Anganwadi	75	80	117	141	102	142	65
Weighted value of major component, $M_{bi} = [\sum^{n}_{i=1} \text{index } s_{b}i] / n$	0.45	0.40	0.25	1.00	0.17	0.72	0.06

Source Census of India (2011a), p 151–183

Table 10 Aggregated weighted values of indicators

Aspects	Name of blocks						
	Terhagachh	Dighalbank	Thakurganj	Pothia	Bahadurganj	Kochadhamin	Kishanganj
Educational Commitment	0.35	0.42	0.51	0.71	0.51	0.47	0.43
Economic Autonomy	0.64	0.66	0.75	0.84	0.88	0.88	0.40
Infrastructure Development	0.18	0.49	0.43	0.53	0.51	0.59	0.61
Access to Technological Advancement	0.33	0.30	0.36	0.31	0.48	0.54	0.88
Livelihood diversity	0.42	0.37	0.44	0.61	0.78	0.51	0.31
Institutional Setups	0.45	0.40	0.25	1.00	0.17	0.72	0.06
Composite Adaptive Index, $ACI_b = [\sum_{i=1}^{6} w_{Mi} * M_{bi}]$	**0.39**	**0.46**	**0.48**	**0.66**	**0.58**	**0.63**	**0.46**

Source Computed by Authors based on data Census of India (2011a), p 57,73–76;151–183, Census of India (2011b), p 26–29; 90–191

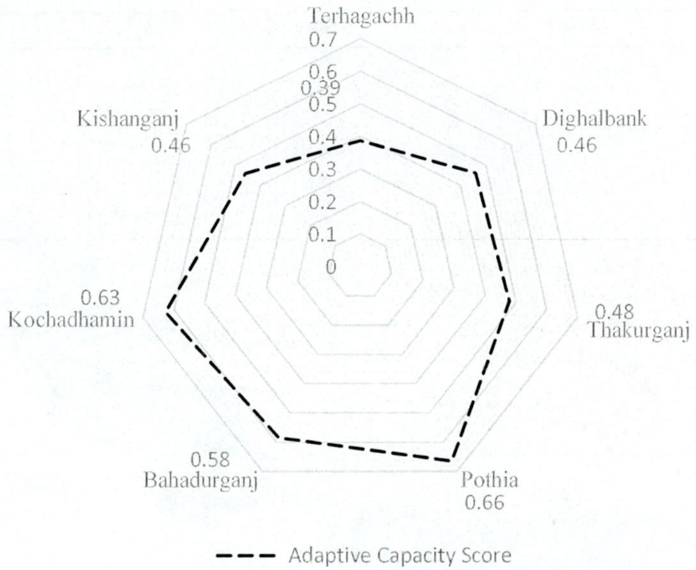

Fig. 8 Spider diagram of components of adaptive capacity index in the CDB of Kishanganj district
Source Prepared by authors

process is to develop a pairwise comparison matrix that gives relative importance to various attributes of the goal. It is followed by checking the consistency of the table and calculation of the largest eigenvalue of the comparison matrix. The consistency is calculated to check whether the intended values are in order and the weighted sum value and criteria index (CI) is calculated using Eq. 4. The foremost task in AHP is to develop a hierarchy structure in which the goal is kept at the top, the criteria on level two, and the alternatives are on level 3. This section aims to develop a ranking system for measuring adaptive capacities. The six aspects of adaptive capacities are chosen criterion and areas of study i.e. 6 blocks of Kishanganj district are the alternatives. Each alternative has different values for different criteria.

The average of the criteria ratio gives the value of λmax. The consistency ratio (CR) verifies the consistency of the judgment. The judgment or the preference is said to be correct only when CR = > 0.10 or 10%. The CR is calculated by dividing CI by a random index (RI) whose value is given by the number of criteria. The number of criteria is 6, so the RI is 1.24 and the value of CR is 0.029 (2.9%). The proportion of inconsistency (CR) is less than 0.10, so the metrics are reasonably consistent and process of decision-making using AHP is finalized and priority value is given based on criteria weights and accordingly, ranking is carried out (Table 12).

The ranking of adaptive capacities through AHP gives the indicator of 'institutional setups' the first rank followed by the 'educational commitment'. Both these indicators prove to be an important characteristic of any society or community to

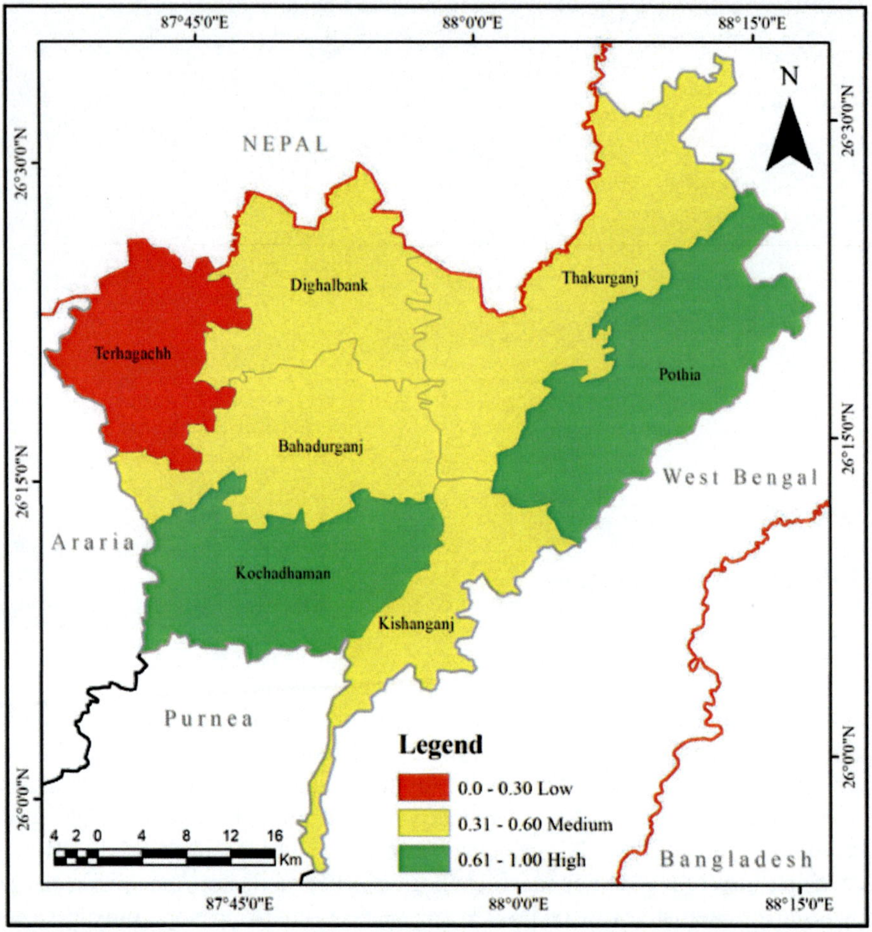

Fig. 9 Spatial variation of components of adaptive capacity index in the CDB of Kishanganj district. *Note* Administrative boundary adopted from Census of India (2011a). *Source* Prepared by authors

effectively respond during disasters and plan for associated disturbances. The presence of different social groups and education facilities makes a region more resilient and less vulnerable.

6 Conclusions and Recommendations

The study calculated block-wise relative adaptive capacity indices to assign vulnerability rank to each block according to its adaptation measures concerning the flood. An integrated approach was adopted in which adaptive capacity has been taken as the function of vulnerability in which adaptive capacity is inversely proportional

Table 11 Normalized Pair-Wise matrix

Criteria	Educational commitment	Access to technological advancement	Economic autonomy	Infrastructure development	Livelihood diversity	Institutional setups
Educational commitment	0.20	0.21	0.20	0.24	0.24	0.67
Access to technological advancement	0.10	0.11	0.10	0.18	0.24	0.07
Economic autonomy	0.20	0.21	0.20	0.18	0.24	0.07
Infrastructure development	0.05	0.04	0.07	0.06	0.05	0.03
Livelihood diversity	0.04	0.11	0.04	0.06	0.05	0.03
Institutional setups	0.40	0.32	0.40	0.29	0.19	0.13

Source Computed by Authors based on data Census of India (2011a), p 57,73–76;151–183, Census of India (2011b), p 26–29; 90–191

Table 12 Weights of the criterion based on the pairwise matrix

Criterion	Priority value (in %)	Rank
Educational commitment	22.2	II
Access to technological advancement	10.9	IV
Economic autonomy	21.3	III
Infrastructure development	5.7	VI
Livelihood diversity	6.5	V
Institutional setups	33.5	I

Source Computed by Authors based on Census of India (2011a) and b

to vulnerability. Several indicators were assigned to calculate the determinants of adaptive capacity. Since the measurement unit for each indicator varies, a normalized procedure was used to convert them into unique values for ease of comparison using the composite index approach.

In the adaptive capacity calculation for the Kishanganj district, Pothia and Kochadhamin blocks were found to have the highest adaptive capacity value making them least vulnerable to floods. Terhagachh, Dighalbank, and Kishanganj blocks have the lowest value of adaptive capacity making them the most vulnerable blocks of the district. The indigenous approach to flood resilience measures adopted by the community also plays an important role in developing adaptive capacity for the flood-affected region. These approaches include flood prediction techniques for livestock and human life-saving strategies. The disaster risk management (DRM) policies and programs are primarily directed toward activities reducing the mortality rate due to floods. The indigenous measures to counter floods need to be acknowledged by the government authorities. The information about early warning systems is disseminated through the use of information technologies like smartphones, radio, tv, etc. Access to these information technologies and the ability to use them plays an important role in receiving the early warning at the right time. The abyss of information and coordination between authority and communities should be minimized through two-way communication from community to authority and vice versa. The effective tool for the reinforcement and application of indigenous knowledge is educational commitment. It will reduce the gap between this indigenous knowledge and modern techniques through institutional setups at the regional and local levels.

The ranking of adaptive measures through AHP concludes that the institutional setup and educational commitment play a crucial role in adapting the external effects to the region. The Government of Bihar adopts the Bihar Disaster Risk Reduction Framework (BDRRF) to achieve the vision of a Disaster Resilient Bihar with ten priority commitments for the period of 2015–2030. The framework should include the institutional assessments and setups in their priority objectives to strengthen the social capital of the affected area. The Panchayati Raj Institutions (PRIs) and Urban Local Bodies (ULBs) need to be empowered with financial and regulatory roles to specifically reinforce the adaptive capacity aspects at lower administrative levels for Disaster Risk Reduction. The constant threat to the agricultural sector due to disaster needs to be taken care of by providing scientific and technological advancements to the farmers. The essential services and critical infrastructure needs to be resilient for the functionality and continuity of the social system, in case of disruptions. The adequate investment in capacity building, knowledge management, and public awareness will strengthen the risk management approach of the government. Flood risk management is a crucial aspect of the disaster resilience framework and the adaptive and transformational capacity of the region should be an integral part of the framework. The DRM program of the district should emphasize developing social institutions and educational institutions to strengthen the coping mechanism of the area. The government shall incorporate the development-based activities in the disaster management policies in addition to disaster-specific risk reduction measures.

Further, development in the research of composite indices can be extended through high-resolution data collection by government agencies. The data availability at a lower scale i.e., municipality and village level needs to be strengthened. The adaptive capacity will improve by the integration of data collection and tracking adaptive capacity from the lower administrative scale to the regional level. In addition, the number of indicators to assess adaptive capacity used may influence the final scores. The application of inductive methods with fewer indicators can be used to determine the adaptive capacity index but once the knowledge regarding the determinants of adaptive capacity is consolidated then the deductive method using multiple indicators will be more suitable for testing the adaptive capacity theory. The indigenous adaptation measures should also include the quantification of the adaptive capacity of a region.

References

Adger WN et al. (2004) New indicators of vulnerability and adaptive capacity. Norway: Tyndall Centre for Climate Change Research. (Technical Report 7)

Adger WN et al (2005) Social-Ecological resilience to coastal disasters. Science 309(5737):1036–1039. https://doi.org/10.4324/9781351201117-18

Bahinipati CS (2014) Assessment of vulnerability to cyclones and floods in Odisha, India: A district-level Analysis. Curr Sci 107(12):1997–2007

Brooks N, Adger WN, Kelly PM (2005) The Determinants of vulnerability and adaptive capacity at the national level and the implications for adaptation. Glob Environ Chang 15(2):151–163. https://doi.org/10.1016/j.gloenvcha.2004.12.006

BSDMA (2018) Bihar Floods 2016. Patna: Bihar State Disaster Management Authority (BSDMA). Government of Bihar.

Cannon T, Schipper L (2014) World disasters report 2014: Focus on culture and risk. Swirzerland: Int Fed Red Cross Red Crescent Soc. https://doi.org/10.1111/j.0361-3666.2005.00327.x.

CARE (2019) Climate vulnerability and capacity analysis (CVCA) Handbook Version 2.0. Switzerland: Climate Change and Resilience Platform (CCRP)

CEDRA (2009) Climate change and Environmental Degradation Risk and Adaptation Assessment: An environmental tool for agencies in developing countries. United Kingdom: Tearfund.

Census of India (2011a) District census handbook. Part A. Kishanganj. Village and Town Directory. Patna: Directorate of Census Operations. Ministry of Home Affairs. Government of India. (Series 11, Part XII A)

Census of India (2011b) District census handbook. Part B. Kishanganj. Primary Census Abstract (PCA). Patna. Directorate of Census Operations. Ministry of Home Affairs. Government of India. (Series 11, Part XII B)

Chakraborty S, Mukhopadhyay S (2019) Assessing Flood Risk Using Analytical Hierarchy Process (AHP) and Geographical Information System (GIS): Application In Coochbehar District of West Bengal, India. Nat Hazards 99(1):247–274. https://doi.org/10.1007/s11069-019-03737-7

Christian Aid (2010) Integrating climate change adaptation into secure livlihoods, Toolkit 1: Framework and Approach. Scotland: Christian Aid.

CWC (2019) State wise Flood Damage Statistics. New Delhi: Central Water Commssion (CWC), Department of Water Resources, River Development & Ganga Rejuvenation, Ministry of Jal Shakti, Government of India

Daszkiewicz C (2020) Ecosystem-based Adaptation Annotation. Available at: https://www.weadapt.org/knowledge-base/ecosystem-based-adaptation/how-to-make-a-floating-vegetable-garden-step-by-step [Accessed: 30 April 2020]

Devadasan PM, Rai R (2019) Women Empowerment through Self Help Groups-Interventions towards Socio-Economic Welfare. International Journal Innovative Practices in IT, Management, Education and Social Sciences. 06(01):424–441

Di Baldassarre G et al. (2016) Drought and Flood in the Anthropocene: Modelling Feedback Mechanisms, Earth System Dynamics Discussions, pp 1–24. https://doi.org/10.5194/esd-2016-65

Disaster Management Department (2017) Damage Flood Report- Cumulative form IX, 2017. Patna: Disaster Management Department. Bihar State Disaster Management Authority (BSDMA). Government of Bihar. Available at: https://state.bihar.gov.in/disastermgmt/SectionInformation.html?editForm&rowId=441 [Accessed 22 Jul. 2021]

Ellis F, Allison EH (2001) The Livelihoods Approach and Management of Small-scale Fisheries. Mar Policy 25(5):377–388

EM-DAT (2021) EM-DAT,CRED / UCLouvain, Brussels, Belgium. Available at: www.emdat.be [Accessed: 9 February 2021]

Few R et al (2017) Transformation, Adaptation and development: Relating concepts to practice. Palgrave communications. 3(1):17092–17099. https://doi.org/10.1057/palcomms.2017.92

Guillén Bolaños T, Máñez Costa M, Nehren U (2016) Development of a Prioritization Tool for Climate Change Adaptation Measures in the Forestry Sector—A Nicaraguan Case Study. Germany: Climate Service Center Germany. (Report 28). https://doi.org/10.1007/978-3-319-99462-8_10.

Hahn MB, Riederer AM, Foster SO (2009) The Livelihood Vulnerability Index: A Pragmatic Approach to Assessing Risks from Climate Variability and Change-A Case study in Mozambique. Glob Environ Chang 19(1):74–88. https://doi.org/10.1016/j.gloenvcha.2008.11.002

Howden SM et al (2007) Adapting agriculture to climate change. Proc Natl Acad Sci 104(50):19691–19696. https://doi.org/10.1073/pnas.0701890104

IPCC (2014) Climate Change 2014 : Synthesis Report. Contribution of Working Groups I, II and III to the Fifth Assessment Report of the Intergovernmental Panel on Climate Change. Edited by R. Pachauri and L.Meyer. Switzerland: The Intergovernmental Panel on Climate Change (IPCC). https://doi.org/10.1016/S0022-0248(00)00575-3.

Kenney MA et al. (2011) Climate change impacts and responses: Societal indicators for the national climate assessment. Washington, D.C: National Climate Assessment (NCA). U.S. Global Change research Programme. (NCA Report Series- Volume 5 C).

Lwasa S (2010) Adapting urban areas in Africa to climate change: The case of Kampala. Current Opinion in Environmental Sustainability. 2(3):166–171. https://doi.org/10.1016/j.cosust.2010.06.009

Maldonado J, Moreno-Sánchez R, del P (2014) Estimating the adaptive capacity of local communities at marine protected areas in latin America: A practical approach. Ecol Soc. 19 (1). https://doi.org/10.5751/ES-05962-190116.

Marshall N et al. (2010) A framework for social adaptation to climate change sustaining tropical coastal communitites and industries. IUCN Climate Change and Coral Reefs Working Group. Gland, Switzerland: International Union for Conservation of Nature (IUCN).

Marzi S, Mysiak J, Santato S (2018) Comparing Adaptive Capacity Index Across Scales: The Case of Italy. J Environ Manage 223(1):1023–1036. https://doi.org/10.1016/j.jenvman.2018.06.060

McLaughlin S, Andrew J, Cooper G (2010) A Multi-Scale Coastal Vulnerability Index: A Tool for Coastal Managers. Environ Hazards 9(3):233–248. https://doi.org/10.3763/ehaz.2010.0052

Metzger MJ, Leemans R, Schröter D (2005) A Multidisciplinary Multi-Scale framework for assessing vulnerabilities to global change. Int J Appl Earth Obs Geoinf 7(4):253–267. https://doi.org/10.1016/j.jag.2005.06.011

Mishra DK (1997) The Bihar Flood Story. Econ Pol Wkly 32(35):2206–2217. https://doi.org/10.2307/4405794

Mishra JN, Kumar R, Kendra YC (2008) Adaptive capacities of community to cope up with flood situations: Flood and livelihood adaptive capacity based compilation. Gorakhpur: Gorakhpur Environmental Action Group and OXFAM Novib

Mishra K, Sinha R (2020) Flood Risk Assessment in the Kosi Megafan using Multi-Criteria Decision Analysis: A Hydro-Geomorphic Approach. Geomorphology. 350(1). https://doi.org/10.1016/j.geomorph.2019.106861.

Morgan CL (2011) Vulnerability Assessments: A Review of Approaches. Switzerland: International Union for Conservation of Nature (IUCN)

Mukuna TE (2015) Vulnerability analysis of the Gender-Differentiated Impact of Flooding in Budalangi Flood Plains, Kenya. J Emerg Trends Educ Res Policy Stud (JETERAPS). 6(2):201–216

Nelson DR, Adger WN, Brown K (2007) Adaptation to Environmental Change: Contributions of a Resilience Framework. Annu Rev Environ Resour 32(1):395–419. https://doi.org/10.1146/annurev.energy.32.051807.090348

NIEPA (2021) School report card, National Institute of Educational Planning and Administration (NIEPA). Available at: http://schoolreportcards.in/SRC-New/SchoolDirectory/Directory.aspx (Accessed: 20 May 2021).

NITI Ayog (2021) SDG India Index & Dashboard 2020–21: Partnership in the Decade of Action. New Delhi: United Nations and NITI Ayog.Available at: https://niti.gov.in/writereaddata/files/SDG_3.0_Final_04.03.2021_Web_Spreads.pdf.

Nguyen T, Nguyen T, Man H (2016) Assessing Adaptive Capacity to Flood in the Downstream Communities of the Lam River. Journal of Geography, Environment and Earth Science International. 5(3):1–13. https://doi.org/10.9734/jgeesi/2016/23522

NRSC (2020). Flood Hazard Atlas of Bihar-A Geospatial Approach. Hyderabdad: National Remote Sensing Centre, Indian Space Research Organization, Dept. of Space, Govt. of India, pp. 53–55.

Reid, H (2016) Ecosystem and Community based Adaptation: Learning from Community-based Natural Resource Management. Climate Dev 8(1):4–9. https://doi.org/10.1080/17565529.2015.1034233

Saaty TL (2008) Decision Making with the Analytic Hierarchy Process. International Journal Services Sciences. 1(1):215–229. https://doi.org/10.1504/ijssci.2008.017590

Saja AMA, Sahid MSL, Sutharshanan M (2020) Implementing Sendai Framework priorities through risk-sensitive development planning – A case study from Sri Lanka. Progress in Disaster Science 5:100051. https://doi.org/10.1016/j.pdisas.2019.100051

Smit B et al. (2001) Adaptation to climate change in the context of sustainable development and equity, Climate Change 2001: Impacts, Adaptation, and Vulnerability. Sustainable Development. New York: Cambridge University Press

Sullivan C (2002) Calculating a Water Poverty Index. World Dev 30(7):1195–1210

Swanson D et al. (2007) Indicators of Adaptive Capacity to Climate Change for Agriculture in the Prairie Region of Canada: An Analysis based on Statistics Canada's Census of Agriculture. Winnipeg: International Institute for Sustainable Development (IISD). https://doi.org/10.1017/CBO9781107415324.004.

Thanvisitthpon N et al (2020) Assessment of Flood Adaptive Capacity of Urban Areas in Thailand. Environ Impact Assess Rev 81(March):106359–106363. https://doi.org/10.1016/j.eiar.2019.106363

The World Bank (2017) Sahel Adaptive Social Protection Program: Addressing the Challenges of Climate Change and Disaster Risk for the Poor and Vulnerable. International Bank for Reconstruction and Development / The World Bank, Washington, D.C.

UNISDR (2009) *UNISDR Terminology on Disaster Risk Reduction.* Geneva: Uniter Nation International Strategy for Disaster Reduction (UNISDR). doi: https://doi.org/10.4324/9781351138444-36.

UNISDR (2015) Sendai framework for disaster risk reduction 2015–30. United Nation. Geneva: Uniter Nation International Strategy for Disaster Reduction (UNISDR).

Thathsarani US, Gunaratne LHP (2018) Constructing and index to measure the adaptive capacity to climate change in Sri Lanka. Procedia Eng 212:278–285. https://doi.org/10.1016/j.proeng.2018.01.036

The World Bank (2010) Mainstreaming adaptation to climate change in agriculture and natural resources management projects. Philippines: The World Bank

WRD (2020) Flood control and drainage: Introduction. Patna: Water Resources Department (WRD) Government of Bihar. Available at: http://wrd.bih.nic.in/ [Accessed: 12 May 2020].

Yadav V, Ibrar Z (2022) Relationship of water stress and flood damage for sustainable development. Water resour manage 36(4):1323–1338. https://doi.org/10.1007/s11269-022-03083-7

Yates, J. S. (2010) A Review of the PVA Methodology in Context :A Comparision with other Frameworks and Experience from the field. Warwickshire: Global Causeway. (Research Report 2010–1)

Yoon DK (2012) Assessment of Social Vulnerability to Natural Disasters: A Comparative Study. Nat Hazards 63(2):823–843. https://doi.org/10.1007/s11069-012-0189-2

Zacarias DA (2019) Understanding Community Vulnerability to Climate Change and Variability at a Coastal Municipality in Southern Mozambique. International Journal of Climate Change Strategies and Management. 11(1):154–176. https://doi.org/10.1108/IJCCSM-07-2017-0145

Perceptions and Representations of Disasters and Climate Change in the Anthropocene

Cruise Tourism, Risk Perception and Public Narratives in Syracuse, Italy

Gaetano Sabato

Abstract In the last few years, the interest in risk-related issues has grown in all social sciences, confirming that individual and collective behaviors and intervention policies are connected to certain perceptions of disasters. Communities are often the starting point as well as the recipients of such studies. In the tourism industry, the issue of safety and risk plays a fundamental role in the dynamics of hospitality and opens up many inquiries. Indeed, one of the risks connected to tourism is the environmental one: the sustainability of tourist flows in destinations or, even better, in host communities is often what determines the success or failure of certain types of tourism. The aim of this chapter is to investigate the ways in which a community, as a destination, perceives specific forms of risk linked to a type of tourism, here specifically the cruise tourism. Therefore, the case of Syracuse, Italy is taken into account, where (precisely near the historic center of Ortigia) in 2020, two cruise ships remained idle for four months, for a long technical stopover. The presence of the ships caused a controversy within the community, among those who considered the ships a potential danger for the environmental pollution and disfigurement of the landscape; and those who, on the other hand, saw opportunities for economic development even in the long term. This study, from a cultural geographical perspective, considered the public narratives produced by some local online newspapers, by analyzing a selection of significant articles about this controversy. The primary goal is to observe the ways in which those narratives are organized, as their "forms" show the (re)production of complex cultural dynamics.

Keywords Cruise tourism · Geography · Conflicts · Tourism · Syracuse

---

G. Sabato (✉)
University of Palermo, Palermo, Italy
e-mail: gaetano.sabato@unipa.it

B. Dahiya et al. (eds.), *Disaster Resilience and Human Settlements*,
Advances in 21st Century Human Settlements,
https://doi.org/10.1007/978-981-99-2248-2_6

1 Introduction

In recent decades, the interest in risk-related issues has grown in all social sciences, making it clear that individual and collective behaviours and intervention policies are connected to certain perceptions of disasters (see at least: Kasperson et al. 1988; Renn et al. 1992; Pidgeon, Kasperson, Slovic 2003). Communities are often the starting point of such studies, besides being the recipients. This is not surprising when one considers on the one hand the effects of any disaster hitting them and, on the other hand, that calamities can be weaker and attenuated only by raising a stronger collective awareness. The cultural and social construction of risk was well highlighted by Douglas (1986, 1992) and Wildavsky (1982) who confirmed that social sciences offered one of the most suitable perspectives for analyzing the symbolic aspects of risk perception. As a matter of fact, from a geographical point of view, the possibility of analyzing the perception of risk in correlation with the spatial dimension is an interesting object of study (see Mercatanti, Sabato 2019; Gómez Cantero et al. 2020; Mercatanti, Montes 2021; Mercatanti, Sabato 2021; De Pascale 2022). Specifically, in the tourism sector, the issue of safety and risk plays a fundamental role in the dynamics of hospitality and opens up many inquiries. Tourism supply and demand, in fact, are organized around the possibility of making and perceiving a trip safe and thus, security is ultimately one of the main aspects to consider when travelling. Due to this consideration, the risk perception related to tourism is a topic dealt with from many points of view in the social sciences (see at least: Fuchs and Reichel 2004, 2006; Lepp and Gibson 2003; Rittichainuwat 2013; Mercatanti 2015; Tarlow 2014). The environmental issue, as represented in recent years, is one of the most typical risks connected to tourism: sustainability of tourist flows, in a certain destination or, even better, in a host community, can often distinguish or determine the quick success or failure of some types of tourism. Certainly, the Covid-19 pandemic has created new problems to global mobility, reverberating huge losses on the tourism industry and even changing the way of traveling and understanding destinations. In terms of security, of course, not only the economic consequences of this new risk must be taken into account, but also a wide-ranging symbolic "construction" strictly linked to the perception of risk and validated by various forms of narration. Here I will take into account a particular form of tourism, the cruising tourism, to observe how the perception of risk can be connected to the environmental issue. In more detail, this study focuses on the case of Syracuse, Italy, where public narratives of cruise tourism were told about the presence of two ships that, due to the Covid-19 pandemic, had stopped for technical reasons in the port and stayed for a few months. For this purpose, I will analyze from a cultural geographical approach some significant texts in articles found in local online newspapers. Indeed, between August and September 2020 the presence of the two cruise ships in the port brought an animated controversy that divided the community of Syracuse into two parts: those inhabitants who considered the presence of the two ships as positive, mainly for the economic return, as the companies paid parking; and those who were more concerned with the negative impact that the two ships were causing on the environment, as the prolonged running of the engines, obviously necessary to keep on-board systems active, would have certainly disfigured the landscape and the waterfall.

2 Cruise Tourism in the Mediterranean Sea and in Syracuse, Italy

For the first time in the last decade, the growth of the cruise market has globally come to a halt due to the Covid-19 pandemic. The optimistic predictions made by market analysts for the year 2020, based on the excellent results of cruise tourism in 2019, were contradicted by an event that was neither probable nor plausible. Indeed, any realistic analysis of the situation before the pandemic could not have taken into account the blocking of all tourist activities all over the world in order to contain the virus. The lockdown imposed a no entry to the routes (prohibited by law in many countries or implemented as a precaution by cruise companies), but obviously, at the same time, the freezing of national and international travels and various tourist services would have caused anyway the drastic reduction of passengers unable to reach most destinations. These two factors caused an interruption of cruising activities for several months, which break which consequently meant a very high costs. In 2019, the total number of passengers predicted for 2020 was around 32 million, instead the total number reached was 5.7 million passengers, that is 80.6% less than the expected figure (CLIA GMR 2020: 2). According to the estimates of the latest report by CLIA (Cruise Lines International Association)—the association that boasts the largest number of registered cruise companies in the world[1]—the global impact of the Covid-19 suspension occurred between mid-March and September 2020 has resulted in a loss of more than 77 billion dollars in the global economic activity, 518,000 jobs and 23 billion dollars in wages (CLIA SCIO 2021: 26). Relevant numbers that reveal how much the cruise sector represents in terms of economic impact, despite the total number of cruise passengers in 2019 (the most consistent recorded) which was around 29.7 million, namely 2% of the total international tourist arrivals (1,460 million) in the same year (CLIA SCIO 2021; CLIA GMR 2020; UNWTO 2020).

Traditionally, the Mediterranean Sea is the second cruise market targeted after the North American and Caribbean: actually, in more recent years, the Asian market has competed with the rest of the European area (the North Sea and the Baltic Sea) for the third place (Risposte Turismo 2020b; CLIA 2020). With reference to Italy, the country was at the eighth place until 2019 worldwide, in terms of passenger volume (CLIA GMR 2020) and this is enough to consider that the number passengers in transit in the Italian ports in 2019 reached 12.2 million for the first time (Risposte Turismo 2020b). This means that Italy alone counts nearly 39% of the 31.5 million passengers travelling in the Mediterranean Sea. A result explained by a combination

[1] According to the CLIA (2021) the association represents the 95% of global cruise capacity with 57 registered companies.

of reasons: attractive destinations, strategic choices of the companies, quality services and, in general, an important inducement for the hosts communities.

Today, cruise tourism represents an important resource for Sicilian tourism. The island, which is one of the 20 administrative regions in Italy, for some years has reached important positions. In the industry indeed, in 2019 it was ranked as first in the country for the number of cruise ship ports. Moreover, in the same year, after Liguria, Lazio, Veneto and Campania it was ranked as fifth for enlivened passengers, that amounted to more than 1.23 million (in 2018 they were over 1.17 million) (see Risposte Turismo 2019, 2020a; Sabato 2018). With reference to the number of cruise ship calls at ports into the Italian context, Sicily was second for several years (after Lazio) with its 11 cruise ports (listed in order of importance) covering the three sides of the island: Palermo, Messina, Catania, Giardini Naxos, Trapani, Lipari, Siracusa, Porto Empedocle, Milazzo, Licata and Pozzallo (Risposte Turismo 2019, 2020b). Taking into consideration the data related to the first three ports in the pre-pandemic period, Palermo, Messina and Catania, a more marked seasonal character can be observed in the months between April and October, with considerable increased numbers in summer, although Palermo is affected by a (reduced) cruise traffic even in the winter months (Risposte Turismo 2020a, b). In 2017 the ports of Palermo and Messina were leading the Sicilian cruise tourism, hosting over 30% of all the companies that have itineraries in Italy, whereas Syracuse hosted about 12% (Risposte Turismo 2017).

Syracuse is located on the southeastern coast of the island, and it is the fourth Sicilian city by population after Palermo, Catania and Messina. The ancient urban site was founded by the Greeks in the eighth century BC, soon becoming one of the most influential cities in the Mediterranean Sea, maintaining a political and administrative centrality until the Middle Ages. In 2005, UNESCO added Syracuse and the Necropolis of Pantalica to the World Heritage List its site and the Necropolis of Pantalica (39 km away from Syracuse). This increased the tourist attractiveness of the city. Incoming tourism in Syracuse grew especially in the last twenty years and it is, above all, a cultural tourism, although not exclusively (the beaches near the city are renowned). Most of its heritage is found on the little island of Ortigia, the oldest urban nucleus and, furthermore, on the archeological park, including the 'Paolo Orsi Museum' considered one of the most important museums in Europe. In addition, the classical performances annually hosted into the ancient Greek Theater, between May and June are important tourist attractions.

As a cruise port, Syracuse mostly developed in the last years. Before 2017, the cruise traffic was hampered by the lack of adequate infrastructures to accommodate ships of standard tonnage: in that year a suitable quay was completed and a second one was realized in 2019 allowing the boarding and landing of passengers without using tenders. Indeed, between 2016 and 2019 the number of passengers welcomed grew by over 33% (Risposte Turismo 2017, 2020a). However, Syracuse ranks mostly as port of call and only sporadically as a home port. The position of the port is strategic for passengers as it allows them to reach the center and many cultural tourist sites quickly: it is located on the border with the island district of Ortigia. Obviously, the contraction in cruise tourism due to the covid-19 pandemic has also had an impact on Syracuse: in terms of handling fleeting, in the first nine months of 2020 the port lost 84% of passengers (−8,600) compared to the same period in 2019.

3 The "Ships of Discord": The Public Narratives of Newspapers

The Italian government, during the first lockdown from March to May 2020, suspended all the non-essential transport, allowing the cruise activities to continue in national ports from mid-August 2020 on. As previously explained in the introduction, between August and September 2020 two cruise ships, the Spirit and the Dawn of the Norwegian Cruise Line, with only the crew on board (about 300 people) to comply with the anti-pandemic rules, were parked in the port for a few months long technical stopover. Indeed, during their stay in Syracuse, the two ships were alternating the roadstead and the quay to be kept running. Their presence caused some polemics, reported by the local media in some online articles: according to a group of inhabitants, the running vessel engines, to allow a correct functioning of on-board systems, actually represented a potential danger to the environment and disfigured the waterfront as they limited the view of the coast. The alleged polluting power of the ships was even dangerous considering the fact that the cruise port is located at the very beginning of Ortigia and, therefore, inside an important part of the historic center. Although at first the departure of both ships was scheduled not earlier than November 2020, in mid-September the Spirit left Syracuse for Brindisi (a city and port in Puglia, in the Southeast of Italy). According to some newspapers, the departure was be connected to the controversies raised by the presence of the ships mentioned earlier, making the company re-deploy one ship to another port in southern Italy. On the other hand, the newspapers also reported other points of view. The ships would respect the environment and pollution safety standards, therefore turning the problem into an advantage for the port and the city economy, as the services provided would be paid regularly by the companies. I have been analyzing residents' perceptions in another study still ongoing, by carrying out fieldwork research and targeted interviews. Here, however, I will consider some texts taken from the most significant articles on both the described positions, in order to observe how the construction of public narratives shapes the perception of environmental risks. Indeed, media texts are interesting and useful objects of study (Carvalho 2000) to understand an important part of the communication process and, even more, to understand the complex interweaving of production and reproduction of how narratives can be mutually influential means through which risk perceptions are expressed and "put into shape". Moreover, from the perspective of Cultural Geography studying media texts allows to study also the communities addressed as well as the links to their representation, culture and spatiality. Here are nine articles are taken from some local online newspapers, two pre-pandemic and seven post-pandemic ones. For each article the headline is analyzed first of all, as it is a particularly "significant" paratext in media communication (Genette 1987; Montes 2006).

(1) The first headline refers to the online newspaper LibertàSicilia.it. The narration aims to enhance the potential of the port of Syracuse in 2020, the following year compared to date of the article. The paratext refers to the future as a present in order to give greater impact to the story:

> LibertaSicilia.it (14 March 2019 - https://bit.ly/3mlcKjn)
>
> Syracuse. Cruises departing and arriving from the big port in 2020
>
> (*Siracusa. Crociere in partenza e arrivi dal porto grande nel 2020*)

The article reports the works on the new quay that will be hosting the cruise ships, it even quotes the new routes followed by some cruise companies that will operate stops in Syracuse. The opportunity to have more ships in port is described as a dream:

> The dream is to be able to host 5 cruise ships at the same time, so as to be a driving force for the growth of tourism also in Ortigia. Last year Syracuse managed to host three ships at the same time.
>
> (*Il sogno è quello di poter ospitare 5 navi da crociera contemporaneamente, così da costituire un volano per la crescita del turismo anche in Ortigia. Lo scorso anno Siracusa è riuscita ad ospitarne tre in contemporanea.*)

(2) The second headline is taken from SiracusaOggi.it and, in its simplicity, describes a very positive re-opening of cruise and pleasure tourism in the Sicilian city at the beginning of the summer season, in 2019. To do this, it uses enumeration, including naval and land vehicles to emphasize the importance of the cruise and recreational volume through the grandeur of the logistics prepared (buses):

> SiracusaOggi.it (1 June 2019 - https://bit.ly/3dTwGGU)
>
> Syracuse. Tourist season at the top: two cruise ships, yachts and dozens of buses
>
> (*Siracusa. Stagione turistica al clou: due navi da crociera, yacht e decine di bus*)

Moreover, the text of the article focuses on the progress of the redevelopment works of the port and the need to build an additional quay for the mooring of cruise ships. The tone is in general optimistic for the new season which starts, and a rotation of ships previously announced is recorded:

> As proposed in recent days, the ships alternate and, in some cases, even arrive at the same time.
>
> (*Come prospettato nei giorni scorsi, le navi si avvicendano e, in alcuni casi, arrivano anche contemporaneamente.*)

(3) The third headline is taken from the SiracusaNews.it newspaper in 2020 and directly mentions the two ships parked in the port of the city. The narration stresses the specialized term "technical stop", put in quotation marks and the clear chronological indication which gives a sense of the time needed to restart (the article, in fact is dated August 2020):

SiracusaNews.it (17 August 2020 - https://bit.ly/34sUHBE)

Cruise ships in the port of Syracuse: "technical stop" until November. The two cruise ships, coming from the United States and in transit through Gibraltar, have no passengers on board, but only the crew members

(*Navi da crociera in porto a Siracusa: "sosta tecnica" fino a novembre. Le due unità da crociera, provenienti dagli Stati Uniti ed in transito per Gibilterra, non hanno passeggeri a bordo, ma solo i componenti dell'equipaggio*)

The article in the first part describes the arrival of the ships, in the second part it focuses on their alternation on the quay specifying that this practice will not hinder the arrival of other cruise ships in the port:

The two ships will alternate on the quay, first one and then the other, but this should not cause any difficulty for the arrival of other boats, a concern that hovers right at the dawn of the cruise season that has restarted today with the landing in Civitavecchia of the Msc Grandiosa.

(*Le due navi si alterneranno in banchina, prima una e poi l'altra, ma questo non dovrebbe comportare alcuna difficoltà per l'arrivo di altre imbarcazioni, una preoccupazione che aleggia proprio all'alba della stagione crocieristica ripartita proprio oggi con l'approdo a Civitavecchia della Msc Grandiosa.*)

(4) The fourth headline refers to the newspaper LibertàSicilia.it and evokes the controversial issue addressing the ships. The paratext highlights the heated debate between the two positions related to the ships in the port and emphasizes one of the negative issues in particular: the hypothesis that the ships operating for reasons other than tourism do not pay for the stopover in the port for stays up to four days. This explains the reason why the ships move alternating from one quay to the bay. The short narration brings together various elements (the controversy, the reason for stopping, the missed-payment), and exposes the position of the opposers toward arguments used to reject the thesis of the advantage of the mooring of ships:

LibertaSicilia.it (18 August 2020 - https://bit.ly/31HYadI)

Syracuse. Controversy on the cruise ships on the technical stop and the "ballet with a trick" [the farce] to avoid paying

(Siracusa. Polemica sulle navi da crociera con la sosta tecnica e il balletto col trucco per non pagare)

The article is a long declaration delivered by a former Municipal Councilor of the town of Syracuse reported by the online newspaper. The text quotes three issues related to the ships at port. First, the negative effect on the landscape of the waterfront: the ships prevent the view of the coast from the island of Ortigia and from the mainland the view of the coast of Ortigia.

The two cruise ships inside the port constitute for four months a barrier of the optical cone that prevents the visual perception of the waterfront

(*Le due navi da crociera dentro il porto costituiscono per quattro mesi una barriera del cono ottico che impedisce la percezione visiva del waterfront*)

Second, the alleged environmental impact of continuous running engines:

The ships remain autonomous thanks to the maintenance of the energy generators active 24 hours a day with considerable emission of exhaust gases into the air.

(Le navi rimangono autonome grazie al mantenimento dei generatori di energia attivi h24 con notevole emissione in aria di gas di scarico.)

Third, the issue quoted in the headline, neglecting the thesis related to the economic benefit of the ships' presence:

If I remember well, I believe that parking on the quay in the port of Syracuse for less than four days is free, so the generous display of paso doble in the harbor offered by the cruise ships this morning, will possibly be repeated periodically in a deafening silence of those who should be watching over and protecting the Syracusan community.

(*Non vorrei ricordare male ma credo che lo stazionamento in banchina nel porto di Siracusa inferiore a quattro giorni sia gratuito, quindi la generosa esibizione di paso doble in rada offerto dalle navi da crociera questa mattina, è possibile che sarà replicato con cadenza periodica tra il silenzio assordante di chi dovrebbe vigilare e tutelare la comunità siracusana.*)

(5) The fifth headline refers to SiracusaNews.it and was published a few days after the previous article. It addresses the safety issue and appeared in the middle of the above mentioned controversy about the alleged negative environmental impact. The title mentions the alderman's official response to the environmental recovery policies and air quality of the Municipality of Syracuse. According to his statement, both the mooring and the level of emissions are within the limits imposed by law. The text reports also the meeting between the municipal councilor and the commander of the Coast Guard of Syracuse (in Italy responsible for safety in port and at sea):

SiracusaNews.it (24 August 2020 - https://bit.ly/3Euc04C)

Cruise ships parked at the port of Syracuse, Gradenigo: "mooring and emissions in order". The new alderman this morning met the commander of the Coast Guard to get enlightened also following the recent controversy

(*Navi da crociera in sosta al porto di Siracusa, Gradenigo: ormeggio ed emissioni nella regola. Il neo assessore stamattina ha incontrato il comandante della Guardia Costiera per avere lumi anche a seguito delle recenti polemiche*)

The article opens by reporting the meeting between the alderman and the commander during which the controversial situation of the ships in port was discussed. Then, a detailed declaration of the alderman clarifies two critical issues: the stop of the ships was authorized according to the regulations in force, since no dangerous conditions were detected. Moreover, the ships do not pay for the simple stop, but for ancillary services necessary for their maintenance (such as bunkering, supplies, and so on). During their stop, the cruise companies adapted anti-pandemic regulations and were testing these services. In this sense, Costa Crociere managed to reach an agreement for future stopovers with the ship Costa Generosa. The text informs that the Municipality and the Harbor Office discarded the opportunity to transfer the two ships in Santa Panagia Bay (in the area North of the city, outside the town) due to the presence of oil tankers in transit and arriving at the pier which was connected to the pipeline of the nearby oil refinery.

With regard to the environmental impact, the text continues reiterating that some tests had already been carried out to verify the possible danger:

> The same Captaincy carried out the surveys on the sulfur contents in reference to the fuel used in the days around August 15th, "verifying the re-entry of values within the limits imposed bylaw - continues Gradenigo. In recent days, the same Harbor Master has asked the ARPA[2] to carry out surveys on water, air and background noise. All data that will be announced shortly by the Commander".
>
> (*La stessa Capitaneria ha eseguito i rilievi sui tenori di zolfo in riferimento al carburante utilizzato nei giorni a cavallo di ferragosto, "appurando il rientro dei valori entro i limiti imposti dalla legge – prosegue Gradenigo -. La stessa Capitaneria ha chiesto nei giorni scorsi all'Arpa di effettuare i rilievi in merito a acqua, aria e rumore di fondo. Tutti dati che verranno a breve resi noti dal Comandante.*)

The declaration of the alderman ends by quoting the opportunity to create an agreed protocol with the Municipality, Harbor Office and ARPA aimed at

> monitoring and mitigating the environmental impact of cruise ships, putting into practice all the actions required by the European directives to reduce marine pollution in ports [...] favoring and encouraging the entry and stop in port to those ship companies that use fuels with a sulfur content of less than 0,1%.
>
> (monitorare e mitigare l'impatto ambientale delle navi da Crociera, mettendo in pratica tutte le azioni previste dalle direttive europee per la riduzione dell'inquinamento navale nei porti [...] favorendo e incentivando l'ingresso e la sosta in porto a quelle compagnie navali che utilizzano carburanti con tenore di zolfo inferiore allo 0,1%.)

(6) The sixth headline, published on the online newspaper SiracusaPost.it (the day after the previous article) deals essentially with the same controversy and reports the declaration in favor of Confcommercio, one of the most important Italian associations of companies operating in trade, services and tourism, with a network of representatives spread throughout the national territory. The paratext reports the controversy by using the declarant's inclusive point of view:

> SiracusaPost.it (25 August 2020 - https://bit.ly/37DNtwB)
>
> Syracuse, cruise ships in port: Confcommercio "Noi favorevoli" (We are in favor)
>
> (*Siracusa, navi da crociera in porto: Confcommercio "Noi favorevoli"*)

The article reports the declaration of the president of the port sector of Confcommercio. According to him and the trade association, the two ships at port are an interesting opportunity to attract other companies in the future. The ship operators have been waiting for years for a suitable dock to be built and the controversy risk to drive the cruise lines away. These initial considerations support three points. First, the permission to dock is granted by the Harbor Office and it was given according to the regulations in force. Second, the operators ask the mayor to welcome ships and crews as most citizens are doing. Moreover,

---

[2] ARPA is the Regional Agency for environmental protection. The regional agencies with ISPRA (Istituto superior per la Protezione e Ricerca Ambientale) in Italy make the National system for environmental protection (Sistema nazionale per la protezione ambientale).

the presence of ships is generating economic repercussions in almost all port categories. Regarding the impact on the environment [...] not everyone knows that the anti-pollution regulations applied to ships are the most stringent that may exist.

(la presenza delle navi sta generando delle ricadute economiche in quasi tutte le categorie portuali. Riguardo all'impatto sull'ambiente [...] non tutti sanno che le norme antinquinamento, applicate alle navi, sono le più stringenti che possano esistere.)

(7) The seventh headline is from the national press agency Ansa.it. It announces the arrival of the first cruise ship in the port of Syracuse after the lockdown and the forced break of tourist activities. The reference to the single ship, the Costa Deliziosa, is synecdoche to bring up the question of the whole sector:

Ansa.it (10 September 2020 - https://bit.ly/2HvrA7U)

Tourism: cruises are back; the 'Costa Deliziosa' in Syracuse

(*Turismo: tornano crociere; 'Costa Deliziosa' a Siracusa*)

The article reports the declarations both by the Mayor of Syracuse and by the Country Manager Italy of Costa Crociere. This latter thanks the city for the welcome:

It is a great pleasure [...] to go back on a cruise with our Syracuse ships and we thank the city for the warm welcome. The arrival of Costa Deliziosa marks the return of cruises to Syracuse and therefore also the recovery of a sector that can have a positive economic impact for the local economy.

(È un grande piacere [...] tornare nuovamente in crociera con le nostre navi Siracusa e ringraziamo la città per la bella accoglienza. L'arrivo di Costa Deliziosa segna il ritorno delle crociere a Siracusa e quindi anche la ripresa di un settore che può avere un impatto economico positivo per l'economia locale.)

On the other hand, the article quotes a new security protocol adopted by the company and suitable for dealing with the pandemic:

Costa developed [...] [a] Safety Protocol, which contains new operational measures adapted to the needs of the COVID-19 situation, perfectly compliant with the provisions on the matter defined by the Italian and European authorities.

(*Costa ha sviluppato un Safety Protocol che contiene nuove misure operative adeguate alle esigenze della situazione COVID-19, perfettamente conformi alle disposizioni in materia definite dalle autorità italiane ed europee.*)

(8) The eighth headline refers to SiracusaOggi.it and reports the departure of ship Spirit heading the port of Brindisi, Puglia (a southeastern region of Italy). The term "preferring" used highlights the company's (the Norwegian) choice with respect to the ports. The underlying argument is the controversy that had arisen in the previous days around the parked ships:

SiracusaOggi.it (16 September 2020 - https://bit.ly/3msKQF5)

Syracuse. The Norwegian Spirit left the Porto Grande, preferring Brindisi

(Siracusa. La Norwegian Spirit ha lasciato il Porto Grande, preferendo Brindisi)

The text focuses on the departure of the Spirit, and tries to explain the reasons for this new:

According to what is learned from industry sources, the large ship opted for Puglia mainly for reasons vacancy on the quay. But also, it is not a secret that the mood created by ships parked in Syracuse played a role in the choice.

(*Secondo quanto si apprende da fonti di settore, la grande nave avrebbe optato per la Puglia principalmente per ragioni di spazio in banchina. Ma non sarebbe stato nascosto che il clima creatosi attorno alle navi in sosta a Siracusa avrebbe avuto un ruolo nella scelta.*)

Furthermore, the article reports news about a new cruise arrival:

In the next few hours the Costa Deliziosa will reach Syracuse again, which has already resumed its activity, albeit in a reduced form and reserved only for Italian tourists.

(*Nelle prossime ore raggiungerà nuovamente Siracusa la Costa Deliziosa che ha già ripreso la sua attività, seppur in forma ridotta e riservata soltanto a turisti italiani.*)

(9) Finally, the ninth headline is offerred by SiracusaOggi.it and informs of a new cruise ship arrival, the Msc Lirica, which in December 2020 replaced the Norwegian ship Dawn (that in the meantime returned to service) at port. According to the text, the Lirica remained idle for three months, until March 2021:

SiracusaOggi.it (8 December 2020 - https://bit.ly/3vYKUPP)

The Msc Lirica in the Porto Grande of Syracuse, idle for three months

(*La Msc Lirica al porto Grande di Siracusa, sosta inoperosa per tre mesi*)

The text mentions the route that the ship should have followed once it arrived at the port, specifying instead that this too would remain idle due to the anti-pandemic measures. In the second part the two Norwegian ships are quoted:

However, the large ship is now docked at the Grand Harbor, where it will remain idle for about three months just like the two Norwegian boats previously.

(*La grande nave è comunque ora in banchina al porto Grande, dove rimarrà per circa tre mesi in sosta inoperosa proprio come in precedenza le due imbarcazioni della Norwegian.*)

If on the one hand, the analysis of the nine articles taken into account so far show a certain interest for cruising in Syracuse, at the same time the texts show a certain conflict between different interests within the same community. A sort of internal crisis coinciding with the normal rotation of ships and routes, due to the Covid-19 pandemic. Only relatively recent the port became suitable for standard cruise ships, thanks to a new quay built which opened new developing opportunities for further cruising destinations. In this regard, the first three articles, all dating back to the pre-pandemic period, addresses cruising in very positive terms, sometimes even enthusiastic (especially article 3). The positive forecasts of the cruise market offer a lot of hope for 2020 and the texts highlight the presence of ships, the number of stops and companies, as well as the efficiency of ground services. The tone of the texts dated August 2020 instead is different: anti-pandemic measures in Italy (and in several countries of the world) appearing more relaxed after the lockdown that had suspended all non-essential transport and tourist services. In that moment the parked ships had become a presence linked to other issues: 'concern' had been expressed (in Italian language "preoccupazione") with the possible docking problems of other cruise ships arriving in the port, even if it appeared a remote hypothesis.

The articles 4, 5 and 6 were published within a few days (respectively 18, 24 and 25 August) and they follow the evolution of a controversy that had developed with the presence of ships. The texts report a shared opinion, part of the community of Syracuse and the debate focuses on three points: the environmental danger caused by the prolonged ignition of the engines; the disfigurement of the waterfront landscape to and from Ortigia; the economic convenience/inconvenience for local businesses. The narratives of these articles are organized around the rhetoric of confrontation between stakeholders: the Municipality (the alderman; the former Councilor), the Harbor Office (even if indirectly), the trade association (the president of the port sector of Confcommercio), the cruise companies (even if indirectly). This is not surprising considering the topics of the debate, however the opinions of ordinary citizens are not present. The public narratives of newspapers prefer to stay on public and official statements: almost the entire article coincides with the statements made by the various stakeholder representatives. This verticality proceeds with a top-down movement, in line with the information on the regulatory situation regarding safety, or on the reasoning regarding economy.

Article 7 is another part of the narrative of the destination, in this case by focusing on the arrival of the Costa Deliziosa ship in September 2020, an event that represents the resumption of tourist activities. The statement reported belongs to the Costa country Manager of the company, who thanks the citizens for the welcome. This is another stakeholder, this time on the side of the companies, which recalls the economic value of the presence of cruising in the Sicilian city. The controversy on ships in technical stops is not mentioned, but is symbolically evoked in absentia, precisely by the reference to the economic discourse.

The narration of article 8 marks the end of the polemic, with the departure of the ship for Brindisi. At the center of the narrative is the company's decision to move one of the two ships that have become the "subject of the dispute". Here another kind of verticality is reproduced: that of the economic relations typical of the tourism industry. Once again, the "voice" of the community is not directly present and the strategy of the company to avoid further conflict is underlined by noticing the decision of the new deployment. Moreover, the article ends with the news of the return at port of the Costa Deliziosa, citing the partial resumption of cruise activities.

The last article, published a few months after the eighth (December 2020), focuses on the arrival of another idle ship, the Msc Lirica, which takes the place of the ship by Norwegian. The text specifies that the cruise ship would remain idle at port for three months, due to the new anti-pandemic measures. At the very end, the narrative uses the rhetorical figure of simile to compare this stop to the one of Norwegian's ships without mentioning the summer controversy.

4 Conclusions

The aim of this study was to investigate the ways in which a community (Syracuse), as a destination, perceives specific forms of risk linked to a type of tourism (in our case cruise tourism). For this purpose, here I considered the public narratives produced by local online newspapers. The analysis shows that the city, as a cruise destination, is opening up to a new phase of development thanks to the quay suitable for standard ships. The texts, by using public narratives with a vertical point of view, reconstruct a controversy created within the community in August 2020, due to the presence of two cruise ships in a technical stop after the most stringent lockdown. Their arrival coincides with the partial resumption of cruising in Italy and in various Mediterranean destinations. However, stationary ships were being perceived by a part of the city as a potential environmental hazard (especially with regard to pollution and disfigurement of the landscape). On the contrary, some citizens saw this presence as an important opportunity for the economic development of local maritime services, as well as for the future development of routes centered on Syracuse. The analyzed newspapers have reported these positions as I have shown in the previous paragraph. During my fieldwork experience, through various interviews carried out for another study not yet published, I was able to ascertain that these positions were indeed those summarized by the articles. However, as I have already stated, the narratives of the observed newspapers are constructed through the statements of the stakeholders, neglecting the inhabitants' point of view. The media debate is ensured by the alternation of the official positions of the Municipality, the Harbor Office (even if indirectly), the trade association and, marginally, of cruise companies. In this sense, the public narratives here studied remain on a top-down model, on a geographical "small scale" that favors the view from above.

In addition, the cruise ship which from a symbolic point of view is usually considered as a 'destination among destinations' (Löfgren 1999; Sabato 2018), here takes on another aspect. Its usual tourist activity is interrupted, although remaining an idle ship, and becomes an element of "foreignism" to a destination community with the dynamism that distinguishes this medium-destination stops. In this sense, its potential divides the perceptions: for some people it continues to be a means for territorial development, for others it becomes a possible danger. It is possible to read, in the various newspaper narratives, the stakeholder attempt to mediate between these two positions without achieving a real result.

Finally, from a geographical and urban space perspective, a number of various spaces can be interpreted, intersecting and overlapping in a "production and reproduction" of spatiality (De Certeau 1990; Foucault 1975, 2004; Lefebvre 1991; Soja 1993) and performativity (Gregory 2004). Indeed, part of the controversy arises from the fact that the ships are located close to the historic center of Ortigia, the heart of the ancient city of Syracuse where currently, one of the districts in which a rapid process of gentrification and turistification is already in place. The perception of cruise ships, therefore, is probably to be connected to the perception that citizens have of this area: a space that needs to be preserved, on land and at sea, or developed in new directions. It is precisely on this level that the perception of risk plays a decisive role, which emerged in various ways within the media narratives previously taken into account. Indeed, it must be also remembered that the arrival of standard ships in the bay is not recent, but the arrival of standard ships at the quay is very recent, a condition that allows the disembarkation/embarkation of a greater number of passengers in a short time. A new condition which also opens up new questions and which will constitute a further challenge for the community.

References

Carvalho A (2000) Discourse analysis and media texts: a critical reading of analytical tools. In: International conference on logic and methodology, vol 33. International Sociology Association, Köln, pp 1–40.

CLIA Cruise Lines International Association (2020) Global Market Report

CLIA Cruise Lines International Association (2021) State of the Cruise Industry Outlook

De Certeau M (1990) L'invention du quotidien, I. Arts de faire. Gallimard, Paris

Douglas M (1986) Risk acceptability according to the social sciences. Russell Sage Foundation, New York

Douglas M (1992) Risk and blame. Essays in cultural theory. Routledge, London-New York

Douglas M, Wildavsky A (1982) Risk and culture. University of California Press, Berkeley-Los Angeles-London

Foucault M (1975) Surveiller et punir: Naissance de la prison. Gallimard, Paris

Foucault M (2004) Sécurité, territoire, population. In: Senellart M (ed). EHESS, Paris

Fuchs G, Reichel A (2004) Cultural differences in tourist destination risk perception: an exploratory study. Tourism (Zagreb) 52(1):21–37

Fuchs G, Reichel A (2006) Tourist destination risk perception: the case of Israel. J Hosp Leis Mark 14(2):81–106

Genette G (1987) Seuils. Éditions du Seuil, Paris

Gómez Cantero J, Morán Martínez C, Losada Gómez J, Carnelli F (eds) (2020) The climate crisis in Mediterranean Europe: crossborder and multidisciplinary issues on climate change. Il Sileno, Lago (CS)

Gregory D (2004) The colonial present. Afghanistan. Palestine. Iraq. Blackwell, Oxford

Kasperson RE, Renn O, Slovic P, Brown HS, Emel J, Goble R, Kasperson JX, Ratick S (1988) The social amplification of risk: a conceptual framework. Risk Anal 8:177–187. https://doi.org/10.1111/j.1539-6924.1988.tb01168.x

Lefebvre H (1991) The production of the space. Blackwell, Oxford-Cambridge, MA

Lepp A, Gibson H (2003) Tourist roles, perceived risk and international tourism. Ann Tour Res 30(3):606–624

Löfgren O (1999) On holiday: a history of vacationing. University of California Press, Berkeley

Mercatanti L (2015) Il declino delle destinazioni turistiche. In: Cusimano G (ed) Il turismo nelle/delle destinazioni. Pàtron, Bologna, pp 183–194

Mercatanti L, Sabato G (2019) Volcanic risk and the role of the media. A case study in the Etna area AIMS Geosciences 5(3):448–460. https://doi.org/10.3934/geosci.2019.3.448

Mercatanti L, Sabato G (eds) (2021) Special Issue: Sustainability and Risk Perception: multidisciplinary approaches. AIMS Geosciences. https://www.aimspress.com/aimsgeo/article/5190/special-articles

Mercatanti L, Montes S (eds) (2021) Global threats in the Anthropocene: from COVID-19 to the future. Il Sileno, Lago (CS)

Montes S (2006) Just a foreword? Malinowski, Geertz and the anthropologist as native. In: Montes S, Taverna L (eds) Sign system studies 34.2: special issue—between semiotics and anthropology: life histories and other methodological issues. Tartu University Press, Tartu, pp 357–386

De Pascale F (2022) Geografie del rischio e della vulnerabilità: approcci teorici ed esperienze didattiche a confronto. Pontecorboli, Firenze

Pidgeon NF, Kasperson RE, Slovic P (2003) The social amplification of risk. Cambridge University Press, Cambridge

Renn O, Burns WJ, Kasperson JX, Kasperson RE, Slovic P (1992) The social amplification of risk: theoretical foundations and empirical applications. J Soc Iss 48(4):137–160
Risposte Turismo (2017) Italian Cruise Watch
Risposte Turismo (2019) Il traffico crocieristico in Italia nel 2018 e le previsioni per il 2019
Risposte Turismo (2020a) Il traffico crocieristico in Italia nel 2019 e le previsioni per il 2020
Risposte Turismo (2020b) Italian Cruise Watch
Rittichainuwat B (2013) Tourists' perceived risks toward overt safety measures. J Hosp Tour Res 37(2):199–216
Sabato G (2018) Crociere e crocieristi. Itinerari, immaginari e narrazioni. Giappichelli, Torino
Soja EW (1993) The spaces that difference makes. Some notes on the geographical margins of the new cultural politics. In: Keith M, Pile S (eds) Place and the politics of identity. Routledge, London-New York, pp 180–202
Tarlow P (2014) Tourism security. Strategies for effectively managing travel risk and safety. Elsevier, Oxford
UNWTO United Nations World Tourism Organization (2020) International tourism highlights

Social Media in Risk Perception and Disaster Management: A Geographical Perspective

Teresa Graziano

Abstract The advent of the last Technological Revolution has completely upset long-entrenched relations among spaces, places and communication flows, by influencing the ways through which information is territorialized and, vice versa, territories are (re)shaped in the virtual dimension. In particular, the internet and the social web have further accelerated the increase in geo-tagged data which are co-created by not professional users. The increasing co-production of space-related information has also influenced risk perception and disaster management. On the one hand, a huge increase in disaster-related data flows diffused by institutional actors is due to a variegated range of purposes (prevention, on-site disaster management, real-time information). On the other hand, a significant amount of data is co-created and shared by users from the bottom-up while the disaster crisis is still unfolding, by integrating official informational sources. Theoretically inserted at the intersection between the Geography of Communication and the Geography of disaster, the chapter provides a theoretical review and a set of descriptive examples demonstrating the growingly pervasive role of ICTs in risk perception and management.

Keywords Disaster geography · Disaster management · Participation · Risk perception · Social media

1 Disaster Geography, Communication and Social Media

Natural hazards can affect different kinds of territories at the global scale to the point that Antronico et al. (2019) argue that "the entire world's population is exposed to natural hazards in different ways and degrees".

Over the last years, along with the growingly impactful anthropic stress, the effects of climate change and the mounting global population, the scale, frequency and impacts of disasters appear to be increasing (UNDRR 2019).

T. Graziano (✉)
University of Catania, Catania, Italy
e-mail: teresa.graziano@unict.it

B. Dahiya et al. (eds.), *Disaster Resilience and Human Settlements*,
Advances in 21st Century Human Settlements,
https://doi.org/10.1007/978-981-99-2248-2_7

In spite of long-lasting human-environmental consequences, disasters are multifaceted events usually concentrated in time and space, which produce a severe damage for the affected communities in terms of human, material or environmental losses.

As Curtis and Mills highlight (2020: 9),

> disasters are inherently spatial—both in terms of the physical processes as well as the human implications. Hurricane tracks, the location of fault lines, how tornadoes are generated—these are patterns or processes that have or leave spatial footprints. Where people live in relation to potential hazards or the societal impact left after a disaster can again be described in terms of spatial patterns. Within these patterns are human places, cultures, and interactions. An earthquake-devastated city is not just a landscape of damage, morbidity, and mortality but it also comprises lost (and recovering) neighborhoods, disrupted social networks, variations in resiliency, and social and environmental (in)justice. As geographers we have the technologies and spatial skills to map, predict, and ultimately understand these landscapes.

Nonetheless, as De Spuches et al. (2020: 506) put it,

> Until the second half of the twentieth century, disasters and risks were subjects exclusive to the hard sciences, analysed with quantitative methods and not considered as possible objects of socio-humanistic interest. Rigidly distinguished between natural and technological, disasters used to be studied in the perspective of a radical determinism as completely independent of human responsibility.
>
> It was around Emilio Quarantelli and Russell Dynes in the Disaster Research Center in Delaware that the first group of social research on disasters emerged. The idea was that disasters are social phenomena, determined not only by physical factors, but by anthropic and cultural ones as well. These pioneering studies claimed for a radical "removal of concepts of naturalness from natural disaster" and shifted the research from the hazard itself to the socio-territorial context involved.

Thus, due to the inherently spatiality of disasters, the geographical perspective can provide useful insights both for the disaster prevention and risk management as well as for post-recovery strategies (Mercatanti and Sabato 2020).

In the specific field of Geography, disaster has become over the years a well-established branch of the discipline, not limited to the most "traditional" analysis of the geo-morphological characteristics of disasters, but also entailing, for instance, the geopolitical implications in terms of transcalar power relations and environmental migratory flows (Donovan 2017).

However, less emphasis is traditionally put on the role of the discipline in interpreting and deconstructing the information and communication flows as well as their territorial implications, which can reshape the spatialities affected by disasters.

A wide range of formal and informal actors such as governments, disaster response agencies, NGOs, volunteers and community members are usually involved in managing disaster process and consequences in order to mitigate against, prepare for and recover from disasters, which need a rapid utilization of information and data from several sources.

As a matter of fact, the advent of the last Technological Revolution has completely upset long-entrenched relations among spaces, places and communication flows, by influencing the ways through which information is territorialized and, vice versa, territories are (re)shaped in the virtual dimension. So, Geography of Information and

Communication can also provide useful insights into the complex tangle of flows, connections and territorial reconfigurations constantly reshaped by ICTs, particularly with the advent of internet and the social web (Albanese and Graziano 2020).

First, new technologies (remote sensing—RS, geographic information systems—GIS, global positioning systems—GPS) are useful tools for mapping, predicting and understanding the pre- and post- disaster landscapes (Curtis and Millis 2010). Over the last years, these tools are growingly used in strict integration with geotagged data provided by local community of not-professionals. On the one hand, a huge increase in disaster-related data flows diffused by institutional actors is due to a variegated range of purposes (prevention, on-site disaster management, real-time information). On the other hand, a significant amount of data is co-created and shared by users from the bottom-up while the disaster crisis is still unfolding, by integrating official informational sources which can result to be very useful in post-disaster management to reach a larger audience, facilitating the exchange of information and improving the speed of communication. The use of social media communication is so much embedded in the disaster management that both private web companies and institutional agencies promote their use: the 2015 World Conference on Disaster Risk Reduction approved the Sendai Framework for Action that for the first time included social media as official tools for natural disaster management. So, while ONU has recommended to use geolocalized standardized hashtag in Twitter during disasters, Facebook has activated the safety check service to communicate one's own condition in case of emergency.

As Laituri and Kodrich (2008) put it, the turning point was the Indian Ocean tsunami (2004) and Hurricane Katrina (2005) which reveal the coming of age of the on-line disaster response community. This integration usually includes both the traditional aspects of top-down disaster management (preparedness, response, recovery, mitigation, and policy) as facilitated by governmental agencies and relief response organizations; and the increasingly huge amount of online data and information growingly co-created by the public, both the one affected by the disaster and a wider public at the global scale.

According to the authors, the role of online disaster response communities includes several actions, such as the possibility to donate money, to disseminate real-time information, often faster than government agencies, and spread interactive communication to seek family members and identifying shelters. What is called online disaster-response community (ODRC) (Laituria and Kodrich 2008: 7) includes "formal and informal networks of people acting as sensors collecting, processing, and delivering information where it is needed".

As a result, online users can act as "sensors", not so much in Goodchild's perspective (2007), as we'll see in the following paragraphs, but rather from a positive and proactive perspective.

Theoretically inserted at the intersection between the Geography of Communication and the Geography of disaster literature, the chapter aims at exploring to what extent the increasing co-production of space-related information has also influenced risk perception and disaster management.

It is organized as follows: the second paragraph deepens the conceptual framework of ICTs geography, by specifically scrutinizing the role of social media in disaster prevention and management; the third paragraph deals with a review of some international web-based experiences and practices of disaster prevention and management, with a particular focus on Italy. The final paragraph includes the discussion in order to highlight potential and critical aspects as well as the final considerations.

2 Geography and the ICTs

The growing pervasiveness of the internet and the web has led many scholars to interpret their complex architecture and, particularly in the domain of geography, their role in reconfiguring real-word spaces and places. Initially regarded as a global network of computers (Dodge and Kitchin 2001), the internet was considered as the most comprehensive information system which produced a bewildering impact on the geographical characteristics of spaces (Brunn 1998). As a result, pioneering geographical researches were focused on theoretical and methodological attempts at "mapping" cyberspace as a new geographical sphere (Dodge 1999), seen as an (apparently) de-territorialized dimension interfering with traditional geographical spaces (Batty 1993) which concurred in overcoming long-entrenched dialectics, such as centre-periphery, near-faraway, here/there.

In particular, in the media geography domain, several researches emphasized how new forms of collective action transcended the traditional views of cultural and spatial fixedness, as far as new social media were regarded as responsible of new forms of social interactions (Jordan 1999; Pickerill 2007). So, in a first phase, cyberspace was largely described in antithesis with real-space owing to the dissolution of traditional spatial boundaries (Mitra and Watts 2002).

However, far from being a de-materialized dimension completely detached from the real-world one, the internet and the web are strictly dependent on, and structured around, the tangible materiality of spaces and places. The supposed a-materiality of virtual spaces does not encompass a process of full de- territorialization of places, nor does it concur in overcoming real-world inequalities and socio-cultural polarization. As underlined by Graham and Marvin's splintering urbanism (2001), the digital dimension is embedded in a deeply territorialized system of cables, networks and technological infrastructures which foster new kinds of inequalities, since it is usually the urban space to be strongly connected in a wider global system of transcalar digital connections.

Kluitenberg (2006: 10) argues that "instead of a strict separation between physical space and informational space, all technological and social trends clearly indicate that these two 'spheres' are becoming more and more closely interwoven".

With the advent of Web 2.0 and Media 2.0 studies, the initial dichotomous distinction between cyberspace and real-space began to dissolve (Gauntlett 2009; Valtysson, 2010) owing to the explosion of user-generated contents and the shift from top-down broadcast media consumption to bottom-up "prosumption" to the point that Meek

(2012a, b) highlight the need to reinterpret the relation between places of relational communication and everyday places (Gauntlett 2009).

Following Wellman's (2001) concept of cyberplace, Meek (2012a, b) focuses on "the spatial hybridity of technologically mediated social life" arguing that.

> Cyberplace, like cyberspace, is defined by interaction; however, its interactions are explicitly embodied in space/time—or events—due to the phenomenology of perception. As such, cyberplace provides us with an embodied engagement with place, whereas cyberspace is disembodied. To stave off preliminary objections, it is not that I'm not arguing that cyberplace is a materiality, or that the internet by nature and function constructs a material place, but that the geographic function of the internet and associated new social media are to create communication linkages; these recursive linkages constitute a contemporary existential spatiality, or cyberplace, that is increasingly intertwined with our lives.

Thus, on one hand, contemporary environments have been transformed into veritable infoscapes, constantly (re)shaped by the increasing amount of big data provided both by institutional actors and (un)aware citizens continuously leaving digital traces (Kitchin 2014). The top-down tech-mediated citizenship, embodied in the strategic discourses about the (neoliberal) efficiency of smart urbanism, mirrors the growingly pervasive vision of a technocratic urban governance where citizens seem to be reduced to human sensors (Goodchild 2007) or sensing nodes (Gabrys 2014).

On the other hand, new technologies have been recently transforming citizens into key players in co-creating and disseminating data and information. Although the internet is not such libertarian, equally accessible or fully democratic as the techno-enthusiasts would depict it, phenomena such as citizen journalism, grassroots information, netizenship, digital campaigning, cyber-dissidence, local scale online activism can only stem from the formerly hacker-shaped architecture of the Web (Graziano 2017).

2.1 Geography and Social Media Communication for Risk Management

While the relations among ICTs, places and spaces have been early investigated from a geographical perspective since the first appearance of the internet and the Web, studies on the role of new technologies in shaping unprecedented communication and information flows during disasters are less numerous in geography and social science.

Nonetheless, as Paradiso puts it (2010), "ICTs deeply affect people-environment relations" over the different stages of disaster prevention and management, ranging from hazard perception and citizen preparedness to relief, recovery and resilience.

A wide array of targets, channels and objectives is included in the multifaceted complex of messages and informational flows which characterize the so-called "risk communication" (Dymon and Boscoe 1996; Reckelhoff-Dangel and Petersen 2007).

Risk communication is one of the most relevant components of the institutional mitigation policies and sustainable hazards mitigation practices developed by private actors in order to tackle the challenges of prevention/disaster and post-disaster communication (Stewart and Rashid 2010; Wang and Ye 2018; Wang et al. 2019).

Houston et al. (2015) even support the view that communication is so much a core element of disaster planning, response and recovery that a truly effective disaster communication may be itself a tool to prevent a disaster or lessen its impact. On the contrary, an ineffective disaster communication may make the effects of a disaster even worse to the point that disasters often are the "result of a crisis in the communication process or a result of a communication breakdown" (Rodríguez et al. 2007: 479).

Over the decades, disaster communication occurred via the mass media (Rodríguez et al. 2007), consisting of warning messages and news coverage generally managed by governmental agencies. This kind of institutional communication, mirroring the government views on disaster response and recovery, significantly influenced the population's perceptions. Thus, it played as a hegemonic tool of power shaping collective knowledge, attitudes and behavior during and after disaster (ib.).

Within the wide range of channels and tools traditionally used by risk communication, social media has growingly become one of the most important platform due to its real-time diffusion, interaction and responsiveness, by emphasizing the space–time dynamics of social responses to disasters (Xiao et al. 2015; Li et al. 2018; Wang and Ye 2018).

As Rashid (2011) puts it, "People are increasingly using social media tools in natural disasters and other emergencies to contact friends and family members as well as to communicate their concerns, fears, and needs" insofar as new smart technologies have a strong potential to increase information capacity, interactivity, and data amounts.

According to several researchers (Lindsay 2011; Curtis and Mills 2010), the two-way social media communication responds to the needs of an effective disaster communication system, which should be low-cost, easy-to-use, scalable, reliable and mobile, including useful data provided by GIS and visualization tools.

In particular, "the use of social media via mobile computing devices such as smartphones and tablets may be particularly helpful during a disaster that occurs without warning or in a remote location. This is because these devices may be used by citizens who are able to document and share information about events as they unfold, even in the absence of professional news organisations and journalists" (ib.).

Houston et al (2015) developed a comprehensive framework to deconstruct the role of social media in risk and disaster management by categorizing disaster social media functions drawing on a literature review (Table 1) in order to provide useful information for future social media applications in disaster management and identify still existing gaps.

What is interesting for the theoretical premises of this chapter—namely the growing interaction between the communication geography and disaster management—is the increasing use of a spatial approach to communicate, based on a wide repertoire of web-based tools and data (Curtis and Mills 2010).

The "spatial turn" occurring in several disciplines, which use methods, tools and theoretical perspectives of geography to spatially visualize and/or interpret a wide

Table 1 Functions and uses of disaster social media according to the disaster phase

Disaster social media uses	Disaster phase
Provide/receive preparedness information	Pre-event
Provide/receive disaster warning	Pre-event
Signal/detect disasters	Pre-event/Event
Send/receive requests for help/assistance	Event
Inform/learn about one's conditions/localization	Event
Document/learn on-going events	Event/Post-Event
Deliver/consume news coverage	Event/Post-Event
Provide/receive response information	Event/Post-Event
Raise/develop awareness; donate/receive; identity/list ways to help	Event/Post-Event
Provide/receive mental health support	Event/Post-Event
Express emotions, concerns; memorialise victims	Event/Post-Event
Provide/receive information about response, recovery, rebuilding; tell/hear stories	Event/Post-Event
Discuss socio-political and scientific causes, responsibility and implications	Post-Event
Reconnect community members	Post-Event
Implement traditional crisis communication activities	Post-Event

Source Houston et al. (2015)

range of phenomena, has been particularly evident during the COVID-19 pandemic. Although the health crisis cannot be strictly considered as a natural disaster in terms of entity and temporal continuity, nonetheless it is useful to think about the pervasively role of web-based visualizing tools, GIS and social media to gather and spread information.

As Boulos and Geraghty (2020: 2) put it,

> As with the original SARS-CoV epidemic of 2002/2003 and with seasonal influenza, geographic information systems (GIS) and methods, including, among other application possibilities, online real- or near-real- time mapping of disease cases and of social media reactions to disease spread, predictive risk mapping using population travel data, and tracing and mapping super- spreader trajectories and contacts across space and time [...], are proving indispensable for our timely understanding of the new disease source, dynamics and epidemiology, and in shaping our effective response to it. [...] Since then we have seen a revolution in applied health geography through Web-based tools. Now, as we deploy these tools to protect human lives, we can ingest big data from their sources and display results in interactive and near-real-time dashboards. ese online dashboards have become a pivotal source of information during the COVID-19 outbreak.

So, it has become increasingly crucial to evaluate the complex tangle of power relations, consistency of informational flows and the territorial impacts of communication due to the growing integration between traditional communication channels and social media when a disaster occurs.

3 A Review of Experiences of Web-Based Disaster Communication

As above-mentioned, over the years several web-based resources, often integrated with other ICT services, have been increasingly used to integrate, complete and support the more traditional forms of disaster communication. For instance, Google Earth has been often used since its origins to map the intensity of post- earthquake tremors or to coordinate the relief actions in the aftermath of Hurricane Katrina in 2005 (Paradiso, 2010).

In particular, in the aftermath of the 2007 California wild fires, Google Earth was extensively used to provide daily updates as to the status of the fires, in terms of roads blocked, evacuation orders and so forth (Curtis and Mills 2010).

Before the upsurge of Web 2.0 social media, it was the blogosphere to play a significant role in diffusing information even in the realm of disaster management, in line with the growing relevance forum and blogs had in 2000's. Their importance has been fully investigated for migrants, cyberdissidents and minority groups (Diminescu 2008; Graziano 2012b),[1] which opens up unprecedented trajectories of theoretical and operational reflection also in the realm of post disaster management if we consider the growing number of migratory flows triggered at the global scale by natural disasters, the so-called "environmental migrants" (Graziano 2012a).

However, less emphasis has been generally put on the functions that the blogosphere can pursue in the field of disaster management, except Zhang et al. (2002) who highlighted their potential in offering family reunion or information in a more flexible way than official websites, whose architecture was more static and less interactive. Blogs after the Tsunami, the Hurricane Katrina or the California fires provided accuracy and speed of information which enormously helped relief and recovery efforts (Perrone 2005).

More recently, with the advent of social media, blogosphere has been gradually replaced insofar social networks such as Facebook and Twitter are largely diffused and provide a two-way interactive communication. Wang et al. (2019) retrieved all geotagged tweets posted between October and November 2012, during and in the aftermath of Hurricane Sandy, one of the most destructive cyclones in the USA since 1900. By evaluating the retrieved tweets, which came from the 14 affected eastern states, the researchers found out that the most physically and socially vulnerable communities had a stronger influence on social response to the Hurricane, by engaging more in disaster-related social media conversations. Thus,

> One may suppose that socially vulnerable groups should be more active in disaster-related social media conversations. On the contrary, our modeling results disclose that these socially vulnerable communities were digitally left behind in the pre-disaster social

[1] The pioneering efforts of mapping global diasporas online were made by the TIC&migrations research group coordinated by Dana Diminescu at the Fondation Maison Sciences de l'Homme of Paris (FMSH). The results of this research can be retrieved from the interactive website (www.e-diasporas.fr) and a dedicated special issue *Social Science Journal*. The author of this chapter collaborating with this project as invited visiting scholar at the FMSH in 2011.

> responses to Hurricane Sandy. This implies that the use of social networking sites for information exchange in natural disaster situations has introduced new inequality where groups with disadvantaged socioeconomic and demographic status were less represented in disaster-related communication on social media (Wang et al. 2019: 2).

3.1 Web-Based Tools and Social Media Disaster Communication in Italy

In order to provide an analytical scheme to synthetize the role of ICTs in disaster geography in Italy, I draw on Laituri and Kodrich (2008) who identified two main tiers of ODRC. According to the authors, the first tier is the formal network that includes all national and international programs developed by response organizations and governmental agencies which use online communication to raise public awareness, support funding campaigns and provide information to prevent disaster. The second tiers include the multi-actor and multi-dimension online disaster response when a disaster occurs, involving a wide array of participants (the mainstream media, the general public, the interested parties such as families, aid groups and so forth). This is strongly dependent on the specific disaster and consequently largely transitory. Furthermore, it is made both of formal networks (institutional actors, broadcast media etc.) and informal ones of online users providing first hand/real-time accounts of the disasters by co-creating online contents on social media (pictures, videos, chatrooms etc.).

Drawing on this framework, I added some elements that I consider crucial to further explore the reciprocal connections among disasters, transcalar flows of communication and territorial implications (Table 2):

Finally, I selected four disasters occurred in Italy over the last years in order to evaluate the role of social media and web-based tools in the different stages of disasters by analyzing the communication channels used and the type of actions according to the disaster stage. I developed the following framework that takes into account the intersections (and even frictions) among type of disasters, actors involved in the pre-ongoing-post event management, the communication channels used and the territorial impacts produced (Fig. 1):

The first stage of content analysis was conducted through key-words (locality name + kind of disaster + year) in the web (websites and blogs) and social media (Facebook and Twitter). It has highlighted the increasing use of social media for disaster communication in Italy, although not in a systematic way as far as the institutional actors are concerned.

In particular, while during the 2009 earthquake in Central Italy social media were not so much diffused and consequently they did not play a relevant role in spreading information, after the 2012 earthquake that affected mainly the Emilia region the amount and entity of social media communication significantly grew. During the Emilia earthquake, the National Institute of Geophysics and Vulcanology (INGV) used the Facebook profile to provide technical information such as localization and

Table 2 Disaster/communication/territory interpreting scheme

Disaster	Earthquake
	Flood
	Eruption
	Drought
Actors	Governmental
	Institutional response/recovery agencies
	ONG/associations
	Journalists
	Citizens
	Affected communities
Traditional communication actions	TV
	Radio
	Newspapers
Online channels	Instant messaging
	Websites
	Blogs/forums
	Facebook
	Twitter
Online communication actions	Advocacy
	Awareness
	Real-time information
	Pro-activeness (organization/mobilization)
	Pro-activeness (fundraising)
Offline/online interaction	Yes
	No
Disaster stages of communication	Pre-disaster
	During
	Post-disaster/recovery

Source Author's personal elaboration drawing on Laituri and Kodrich (2008)

magnitude, without any other indication targeted to affected population. However, this was the first natural disaster to be greatly commented online, with the hashtag #terremoto (earthquake) being among the trending topics in the social platform. The limited use of social media by institutional actors characterized also the 2013 flood in Sardinia, whilst on the contrary web users used them in a growingly pervasive way. While Twitter was mainly used to provide real-time information and Facebook to share more emotional narratives of the events, during the Sardinia floods the users were mobilized from the bottom-up through a series of online actions through

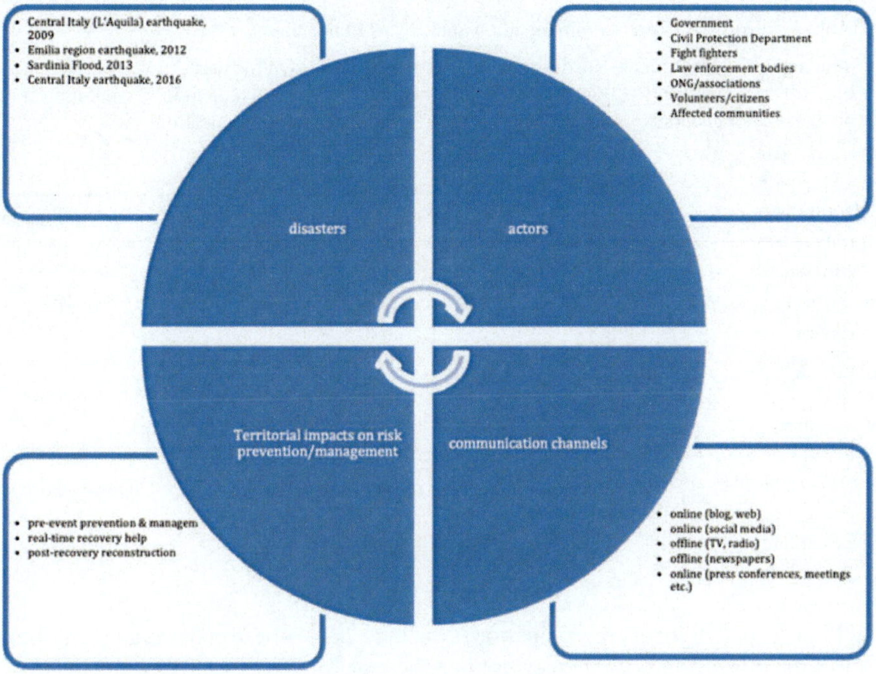

Fig. 1 Disaster-territory-communication framework. *Source* Author's elaboration

the hashtag #allertameteoSAR, finalised to operatively support relief efforts and visualized in a collaborative webGis map (SardSOS).

Some years later, the Civil Protection Department created the #SocialProCiv digital community open to all actors variously involved in risk prevention and disaster management. What is more, the Civil Protection has also launched a campaign "Io non rischio" (I don't take a risk) in order to spread awareness about the Italian fragile hydro-geological heritage, by utilising social media also in the pre-event stage.

However, during the 2016 earthquake that affected Central Italy, the Civil Protection had not a Twitter account where updates about the event could be collected, so that even in this case social media communication was mainly supported by private users from the bottom-up, some of them organized in collective accounts such as @terremotocentro, as well as by the online channels of newspapers and journalists (Table 3).

4 Discussion and Conclusions

Over the last decades the extensive use of technology-mediated communication in territorial governance has encompassed a wide repertoire of tools and practices, ranging from Volunteered Geographic Information to citizens' e-planning. Embodied

Table 3 Territory-disaster-communication analysis of Italian cases

Year and Place of disaster	Traditional media channels	Institutional actors involved in communication flows	Use of social media by institutional actors	Online users' involvement in communication flows	Disaster stage communication
2009 Central Italy earthquake	Yes	Yes	No	Extremely limited	During/Post
2012 Emilia Region Earthquake	Yes	Yes	Extremely Limited	Yes	During/Post
2013 Sardinia Flood	Yes	Yes	Limited	Yes	During/Post
2016 Central Italy earthquake	Yes	Yes	Limited	Yes	During/Post

within the paradigm of e-participation (Graziano 2017), these patterns and tools imply citizens' increasing self-engagement in policy making processes, by referring to new forms of citizens' self-involvement in public decision making that overlap, rather than replacing, more traditional modes of engagement. So, it is crucial to critically explore to what extent e-participation through social media communication can influence the effectiveness of relief efforts during and after a natural disaster. As above-mentioned, the UN Sendai Framework for Disaster Reduction 2015–2030 highlighted the role of social media communication in raising awareness, providing real-time information, sharing updates, and support recovery in the different stages of a natural disaster.

The brief exploratory review of experiences of disaster social media communication in Italy have underlined on the one hand the huge potential of ICTs in the whole disaster cycle (from pre-event awareness to post-event recovery), although a still existing communication gap can be noticed as far as the institutional communication is concerned. What is more, in Italy the complexity of managing informational flows during disasters is further complicated by the overlapping competencies and functions which characterize the multi-level territorial governance, so that the role of communicating is shared among local municipalities, provinces, regions and national government, apart from the recovery/relief agency and actors. However, the experience gathered during the pandemic in terms of online communication will probably represent a set of practices which can be used by institutional actors in case of future natural disasters.

Despite the huge potential, some limitations must be underlined. First, in some cases of earthquakes or eruption the electromagnetism can affect communication systems. Second, social media data can lack accuracy, both in terms of geolocalization (only a tiny percent of Twits have precise longitudes and latitudes, according to Wang et al. 2019) and in terms of information correctness. This has been particularly evident during the Covid-19 pandemic and the related "infodemic":

As Boulos et al. 2020, 9 highlight,

> During infectious disease outbreaks and epidemics, social media play an important role in communicating verified facts and correct prevention tips to the masses, but also carry the risk of 'virally' spreading misinformation, confusion and fear among the general public. In the case of COVID-19, false or misleading information, (such as 'eating sesame oil or garlic can help prevent and cure coronavirus' and a decade-old map showing global air travel), rumours and panic have been spreading globally on social media much faster than the virus. (...) In fact, it has been said that the WHO is fighting a parallel pandemic (or 'infodemic') of misinformation besides COVID-19.

As a matter of fact, Geography of communication has deeply investigated the impacts of new technologies and web-based informational flows in terms of (re)territorializations, new dialectics (centre-periphery) and new forms of socio-economic polarization fostered by digital divide.

Paradiso (2010) holds the view that

> Traditional GIT, in particular ICT and the Internet, become more and more intertwined with the 'real' world so that the resulting geocyberspace has many different uses. ICT can significantly support daily routine management and new operations also in all phases of the disaster cycle. [...].
>
> In order to learn from data and experiences, and to glean indications for developing ICT systems for hazard awareness, citizen preparedness and crisis mitigation can be set up through a more complete 'disaster' cycle than the mere emergency relief response, consisting of the following phases: preparedness, warning, response and relief, recovery and resilience.

As this chapter has demonstrated, the geographical perspective can provide a theoretical and analytical framework capable to critically explore social media communication flows within the whole "disaster cycle"—ranging from preparedness to recovery and resilience—as well as providing institutional actors with useful operational indications and guidelines.

As ICTs shape the relations among environments, territories and local communities both in terms of perceived imageries as well as of tangible effects which reconfigure the concrete materiality of places (Albanese and Graziano 2020), this is even more evident in cases of natural disasters insofar as new technologies and particularly social media play a crucial informational role useful for relief efforts.

While the relevance of GIS of disaster simulation, planning, risk assessment and disaster management is widely recognized (Curtis and Mills 2010), less emphasis has been put hitherto to the opportunity of spatializing data retrieved from social media co-created by a variegated repertoire of actors, ranging from activists and local communities to technologically savvy users and citizen journalists.

Despite still entailing the above-mentioned limitations, the use of social media communication during the "disaster cycle" is a deeply geographical issue, for the implications in terms of overlapping ontic dimensions (the real-world and the virtual sphere) and the tangible effects produced on places and spaces by the hybrid spatialities halfway between the real and the digital. Thus, the controversial notion of "people as sensors" (Goodchild 2007)—namely people collecting data mostly in an unware way—is mobilized with a positive meaning in case of natural disasters since geo-spatialized ICTs can provide bottom-up territorialized information beyond mainstream channels of emergency response.

References

Albanese V, Graziano T (2020) Place cyberplace e le nuove geografie della comunicazione. Bononia University Press, Bologna

Antronico L, Coscarelli R, De Pascale F, Condino F (2019) Social perception of geo-hydrological risk in the context of urban disaster risk reduction: a comparison between experts and population in an area of Southern Italy. Sustainability 11:2061. https://doi.org/10.3390/su11072061

Batty M (1993) The geography of cyberspace. Environ Plann B 20(6):615–616

Boulos K, Geraghty EM (2020) Geographical tracking and mapping of coronavirus disease COVID-19/severe acute respiratory syndrome coronavirus 2 (SARS-CoV-2) epidemic and associated events around the world: how 21st century GIS technologies are supporting the global fight against outbreaks and epidemics. Int J Health Geogr 19:8. https://doi.org/10.1186/s12942-020-00202-8

Curtis A, Mills JW (2010) GIS, human geography, and disasters. Cognella, a division of University Readers, Inc.

De Spuches G, Sabatini F, Palermo G, Caravello E (2020) Risk narrations and perceptions in the COVID-19 time. A discourse analysis through the Italian press. AIMS Geosci 6(4):504–514

Diminescu D (2008) The connected migrant: an epistemological manifesto. Soc Sci Inf 47(4):565–579

Dodge M (1999) Measuring and mapping the geographies of cyberspace: a research note. Netcom 1(2):53–66

Dodge M, Kitchin R (2001) Mapping cyberspace. Routledge, London

Donovan A (2017) Geopower: reflections on the critical geography of disasters. Prog Hum Geogr 41(1):44–67

Dymon UJ, Boscoe FP (1996) Newspaper reporting in wake of the spring flood in Northern California. Natural Hazards Research and Application Information Center, Quick Response Report #81, University of Colorado at Boulder, Boulder, CO

Gabrys G (2014) Programming environments: environmentality and citizen sensing in the smart city. Environ Plann Soc Space 32:30–48

Gauntlett D (2009) Media studies 2.0: a response. Interact Stud Commun Cult 1(1):147–157

Goodchild MF (2007) Citizens as sensors: the world of volunteered geography. Geo J 69(4):211–221

Graham S, Marvin S (2001) Splintering urbanism: networked infrastructures, technological mobilities and the urban condition. London and New York, Routledge

Graziano T (2012a) "Il migrante "ambientale", tra nuove configurazioni delle migrazioni transnazionali ed evoluzione delle dinamiche demografiche. Bollettino della Società Geografia Italiana 2: 223–241

Graziano T (2012b) The Tunisian diaspora: between 'digital riots' and web activism. J Soc Sci Inform 51(4):535–551

Graziano T (2017) Citizen e-participation in urban planning: achievements and future challenges in a Mediterranean city. Int J E-Plann Res 6:3

Houston JB, Hawthorne J, Perreault MF, Park EH, Goldstein Hode M, Halliwell MR, Turner McGowen SE, Davis R, Vaid S, McElderry JA, Grif SA (2015) Social media and disasters: a functional framework for social media use in disaster planning, response, and research. Disasters 39(1):1–22

Jordan T (1999) Cyberpower: the culture and politics of cyberspace and the internet. Routledge, New York

Kitchin R (2014) Thee data revolution: big data, open data, data infrastructures and their consequences. SAGE, London, p 2014
Kluitenberg E (2006) The network of waves: living and acting in a hybrid space. Open 11:6–16
Laituri M, Kodrich K (2008) On line disaster response community: people as sensors of high magnitude disasters using internet GIS. Sensors 8:3037–3055
Li Z, Wang C, Emrich CT, Guo D (2018) A novel approach to leveraging social media for rapid flood mapping: a case study of the 2015 South Carolina floods. Cartogr Geogr Inf Sci 45(2):97–110
Lindsay BR (2011) Social media and disasters: current uses, future options, and policy considerations. Congressional Research Service, Washington, DC
Meek D (2012a) YouTube and Social movements: a phenomenological analysis of participation, events and cyberplace
Meek D (2012b) YouTube and social movements: a phenomenological analysis of participation, events. Antipode 44(4):1429–1448
Mercatanti L, Sabato G (2020) Sustainability and risk perception: multidisciplinary approaches. AIMS Geosci 7(2):219–223
Mitra A, Watts M (2002) Theorizing cyberspace: the idea of voice applied to the internet discourse. New Media Soc 4(4):479–498
Paradiso M (2010) Communications mode(l)s and disasters: from word of mouth to ICTs. Netcom 24–3(4):181–200
Perrone J (2005) The coming of age of citizen media. Natural Disasters, December 26, The Guardian
Pickerill J (2007) Autonomy on-line: Indymedia and practices of alter-globalisation. Environ Plan A 39(11):2668–2684
Rashid H (2011) Interpreting flood disasters and flood hazard perceptions from newspaper discourse: tale of two floods in the Red River valley, Manitoba, Canada. Appl Geogr 31(2011):35–45
Reckelhoff-Dangel C, Petersen D (2007). Risk communication in action: The risk communication workbook. EPA Office of Research and Development, National Risk Management Research Laboratory, Cincinnati, OH
Rodríguez H, Quarantelli EL, Dynes RR (2007) Handbook of disaster research. Springer, New York, NY
Stewart RM, Rashid H. (2010) Community strategies to improve risk communication and reduce social vulnerability: a case study from the Red River Basin. Manitoba Heritage Thesis, Retrieved from https://mspace.lib.umanitoba.ca/xmlui/handle/1993/8023 on 5 June 2021
UNDRR (2019) Global assessment report on disaster risk reduction. Switzerland, United Nations Office for Disaster Risk Reduction (UNDRR), Geneva
Valtysson B (2010) Access culture: Web 2.0 and cultural participation. Int J Cult Policy 16(2):200–214.
Wang Z, Lam NSN, Obradovich N, Ye X (2019) Are vulnerable communities digitally left behind in social responses to natural disasters? An evidence from Hurricane Sandy with Twitter data. Appl Gcogr 108:1–8
Wang Z, Ye X (2018) Social media analytics for natural disaster management. Int J Geogr Inf Sci 32(1):49–72
Wellman B (2001) Physical place and cyberplace: the rise of personalized networking. Int J Urban Reg Res 25(2):227–252
Xiao Y, Huang Q, Wu K (2015) Understanding social media data for disaster management. Nat Hazards 79(3):1663–1679
Zhang D, Zhou L, Nunamaker Jay Jr F (2002), A knowledge management framework for the support of decision making in humanitarian assistance/disaster relief. Knowl Inform Syst 4(3):370–385

Environment as a Weapon: History, and the Hazards of War

Charles Travis

Abstract This chapter will discuss how environmental landscapes, and resources, have been historically weaponized and deployed as human induced hazards in the practice of warfare. The *Integrated History and Future of People on Earth* (IHOPE) project notes that "human history has traditionally been cast in terms of the rise and fall of great civilizations, wars, and specific human achievements. This history leaves out the important ecological and climate contexts that shaped and mediated these events. Human history and earth system history have traditionally been developed independently, with little interaction [...] Therefore, separate methods of describing these histories have been developed, and there have been few attempts to integrate these histories... across these fields of study." Geography, in theory if not in practice, does incorporate and synthesize perspectives from the human and earth sciences. This chapter by employing the lenses of literary, cultural, and historical geography will explore the landscapes of warfare, the human use of environment and resources as weapons, and speculate on the *Gaian* planetary 'counter-insurgency' against *homo-sapiens.* If as Anthropocene discourses speculate—humans have indeed become 'a geological force'—then the classical definition of geo-hazard needs to be modified from traditional precepts to include human induced environmental phenomena. This chapter will provide a selected history on the framing and use of nature as a weapon; secondly it will discuss the industrial and ecological development of war within Anthropocene discourses, and thirdly, reflect upon whether global warming and related phenomena constitute in chemist and biophysicist James Lovelock and micro-biologist Lynn Margulis' *Gaian* geo-biological reframing of Earth System Sciences, a sign that the "earth-goddess" is waging a counter insurgency against the millennial onslaughts of human civilization.

Keywords Environment · War · Historical-literary-geography · Anthropocene · Gaia

---

C. Travis (✉)
Department of History, University of Texas, Arlington, USA
e-mail: charles.travis@uta.edu

Trinity Centre for Environmental Humanities, Trinity College Dublin, Dublin, Ireland

B. Dahiya et al. (eds.), *Disaster Resilience and Human Settlements*,
Advances in 21st Century Human Settlements,
https://doi.org/10.1007/978-981-99-2248-2_8

1 Introduction

On Vladimir Putin's order, the Russian Federation Army's tanks and troop carriers emblazoned with the dripping, spray-painted letter Z crossed the Ukrainian border on 24 February 2022. Military experts had predicted such a move before the spring thaw, as higher temperatures would render the landscape, muddy and intractable for the movement of heavy combat vehicles an artillery. In such a manner, Putin deployed Eastern Europe's climate and landscape as weapons. The military theorist Carl von Clausewitz famously notes in *On War* (1832) that human combat is an extension of political discourse, and though the strategies of war may be drafted on paper, they are executed within the physiography of battle fields where the natural environment and its factors must be considered. Soils, vegetation, geology, topography, watersheds, lakes, seas, oceans and climate become part and parcel in the waging and outcome of wars. "Take the weather," he writes, "Here the fog prevents the enemy from being discovered in time, a battery from firing at the right moment, a report from reaching the general; there, the rain prevents a battalion from arriving [...] the cavalry from charging effectively because it is stuck fast in heavy ground."[1] History provides many accounts where environments have become transformed into weapons, rather than just serving as the passive staging grounds, or neutral theatre of battle.

If as Anthropocene discourses speculate—humans have indeed become 'a geological force'—then the classical definition of geo-hazard, should be modified from its traditional precept, to include human induced phenomena. This chapter will reflect on how environments and natural resources have been weaponized, and how human agency has transformed hazards into tools of warfare. The *Integrated History and Future of People on Earth* (IHOPE) project observes "human history has traditionally been cast in terms of the rise and fall of great civilizations, wars, and specific human achievements. This history leaves out the important ecological and climate contexts that shaped and mediated these events. Human history and earth system history have traditionally been developed independently, with little interaction [...] Therefore, separate methods of describing these histories have been developed, and there have been few attempts to integrate these histories... across these fields of study."[2]

Geography, in theory if not in practice, synthesizes the study of human history, chorology and the earth and environmental sciences. By employing the lenses of literary, cultural, and historical geography, this chapter will provide a selection of examples where the natural environment was historically deployed as a weapon and ally; secondly it will discuss the industrial and ecological development of war within Anthropocene discourses, and thirdly, reflect upon whether global warming and related phenomena constitute in chemist and biophysicist James Lovelock and microbiologist Lynn Margulis' *Gaian* geo-biological reframing of Earth System Sciences, a sign that the "earth-goddess" is waging a counter insurgency against the millennial onslaught of human civilization.

---

[1] Clausewitz (1873), 40.

[2] IHOPE Science Plan v4 6.9, 2.

Fig. 1 Aurochs, Horses and Deer, Lascaux Cave, France. (Wikimedia Commons & Prof. Saxx) Public Domain

2 Environment and War

Arguably, from the time *homo-sapiens* first harnessed fire, the environment has served subsequent civilizations as an arsenal. Both as a means to ward off the climatic and predatory elements of nature (cold and carnivores) and as a tool of internecine combat and subsistence. J. Donald Hughes notes that "hunting was often regarded as a form of warfare, and art often portrayed humans in battle with animals. The idea that hunting and warfare are similar [...] is much older than the classical period,"[3] illustrated by the pre-literate paintings adorning the walls of a warren of caves at Lascaux, France (Fig. 1).

Created by the Magdalenian (c.a. 17,000–12,000 BCE) cultures of the present-day French region of Dordogne, these firelit pre-historic "art galleries" were given the names the *Hall of Bulls*, *Chamber of Felines* and the *Passageway*, after their discovery in 1958. In the Mesopotamian *Epic of Gilgamesh*, (1600–1155 BCE) an ancient saga first inscribed on clay tablets, the eponymous hero and his companion Enkidu hunt and kill the animal-god Humbaba, guardian of the cedar forests of Lebanon. It has been speculated that Enkidu's objective was to harvest and bring timber to construct dwellings and temples on the treeless plains between the Tigris and Euphrates rivers.[4]

[3] Hughes (2013).

[4] Ibid., 129.

2.1 Poisons, Plagues, Pestilence and Sieges

For a millennia, indigenous South American peoples deployed plant and animal derived toxins such as curare and the secretions of poison dart frogs for combat and hunting. In 6 BCE Assyrian soldiers poured ergot contaminated rye into Greek wells to poison the water, and in 429 BCE Spartans at the Siege of Plataea, during the Peloponnesian War, encircled the city with bonfires kindled from resinous pine and placed lumps of sulphur on the flames to create plumes of greasy, toxic smoke. The poisonous fumes were then wafted across Plataea with deadly effect.[5] In the wider classical and imperial Mediterranean, Greco-Roman methods involved dropping bodies of cholera and plague victims over city walls, or leaving human remains "on the ground the enemy was expected to occupy... to poison water sources."[6] In his epic chronicle of the siege of ancient Troy (in present day Turkey), Homer laments in the *Iliad* (8 BCE) how:

> He in wrath
> Against the king had stirred throughout the best
> Fell plague, whereby the troops lay perishing.[7]

Pantelis Michelakis notes that in the *Iliad*, Sophocles' *Oedipus the King*, and Thucydides' *History of the Peloponnesian War* the disease vectors of natural and intentional epidemics depicted in ancient Greek literature:

> ... travel[ed] along the vertical routes followed by gods, prayers, smoke, and smells, and along the horizontal networks followed by sharp arrows, the traffic in prizes of war, and information. The plague exposes and hijacks the porosity of spaces whose physical boundaries and symbolic identities are conventionally perceived to be stable and autonomous.[8]

In the Levant, environmental restrictions on the practice of siege warfare were outlined in the *Book of Deuteronomy*: "When in your war against a city you have to besiege it a long time in order to capture it, you must not destroy its trees, wielding the ax against them. You may eat of them, but you must not cut them down (Deut. 20:19)." Both the Judaic *Torah* and Christian *Book of Exodus* tell of hornet (הָעִרְצ) swarms sent to drive out "the Hivites, the Canaanites and the Hittites," from the promised land of Israel.[9] Locust plagues often invoked in biblical accounts, seem to be a byproduct off ancient, as well as modern Levantine and South Asian warfare. Jacob Wright notes that "during a time of extended military conflict, fields cannot be plowed. In western Asia this situation allows grasshopper egg pods, deposited by the female in the topsoil, to survive and reach maturity. As a consequence, there is a high chance that the region will face a locust plague in the following year. Such

5 Mayor (2008), Madsen (2005).

6 Neufeld (1980), 38.

7 Homer (1884).

8 Michelakis (2020).

9 Neufeld (1980).

was the case, for example, in Afghanistan after the war in 2002. For ancient western Asia, locust plagues after lengthy periods of war are evidenced by various letters."[10]

Bio-weapons have their roots in plagues, such as small-pox which is speculated to have infected Athens in 430 BCE after spreading from Egypt. In the 1500s, smallpox was carried into Mesoamerica by the Cortez and his Spanish armies, resulting in the population of Mexico being decimated in only 50 years from 30 million to less than 2 million, contributing to the downfall of the Aztec, and Mayan empires.[11] In early modern history, the first documented record of the smallpox virus deployed as a weapon is from 1763, when English general, Sir Jeffrey Amherst, donated blankets from a smallpox hospital to indigenous tribes who had allied with the French against the British in the Seven Years War (1756–1763).[12]

During the American Civil War (1861–1865) earthworks, combat trenches, and prisoner of war camps- such as the infamous Confederate camp at Andersonville, in which thousands of Union soldiers perished, due to starvation and disease, illustrate how "the manipulation of the built environment had been done deliberately to engender disease."[13] Antebellum physicians cognizant of "benign" and "severe" malarial disease, knew that its vectors could be mapped geographically," and that it was "a disease that flourished mostly in the lower [American] South with occasional forays into [the states of] North Carolina or Kentucky during especially hot summers."[14] Southern Confederates hoped that malarial and yellow fever could be weaponized to repel an invasion from the northern by Union forces. Subsequently, northern troops dealt with the marshy and fetid fighting environments of Dixie's southern landscape with two strategies: the use of quinine dissolved in whiskey and the marshalling of black soldiers to duty in environments susceptible to malarial conditions.

In addition, charges and suspicions of bio-warfare were levelled against the Union in 1862 after three rebel soldiers returned to the Confederacy in a prisoner exchange presented with the symptoms of small-pox. A full-blown epidemic then broke out in the autumn: "from October 1862 to January 1864 there were 2,513 cases of smallpox among Confederate troops in the East, with 1,020 deaths. While some charged that the Union sent the Fort Delaware cases to Richmond in order to spread disease, others argued that the outbreak occurred because Confederate troops were in areas of Maryland where smallpox prevailed."[15] From ancient Mesopotamia to nineteenth century North America, environments modified by human warfare acted as loci for disease incubation, and transmission. It is a conjecture at best to say whether plagues of insects directed by Moses against Pharoah to free Hebrew slaves could have resulted from such a phenomenon. However the weapon of environmental pestilence described in the *Book of Exodus* anticipates the *Book of Revelation's* four horsemen

[10] Wright (2008).

[11] Geddes (2006).

[12] Ibid.

[13] Humphreys (2016). Ref. 12.

[14] Ibid., 15.

[15] Ibid., 26.

Fig. 2 *Battle of Thermopylae*, Engraving by Jacob Abbott (1803–1879) (*Wikimedia Commons*) Public Domain

of the apocalypse—sword, famine, wild beasts, and plague—the latter three ecocides directed figuratively against the Roman Empire, by John in exile on the Island of Patmos during the reign of Domitian (81–96 CE). Tina Pippin notes that: "Rome is portrayed as burning forever (Rev 19:3) as a woman on the eschatological funeral pyre,"[16] perhaps a biblical foreshadowing of current Anthropocene discourses and the planetary model of *Gaia*—with the earth-goddess in flames due to the ravages of human induced global warming (Fig. 2).

2.2 *Terrain, Wood, Weather and Water*

In 480 BCE the Spartan and Greek armies employed topography to entice a massive Persian invading force led by King Xerxes I into a narrow mountain pass. In dictating the location of the battle, a cohort of 300 Spartan soldiers led by Leonidas, weaponized a ravine at the Battle of Thermopylae.[17] The Spartans fought a pitched battle for seven days before being utterly defeated, but in their loss, they stalled the larger army. This gave the Greeks time to retreat, re-organize, and fight again in a naval clash near the island of Salamis and then win a decisive victory in 479 BCE at Battle of Platae, where saved by their Spartan allies at Thermopylae, the Greeks ended, once and for all the Persian invasion. During the Roman Republic

[16] Pippin (2018). Ref. 182.

[17] Haning (2009).

(509-27 BCE) the frontiers between the 'Eternal City' and mainland Europe were marked by the Alps and landscape features such as rivers, forests, and ridges, which were defended by Roman legions, situated in garrisons, strung out in nodes along military perimeter lines. From the first century BCE to fourth century CE, the Rhine and Danube rivers formed the principal boundaries of such a Roman frontier line. Acting as a border, the Rhine designated the imperial province of Gaul on its western and southern banks from the forest dwelling Germanic tribes. To Roman eyes, the West bank of the Rhine stood for "order"; the East for cultural and environmental "chaos."[18]

In June of 56 BCE Julius Caesar, preparing for the invasion of Britannia, attempted to secure his flank on the Rhine against the Germanic tribes, by constructing a massive wooden bridge to carry his legions over the 300-foot span of the river. A technological marvel of period military engineering, the bridge was seemingly flung across—in the eyes of the Germanic tribes on the eastern bank of the Rhine—by Roman gods. Its presence stunned the tribes of into sending peace hostages or fleeing into the forest. Only the Germanic Suevi people stood their ground and Julius Caesar, after eighteen days of fighting the tribe, recognized their limitless ferocity, and marched his legions back across the Rhine, burning the bridge in his wake. With the consolidation of the Roman Empire under Caesar Augustus (63 BCE-CE 14) limites (paths) across mainland Europe were manned by garrisons hewn from trees culled from the heavily forested regions. As John McNeill notes:

> For about a century and half in northern Europe and for half a century along the Danube frontier, Roman authority rested on these wooden strongholds. A beautiful frieze on Trajan's column in Rome, depicting scenes from the campaigns late in the first century A.D. [CE] along the Danube, testifies to the importance of timber-mostly oaks in this case-in Roman fortification of the early empire.[19]

In 9 CE the prefect-general Quinctilius Varus, under a mandate from Caesar Augustus, led his XVII, XVIII and XIX Legions across the Rhine to pacify Germania and established a Roman province in its eastern forestlands. Arminius, a Germanic chieftain and his tribesmen ambushed and slaughtered Varus' three legions as they marched by an outcrop named the Kalkrieser Berg in the Teutoburg Forest. Arminius, a Cherusci tribesman granted Roman citizenship in the Equestrian Order for having served in the legions, understood the imperial army's strengths and weaknesses. Realizing a frontal assault would be suicidal, Arminius through cunning and deceit crafted an environmental snare, by baiting Varus to send troops to quell a fictitious rebellion in north Germania (Fig. 3).

Luring the three legions into a column that ran for miles along a narrow and boggy track through the dark and rain sodden Teutoburg Forest, Arminius and his tribespeople sprung a trap on 20,000 hapless, and mired Roman soldiers.[20] In the hand-to-hand combat that ensued, Varus' three legions were destroyed, with surviving legionnaires beheaded and sacrificed on bon-fires to the Germanic forest-gods.

[18] Abrecht (2020), 5.

[19] McNeill (2004).

[20] Goulding (2000).

Fig. 3 *Furor Teutonicus* at Kalkrieser Berg, Teutoburg Forest by Paja Jovanović (1899) (*Wikimedia Commons*) Public Domain

Roman historian Tacitus observed in *Germania* (98 CE) how the *Songs of Arminius* came to celebrate the insurgency of the forest tribes against Caesar Augustus's imperial thrust east of the Rhine.[21] The region comprised an environment that Germanic tribes imbued with sacredness: "their holy places are woods and groves, and they apply the names of deities to that hidden presence which is seen only by the eye of reverence."[22] Romans became wise in defeat; six years later in 15 CE, Germanicus Tiberius Caesar, having reconnoitered "obscure forest-passes," and raising "bridges and causeways over watery swamps and treacherous plains," led legions back into the forest where the remains of Varus' soldiers lay. Using the Roman *gladius*, a short-sword suited for close combat, Germanicus countered the attacks of the forest tribes.[23] Visiting the site of Arminius' victory the Romans discovered,

> ... the whitening bones of men, as they had fled, or stood their ground, strewn everywhere or piled in heaps. Near lay fragments of weapons and limbs of horses, and also human heads, prominently nailed to trunks of trees. In the adjacent groves were the barbarous altars, on which they had immolated tribunes and first-rank centurions. Some survivors of the disaster who had escaped from the battle or from captivity,... pointed out too, the raised ground from

---

[21] Tacitus (1999), Byock (2012).

[22] Tacitus (1993).

[23] Goulding (2000).

which Arminius had harangued his army, the number of gibbets for the captives, the pits for the living, and how in his exultation he insulted the standards and eagles.[24]

In addition to woodland topography and terrain, fire was a weapon used by imperial forces and tribal warriors. In the first century BCE a Roman General describes "a massive forest fire set by Germanic Barbarian tribes that burned every bit of ground cover and scorched the soil down to the roots of the trees."[25] In turn, Roman legions deployed the tactic against the forest tribes of Gaul and Germania.

Between the thirteenth and fourteenth centuries, the Mongol Empire (1206–1368 CE) emerged from East Asia, as the most powerful, equestrian military force in world history.[26] It was sustained by a nomadic culture that resourced one of the largest biomes on the face of the earth—the Eurasian Steppe. However, it is not surprising that Mongol Conquests did not penetrate beyond the Volga River and halted in swampy marshes Hungary, Poland and thick forests of Western Europe. Mongol methods of warfare were better suited to open plains, not dense woodlands. Indeed, Khan was forced to limit the size of his army based on the availability and amount of pasture on which his horses could subsist. But a judicious use of the environment and delegation of powers to local leaders allowed successive Mongol Khans to establish a Eurasian Steppe empire that spanned from the Volga to the Amur rivers (the present-day boundary between Russia and China).[27] The *Secret History of the Mongols*, a poetic text written in 1240 CE depicts the environmental dimensions of the long-range sweeps of Genghis Khan's (Temüjin) 'Mongol Hordes':

> The Mongyol which have revolted and are gone out with Temüjin's, whither will those go? Those are,
>
> Ones which have beasts to ride [which are] horses;
>
> Ones which have shelters [which are] trees.
>
> …
>
> Eating remnants [of blades] of grass [for food],
>
> Burning dried dung [for fuel],
>
> …
>
> Passing high passes,
>
> Passing over wide rivers,
>
> Campaigning a long campaign,
>
> Hath thought of setting in order his many peoples.[28]

One result of Genghis Khan's conquest of present-day Eastern Europe, was the consolidation of the Rus and Slav peoples near present day Moscow, leading

---

[24] Tacitus on the Teutoburg Forest, Sect. 1.60–62 of *Annals*, Trans. Alfred John Church and William Jackson Brodribb, 1995–2021, *Livius.org*. https://www.livius.org/sources/content/tacitus/tacitus-on-the-teutoburg-forest/.

[25] Hupy (2008). Ref. 408.

[26] Mongol derives from the toponym *Onongol* for the Onon River that flows through present day Mongolia and the Russian Federation, (*gol* the Uighuro-Mongol word for river).

[27] O'Sullivan and Miller (2015).

[28] Cleaves (1982).

Fig. 4 Napoleon's retreat from Moscow. J. Rousset. 1913. (*Wikimedia Commons*) Public Domain

to the emergence of the Russian heartland between the Ural Mountains and the North European Plain. This landscape became a natural medieval and early modern battlefield and the main theatre of combat in the First and Second World Wars of the twentieth century (Fig. 4).

The processes of Pleistocene glaciation and the Scandinavian ice-sheet that sculpted this landscape created the route along which Charles XII of Sweden in 1708, Napoleon Bonaparte in 1812 and Adolf Hitler's Operation Barbarossa in 1941 invaded Russia. The main thrusts of each campaign followed a glacially inscribed continental drainage divides between Minsk and Smolensk to enter Moscow through its 'River Gate' to the east. However clear the path may have been, the seasons intervened. Respective military expeditions collapsed, in part due attrition, broken supply lines and disease brought about by the summer heat. With the seasonal change to winter, the region's bitterly cold climate and extreme sub-zero temperatures, picked clean the carcasses of the retreating Swedish, French and German armies.[29] Russian poet Mikhail Lermontov's *The Fields of Borodino*, describes the atmosphere following the conclusion of the major battle in Napoleon's invasion of Russia:

The living equaled the dead;
And a cold night came,
Separating those left,
With its black darkness.
Field batteries stopped talking,

[29] Winters et al. (1988).

Fig. 5 Bird's eye view of The Hague, Leiden, Delft. The illustration shows the siege of Leiden, on 3 October 1574. Frans Hogenberg (1540–1590) (*Wikimedia Commons*) Public Domain

The enemy retreated:
But the cost of the day was dear!
Saying in spirit: Miserere!
I reclined my head
On a corpse, hard as if on a bed.[30]

While the North European Plain and its climate seemed to devour any army intent on invading Russia, for the Dutch, the North Sea became a defensive weapon to deploy in the low coastal country of the Netherlands. In the long Dutch quest for independence from the Spanish Habsburg monarchy over the course of the Eighty Years' War (1568–1648) the sea and the Meuse, Rhine and Scheldt rivers, created aqueous garrisons protecting the Netherlands from more powerful neighbors and their armies. During the second Spanish Siege of the city of Leiden in 1574, the Dutch destroyed dikes and flooded the low-lying land with sea-water. Between August and October 1574 Dutch armies engaged in amphibious warfare to drive Spanish troops away (Fig. 5).

In 1629, Prince Frederick Henry engineered flood levels to prevent deep hulled Spanish ships from entering Netherland's harbors, which in turn allowed flat-bottomed Dutch gun boats to strike enemy fleets anchored at sea and then retreat to a

[30] Lermontov (1979), Bagno (2013). Ref. 239.

safe warren of shallow inland waterways. During attacks on its southern flank, Dutch armies retreated north of the rivers, opened sea dikes and flooded the low country southeast of Amsterdam. This defensive system "relied on a high-maintenance hydraulic system of dike, drains, sluices and windmills to keep the seas at bay," but was also "highly vulnerable to human-made and natural disasters."[31] Dutch Golden Age poet Joost van den Vondel referred to Haarlem Lake as the "Water Wolf." Though a successful 'natural' weapon for "chasing foreign enemies," Vodel advocated draining the lake. For the poet, the system stood as the heart and symbol of a human induced natural hazard that occurred when the lake's dikes were broken to fend off foreign armies. But in the process the defensive measure destroyed valuable agricultural land in central Holland, while protecting the coastal enclaves of Dutch burghers and their trading houses:

> Enclose with a dike this animal that plagues you
> That the prince of the winds flies there on his [wind]mill's wings
> The fast prince of winds knows how to chase the Water Wolf
> To the sea, whence he came, ever voracious.[32]

During the French invasion of the *Rampjaar* ('Disaster Year') in 1672, torrents of water and mud halted a French invasion by miring their troops in muck. Despite the success of the "Water Wolf," the period signaled the decline of the Dutch Golden Age. During the Flanders Campaign in the winter of 1794–1795 the "Water Line" froze, allowing French troops to easily cross into the low country.

2.3 *Eighteenth Century Resource Battles and Wars*

The 'discovery' of a 'New World' in the northern and southern hemispheres of the Americas sparked a race between Spanish, French and English empires to control and harness the continental territories and their resources. Gold and silver in Meso and South America, animal pelts, agricultural lands to cultivate cotton, tobacco and sugarcane and seemingly endless forests from which to harvest timber, instigated imperial conflict. In particular, the discovery of the Grand Banks Cod fishery off the shores of North America in 1497 heralded a rush for oceanic protein fostered and European fishery and naval fleet building boom contributing to deforestation, and the emergence of world's oceans as new, global theater for war.[33] Between the European early modern period and the Enlightenment age, increases in population, the multiplication of nation states, and wars pitting bourgeoning empires against each over territories in the 'Old' and 'New' Worlds led to a colossal demand for wood:

> The Atlantic powers-the Netherlands, Britain, France, Portugal, and Spain- all needed navies to maintain their imperial might, and their navies in turn needed ship timber. By the late

[31] Kreike (2021).

[32] Ibid., 24.

[33] Travis et al. (2020).

> sixteenth century, European warships had become large, heavy cannon platforms, built to withstand enemy fire more than to evade it. They were as large as any wooden ships ever built, as much as two thousand tons burden [...] at the beginning of the eighteenth century the building of a warship could consume 4,000 trees- and for smelting iron. Forests were depleted faster than they could regrow.[34]

In May 1762, as the Seven Year's War (French-Indian War) was nearing its close, a flotilla of French warships slipped from the port of Brest on the coast of Brittany past a British naval blockade and set sail across the Atlantic Ocean to Newfoundland. Making landfall, French troops captured St. John's, the capital of the English Grand Banks Cod colony in North America. Whilst the Atlantic Ocean Slave, Tobacco and Sugar Triangle has received just and warranted attention, a Cod, Timber and Sack (wine) Triangle also existed, and its trade infused massive amounts of inexpensive protein into the diets of common folk in Europe. In addition, the capital raised by this North Atlantic trade triangle prompted the Jesuit historian Pierre de Charlevoix to chronicle that the Cod fisheries of the Grand Banks were the "true mines, which are the more valuable, and require much less expense [sic] than those of Peru and Mexico."[35] By taking control of the Cod 'gold mine, and Newfoundland's plentiful spruce forests, the French planned to use these natural resources as bargaining chips in future peace negotiations with the British and to rebuild their greatly diminished navy. Utilizing Newfoundland's hilly and rocky landscape, the French placed artillery on Signal Hill, a promontory that rose above the island providing an outlook over the North Atlantic (Fig. 6).

However, this territorial advantage was turned against the French on 15 September 1762, when amphibious English troops, landing at Halifax Harbour captured Signal Hill, bombarded the French position at Fort William into submission, and recaptured St. John's, the capital of the British Cod Empire. Thirteen years later, the issue of 'Taxation Without Representation' acted as the spark that ignited the American War of Revolution (1775–1778). While taxation was the political 'flintlock,' it was the enormous wealth to be derived from North American natural resources that fired the flame of freedom lurking behind the conventional façade and idealism of United States historiography of the War of Independence. The taxes rebel colonists refused to pay had been levied on staples of tea and sugar, but, it was the royal monopoly on external trade favoring the powerful East India Company that infuriated Boston merchants, and the emerging American oligarchy. This class possessed access to an infinite reservoir of natural resources, and the British crown and the company stood in the way of the accumulation of vast 'American' riches. In February 1775, King George III of Britain asked the House of Commons to declare the colony of Massachusetts in rebellion. Initially perceived as an English Civil War, the transatlantic conflict drew the French navy, and Hessian mercenaries to the home ground of the colonials. The small and hastily assembled rebel armies "utilized their familiarity with the topography" of the Atlantic seaboard and wilderness frontiers of New York, the Ohio Valley, the Appalachians and swampy low countries of the Carolinas

[34] McNeill (2004), Lovelock (2019).

[35] Travis et al. (2020), p. 3.

Fig. 6 1762 Perspective view of the French attack on Newfoundland near St. John's. (Wikimedia Commons) Public Domain

to their advantage. Continental armies, drawn from militias formed as early as the 1620s to protect colonial settlements from indigenous tribes, had been blooded and wizened by the Seven Years Wars (1754–1763). Generals like George Washington, and Nathanael Green adopted proven methods of wilderness warfare practiced by native peoples. In the 1760s, Swiss born Henri Boquet, a British Army brigadier general, warned soldiers trained in conventional English army tactics that they would "receive incessant fire," "from every tree, log or bush," and find themselves "surrounded by a circle of fire,... like an artificial horizon."[36] Bouquet's advice was largely ignored by the military strategists schooled in the orthodox tactics of early modern European warfare.

The climatic conditions of the Little Ice Age (1300–1860 BCE) also impacted the course of the American Revolution.[37] Generals in the eighteenth century rarely led armies into battle during the winter season. Snow, sleet and freezing rain hampered fighting conditions, turning dirt roads into muddy tracks, bogging down marching columns, and hindering the transport of artillery pieces and supply wagons. Subzero temperatures exposed soldiers to hypothermia and made troops vulnerable to illnesses. Generally early modern armies assembled in winter quarters to stay warm,

36 Tierney (2006), 26.

37 Glenn (1996).

preserve strength, numbers, and wait for spring thaws in order to resume military campaigns.

In contrast, General Washington and his de-moralized army's 1776–1777 winter campaign in New Jersey endured weather conditions to surprise a Hessian garrison at Trenton on 26th December and claim a victory over the British at the Battle of Princeton on the 3rd of January. The capture of the Hessian garrison at Trenton followed Washington's famous crossing from Pennsylvania to the New Jersey shore across the ice chocked waters of the Delaware River on flat-bottomed Durham Boats, designed to carry steel processed from regional iron deposits, discovered in the 1600s. On Christmas morning 1776, Washington received intelligence that British forces were planning to cross the west bank of the Delaware once the river was frozen over. A 32-degree fahrenheit isotherm (freezing) line fluctuated with conditions of freezing rain and sleet upstream and showers of rain and ice pellets falling downstream, with Trenton in the center of this zone of mixed precipitation.[38] Crossings were attempted twenty miles south of Trenton, across from the garrison itself, and three miles north of the town at McKonkey's Ferry as immortalized in Emanuel Leutze's dramatic 1851 painting The latter, led by Washington was the largest, and most successful; the first two were hindered by heavy ice—which troops but not heavy artillery could traverse. Crossings were underway at 3.00 A.M. on the 26th of December, with Washington's staff recording that "the storm is changing to sleet and cuts like a knife."[39] An anonymous poem, titled *The Battle of Trenton*, composed by a soldier who participated in Washington's attack, describes the weather conditions facing his troops (Fig. 7):

On Christmas-day in seventy-six,
Our ragged troops, with bayonets fixed,
For Trenton marched away.
The Delaware see! the boats below!
The light obscured by hail and snow![40]

After the crossing, Washington's forces split into two groups and marched south towards Trenton, with the wind and sleet blading their backs. With their flintlocks rendered useless, the Americans deployed bayonets, and the stinging rain and ice pellets of a forceful north-western gale that blinded the Hessians from seeing the rebel forces descending upon them.[41] The attack commenced at 8.00 AM, and lasted but an hour, with the garrison surrounded and two-thirds of its Hessians killed, wounded, or captured.[42] On New Year's Day 1777, the British under General Lord Charles Cornwallis learned of the capture of Trenton. Deploying forces from Princeton on 2 January during a thaw, troops marched along roads, that were "very muddy, almost

[38] Ludlum (1976).

[39] Ibid., 79.

[40] Anonymous, 'The Battle of Trenton' in Henry Wadsworth Longfellow (1876–79).

[41] Ibid.

[42] Ibid.

Fig. 7 Washington Crossing the Delaware, by Emanuel Leutze (1851). (Wikimedia Commons) Public Domain

over our shoe tops."[43] At Trenton, Washington found himself encroached between the large approaching British force, muddy roads and an ice-choked Delaware River. Cornwallis retired for the night, assuming he "would bag the fox," the next day.[44] However Washington, a Virginia planter, attuned to reading the weather, perceived the tell-tale signs of an approaching cold front. Nightfall on 2 January found temperatures plummeting, freezing the muddy roads. Muffling wagon and cannon wheels with cloth, Washington took advantage of a 15-hour freeze to slip his troops past Cornwallis' flank and march 18 miles northeast over frozen roads to surround and defeat British forces at Princeton on 3 January.[45] Though the British viewed the loss of these two winter battles as but minor defeats, American morale was galvanized. The victories boosted the confidence of Washington's continental and militia forces into actually believing the war against King George's army could be won. As a consequence, many soldiers whose enlistments had expired, joined up again, much to the chagrin of the British generals who had counted on the mass demobilization of colonial soldiers after the brutal winter slog to quell the rebellion.

The weather however, proved to be a double-edged sword. In the winter of 1777–1778 temperatures at Valley Forge, Pennsylvania, almost froze the revolution in its tracks. Nearly a quarter of the American army—2500—died from cold and disease. General Washington wrote to the Continental Congress that his troops were not

[43] Ibid., 80.

[44] Ibid., 81.

[45] Ibid., 79.

"made of sticks and stones and equally insensible of frost and snow," lamenting that "three or four days of bad weather would prove to be our destruction."[46] As history records, his army survived, and Washington's experiences of combat in the Ohio Valley with the Iroquois, Shawnee and Delaware during the French-Indian War, were brought to bear. In 1779, to end the Iroquois military alliance with the British army, American troops turned to methods of environmental warfare and targeted the warrior bands' "means of subsistence: Thousands of Iroquois dwelling houses were destroyed in a scorched-earth manner and whole crops were uprooted," forcing the Iroquois to break its pact with the British.[47] This established a precedent for the U.S. Army—Indigenous American Wars of the nineteenth century, by deploying the strategy of the ecological targeting of tribal subsistence, as in the case of the Buffalo War to destroy the major food source of the Great Plains Lakota Sioux. Like Arminius' Germanic tribespeople facing larger Roman legions, forestlands, swamps, valleys, and mountain ranges provided places from which American generals could launch strategic attacks, retreats, and coordinate with French naval forces, such as the land and sea campaign that dealt a fatal blow to the British with the capitulation of Lord Cornwallis after the Siege of Yorktown in 1781. A poem *On the Departure of the British from Charleston* published in 1782 celebrates American General Nathaniel Greene's unconventional use of the terrain:

> Through barren wastes and ravaged lands,
> He led his bold undaunted bands;
> Through sickly climes his standard bore
> Where never army marched before.[48]

The larger, but slower British troops, contending with French naval forces at sea, and rebel colonists who could fade invisibly into the wilderness, could not penetrate into the back-country effectively to destroy American forces.

3 The Anthropocene and Ecological Warfare

The American Civil War (1861–1865) as arguably the world's first Anthropocene war, can be framed as an internecine conflict between an industrialized, mechanized North, and the feudal, agricultural slaveocracy of the South. As a battle for the nation's soul in which preservation of the Union, and freedom for the enslaved was pitted against the specious issue of State's rights, war manifested as "an ecological event that not only affected people but also altered natural systems and reshaped the already complex interaction between humans, other organisms, and the physical environment."[49] As Lisa M. Brady notes the, "landscape was not simply a backdrop

---

[46] Winters et al. (1988), p. 91.

[47] Tierney (2006), 6.

[48] Freneau (1809).

[49] Browning and Silver (2020), p. 4.

to the events of the war—a place where battles took place—but a powerful military resource and an important factor in military decision making."[50] Strategies blending tactics, trenches and terrain, evolved as climate, disease, aqueous and geological features, physiography, and topography shaped the various theatres of battle from the Potomac River, to Virginia, Georgia, Tennessee and Kentucky and across the Mississippi Valley and into Texas and California.[51] For example, in 1862 severe meteorological conditions during the Battle of Antietam (Sharpsburg) and the siege of Vicksburg, Mississippi exemplify how weather, crops, game and food staples were weaponized during the war (Fig. 8).[52]

Geographically, the scale and extent of warfare increased during the war in size because of greater number of rifles and ammunition supplied by industrial revolution armament factories.[53] For instance, Lee's 1862 invasion of the northern states of Maryland and Pennsylvania culminated in the Battle of Antietam on 17 September recorded the highest death toll experienced to that point in United States history, prompting the poet Emily Dickinson to write:

The name—of it—is "Autumn"—
The hue—of it—is Blood—
An Artery—upon the Hill—
A Vein—along the Road—

Great Globules—in the Alleys—
And Oh, the Shower of Stain—
When Winds—upset the Basin—
And spill the Scarlet Rain—[54]

The disemboweled landscape which Dickinson's poem "depicts is a strikingly accurate transcription of the terrain around Sharpsburg, the town through which Antietam Creek flows,"[55] and the poetess perceived the Civil War as an ecological catastrophe entangling the human body and the environment on a savage and intimate scale. Evocations of "arteries upon the hills, bodies of stone, winds that make as to devour" in her war poems indicate Dickinson's ecological imagery "may be far less metaphoric than we assume."[56] The conflict between an industrial north and an agricultural south juxtaposes medieval and industrial age ideologies and infrastructures (feudal, slave plantations protected by Confederate calvary raids and entrenchments versus Union troops transported to battlefields by superior railroad and steamship systems, coordinated by generals across telegraph lines), which Cody Marrs argues signals the emergence of the Anthropocene: "To feed the war machine,

[50] Brady (2005).
[51] Winters (2001), Hess (2018).
[52] Browning and Silver (2020).
[53] Hupy (2008). Ref. 408.
[54] Dickinson (1983).
[55] Hoffman (1994). Ref. 10.
[56] Marrs (2017). Ref. 218–219.

Fig. 8 Battle of Antietam. 19 September 1862 (*Wikimedia Commons*) Public Domain

foundries were erected, forests destroyed, animals slaughtered, mines depleted, railroads extended, and guns and shells and bullets produced *en masse*—all to sustain massive, locust-like armies that left little in their wake and navies forged for the first time out of iron and powered by steam."[57]

Despite being powered by the industrial might of the North, until the Battle of Gettysburg in 1863, Confederate Generals Robert E. Lee and "Stonewall" Jackson were able to deploy topography and terrain as weapons so skillfully, that the plodding Union army under a succession of mediocre generals was left thwarted and demoralized. In response, President Abraham Lincoln, promoted the pugnacious Ulysses S. Grant to the position of Lieutenant-General of the Union Army. Unlike his predecessors Grant, now holding the highest rank ever achieved by a soldier, did not retreat in the face of Confederate advances and he and his Major General William Tecumseh Sherman, would go on to wage 'total war' against the Confederate army and the plantation economy it was fighting to preserve (Fig. 9).

The Battle of the Wilderness in 1864, prior to the capture of the Confederate capital of Richmond, Virginia, signaled the intent of Grant's war of annihilation. The dense woodland near the Rapidan River, known as the Wilderness, seemed to suit Lee's deployment of environment as a weapon, as its rough terrain would shelter Confederate forces and hinder Union artillery. But gunpowder smoke from close fighting blinded soldiers and exploding Union shells sparked a conflagration in the dry forest. The Wilderness flared into an inferno consuming both Union and Confederate troops. Approximately 30,000 men on both sides were killed or wounded, but as the landscape burned and his men died, Grant refused to retreat.[58] In November and December of 1864, Sherman's campaign in Georgia enforced Abraham Lincoln's Emancipation Proclamation and left the burning ruins of cotton plantations in the wake of

[57] Ibid., 220.

[58] McPherson (1988).

Fig. 9 Human Skulls at the site of the Battle of the Wilderness, 1864. (Wikimedia Commons) Public Domain

the northern army as it marched to the sea. Sherman's scorched earth warfare was deployed effectively to decimate the labor, capital and resource base of the Confederacy, leading eventually to Lee's surrender to Grant at Appomattox Courthouse, Virginia in 1865.[59] Union strategy destroyed the "enemy's primary relationship with the natural world," by exposing "the tenuous nature of southerner's control"[60] over an environment dominated by a slaveocracy. In the final analysis, the prosecution of the Civil War unleashed and illustrated how industrial scale slaughter could be achieved by deploying the environment as a weapon. At the height of the war in 1864, George Perkins Marsh, perhaps taking note of the ecological upheaval caused by the war, published *Man and Nature*, arguing that industrialized civilization had commenced "an almost indiscriminate war-fare upon all the forms of animal and vegetable existence" as it advanced towards the chimera of a technological utopia.[61]

3.1 Global Scales

The conflicts of the First (1914–1919) and the Second (1939–1945) World Wars marked the global industrialization of war and weaponization of the environment on a planetary scale. The German School of Geopolitik appropriated the bio-geographical arguments of Friedrich Ratzel's *Der Lebensraum* (1904) which equated national territorial aspirations to an organism's search for "living-space". This manifested in Hitler's commands to the German army to occupy by stealth, and perform blitzkrieg,

[59] Ibid.

[60] Brady (2005).

[61] Marsh (2003).

Fig. 10 Huang He (Yellow River) flooding, June 1938, Source: 中华民国史画 (Wikimedia Commons) Public Domain

lighting invasions to gain territory in Poland and the Ukraine in order to acquire raw resources and expand Nazi Germany's *lebensraum.*[62] The Vietnam (1955–1975) and first Persian Gulf (1991–1992) Wars serve as examples of how mechanical, chemical, and petrological industries came to harness and employ environmental elements as weapons of destruction (Fig. 10).

Second to the atomic bombings of Hiroshima and Nagasaki in 1945, the largest act of environmental warfare took place during the Second Sino-Japanese War in June 1938. To stop advancing Imperial Japanese forces, the Chinese Nationalist army dynamited the Huayuankow Dike on the Huang-He (Yellow River). The tremendous torrent unleashed near Chengchow cascaded down the Yellow River watershed sweeping through Honon, Anhhwei and Kiangsu provinces. Several hundred thousand people drowned and many more million were left homeless. The flood destroyed eleven cities, four-thousand villages and washed away several million hectares of crops and topsoil.[63]

The *New York Times* reported that the flooding may have benefited the Japanese nearly as much as the Chinese, as it allowed imperial troops to safely re-group and mobilize south along the east side of the flood zone without fearing a Chinese attack on their flanks. However, after the waters receded, the Yellow River settled into a new channel, and the Japanese lost control of the flooded provinces to guerrilla and

62 Travis (2001–2002).

63 Westing (1980).

bandit groups. Later, during the Chinese Civil War of 1945 to 1959, nationalists experienced a political backlash over their war-time flooding strategy as people's anger in the inundated regions of the Huang He caused many to join the Communists and march under Mao's 'Red Banner.'[64]

The Vietnam War (1955–1975) and first Persian Gulf (1991–1992) War serve as examples of how chemical and petrology industries transformed environments into weapons of destruction. During the 1960s and early 1970s in Southeastern Asia, tactics employed by soldiers in the American War of Independence (1775–1782) were turned against U.S. forces in the tropical lowlands, hills and highlands of the Indochina Peninsula. The Vietcong built "Ho Chi Minh" trails winding north and south through the dense tropical forests of Vietnam and kept open supply lifelines for their "hit and run" guerrilla fighters. American soldiers, like British redcoats before them, rarely glimpsed the elusive enemy and were often ambushed in the thick foliage of Vietnam's jungles that cloaked Vietcong mines and snipers. Atrocities like the massacre of unarmed villagers at My-Lai in 1968 echoed George Washington's tactics against the Iroquois. Subsequently, increased efforts were made by U.S. forces to defoliate the dense tropical Selva covering the trail and tunnel complexes supplying and garrisoning North Vietnamese fighters. Explosive munitions, herbicides (such as the dioxin Agent Orange), the use of Rome Plows (militarized bulldozers with stinger blades) and massive aerial bombardments from B52 bombers inflicted industrial scale damage to the Mekong River watershed, but failed to dislodge a guerrilla army embedded in its own landscape (Fig. 11).[65] The die for American military engagements in the two Persian Gulf Wars was perhaps cast in the 1930s when Standard Oil of California geologists discovered industrial scale reserves of oil on the eastern shores of the Saudi Arabian peninsula. For most of the twentieth century, preserving the flow of petroleum and securing Saudi-Arabia and Gulf Arab States were among the highest priorities of the United States' geo-political strategy (Fig. 12).

In 1990 a dispute over slant-petroleum drilling in the giant Rumaila oil field precipitated Iraq's invasion of Kuwait, and launched the first Persian Gulf War (1991–1992), as U.S. President George H.W. Bush, a former Texas oil-man, led Coalition forces in Operation Desert Shield to oust Saddam Hussein's army. In retreat, Hussein ordered Iraqi troops to set Kuwait's oil refineries and fields ablaze. Dialogue from William Broyle's screen play for *Jarhead* (2005) a film by Sam Mendes, depicts U.S. Marines wandering the toxic, apocalyptic landscape under a rain of crude oil spewing from burning Kuwaiti well heads:

> **Private Chris Kruger**: They lit up the oil wells. It's raining oil. It's raining oil, fellas. You ever see that movie *Giant*? You've seen the movie *Giant*. James Dean, man. "My well came in, Bick. I'm rich, Bick. Richer than you."
>
> **Corporal Alan Troy**: Come on, shut your trap. That shit's poison. Come on.
>
> **Private Anthony Swofford**: The earth is bleeding.
>
> **Staff Sgt. Sykes**: Well, you'd better get used to it, 'cause we're gonna be living in it.[66]

---

64 Dutch (2009).

65 Hupy (2008). Ref. 408.

66 Broyles (2005).

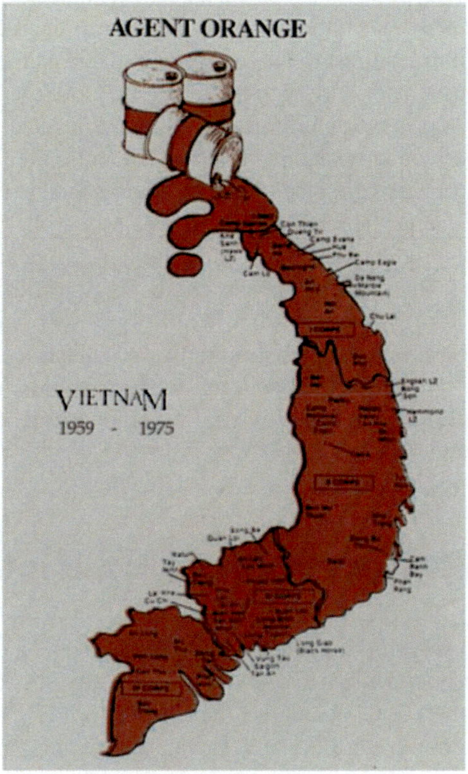

Fig. 11 Agent Orange Poster (*Wikimedia commons*) Public Domain

Fig. 12 Kuwaiti Oilfields on Fire. Jonas Jordan, United States Army Corps of Engineers (*Wikimedia Commons*) Public Domain

The burning of an estimated four to six million barrels of crude oil created a human induced geo-hazard. In contrast to the Eurasian Steppe pasturelands that determined the extent of Genghis Khan's thirteenth century empire, oil reserves, particularly since the Second World War, have determined the geographies and scales of mechanized combat. The U.S. oil embargo against Imperial Japan in 1942 instigated the latter's attack on Pearl Harbor, bringing the 'sleeping giant' of industrial America into wars in Europe and the Pacific. The arrival of America as a global power on the world's stage, is today countered by the People's Republic of China and the island building strategy of its navy in the South China Sea, another region with purportedly untapped oil reserves.

The ravenous consumption and burning of fossil fuels accelerated from the mid-nineteenth century and the Green-House gas effect would come to symbolize the monstrous potential of a new and Industrial Age human agency to unleash itself as a geological force.[67]

4 *Gaia* at War: The Nuclear Age and Global Warming

On 16 July 1945, a polychromic mushroom cloud rose over a desert in New Mexico named the *Jornada del Muerto* (*Journey of Death*) by sixteenth century Spanish conquistadores. Robert G. Oppenheimer the physicist who led the Manhattan Project that assembled the bomb recalled in a television interview in the 1960s that after viewing the event in the desert landscape:

> We knew the world would not be the same. A few people laughed, a few people cried. Most people were silent. I remembered the line from the Hindu scripture, the *Bhagavad Gita*; Vishnu is trying to persuade the Prince that he should do his duty and, to impress him, takes on his multi-armed form and says, "*Now I am become Death, the destroyer of worlds*." I suppose we all thought that, one way or another.[68]

The *Jornada del Muerto* where Trinity Base is located, is the homeland of the indigenous Pueblo and Apache peoples. Its physical geography evoked a "sense of wonder" among the Los Alamos physicists, with "its striking and vexed status as 'nature'" bearing "heavily upon how the Manhattan Project and how science in general-purports to be about the manipulation of nature."[69] The detonation, shock-wave and nuclear signal generated by the fission bomb at the Trinity test site inaugurated the Atomic age, just as James Watt's steam engine kick-started the Industrial Age, and the pre-historic harnessing of fire created a lit the path to emergence of human civilization (Fig. 13).

These milestones in human technical history contributed to the concept of the Anthropocene. Coined by Paul Crutzen in 2002, there have been various arguments about identifying exactly what carbon or man-made nuclear stratigraphic signal

[67] Hupy (2008). Ref. 408.

[68] Pontin (2007).

[69] Banco (2012). Ref. 494.

Fig. 13 Trinity test base camp, edge of *Jornada del Muerto* desert, New Mexico, May 1945. (*Wikimedia Commons*) Public Domain

embedded in the earth's sedimentary layers, ice cores, or tree rings defines this new geological era.[70] The debate to locate the origins of an epoch to succeed the Holocene, is situated at an ontological/epistemological borderland across which civilization, technology and environment converge, intersect and diverge in endless loops. Oppenheimer, in an instance that illustrates Bruno Latour's definition of ecological crises (which "have no clear boundaries, no sharp separation between their own hard kernel and their environment") explained the impetus behind the construction of the atomic bomb:

> The reason we did this job is because it is an organic necessity. If you are a scientist you cannot stop such a thing. If you are a scientist you believe that it is good to find out how the world works; that it is good to find out what the realities are.[71]

The Manhattan Project, a hybridization of culture and environment, harnessed the forces of the nature to create an unprecedented weapon of mass destruction. Fission bomb detonations over Hiroshima, and Nagasaki illustrate the full monstrosity of remediating the environment as a weapon. William Golding's *Lord of*

[70] Crutzen (2002), Zalasiewicz et al. (2011), Anthropocene Working Group (2019).

[71] Marzec (2015), 147.

the Flies (1954) tells the story of a group of British schoolboys marooned on a tropical island as the result of a nuclear war; they descended into barbarity, engaging in tribalism, bloodletting and hunting their own, before being rescued by a warship, itself an object involved in savage warfare. It is ironic that the power of a nuclear conflagration contextualizes the novel's tale of human regression to a more primitive state where boundary between hunting and warfare are blurred. And it was Golding who suggested the name *Gaia*—the earth goddess—to geophysicist James Lovelock and his re-modelling of Earth Systems Sciences in 1972, which argues that the earth acts like a self-regulating 'organism': "able to homeostat the planet for an optimum physical and chemical state appropriate to its current biosphere."[72] Leah Aronowsky notes that elements of Lovelock's idea derived from research he conducted under the sponsorship of the multi-national oil and gas company Royal Dutch Shell "to identify organisms whose biological activities might double as climate-stabilizing mechanisms."[73] Furthermore she claims that *Gaia,* as a *homo-sapiens* de-centering model constitutes an environmental Rorschach test that is "equally appealing to free-market evangelists, Earth-systems scientists, ecofeminists, and science-studies scholars alike," and that "such a displacement of human exceptionalism can be leveraged equally for a doctrine of neoliberal environmental governance or for an embrace of radical biological alterity."[74]

In contrast, Bruno Latour in his *Gifford Lectures*, observed that understanding *Gaia's* at times unpredictable bio-geographical rhythms "depends upon constant observation" and likened Lovelock's earth goddess model to "an opera... that has neither score nor an ending, and it is never performed twice on the same stage."[75] Furthermore, Lovelock's perspective, enhanced by micro-biologist Lynn Margulis' work, recognizes the inadequacy of the disciplinary silos of modern political-economics, Victorian biology and classical physics in addressing the planetary scale "wicked problem" of global warming.[76] Lovelock observed that "*Gaia* theory adds to Darwin's great vision. There is no longer any need to consider the evolution of the species separately from the evolution of their environment. The two processes are tightly coupled as a single indivisible process."[77] As a self-regulating process, *Gaia,* like one of its smallest components, the cell, will protect its homeostasis and is "weaponizing" itself against anthropomorphic threats. Reflecting on *Gaia,* Latour employed a "war of worlds" environmental heuristic, of two opposing camps: the 'moderns' who believe it is possible to go on living in the rational, united and apathetic nature of the Holocene, and the 'terrans' people of *Gaia.*

Is the COVID-19 virus that spread from Wuhan, China in 2019, to a global pandemic in 2020 an example of a *Gaian* weapon, or a further instance of the human weaponization of nature? As Francesco de Pascale and Jean-Claude Roger observe

---

72 Lovelock (1972), 579.

73 Aronowsky (2021), 308.

74 Ibid.

75 Latour (2017), 107.

76 Lovelock (1989).

77 Ibid., 222.

the: "Coronavirus, in fact, is nothing more than a Latourian hybrid an object belonging to both nature and culture, a 'product' of the Anthropocene,"[78] and one imbricated in the cascading effects of global warning on biological, geological, oceanic and atmospheric scales. In 2014 the U.S. Secretary of State John Kerry was ridiculed in some quarters for stating that "in a sense, climate change can now be considered another weapon of mass destruction, perhaps even the world's most fearsome weapon of mass destruction."[79] However, Donna Haraway, drawing on Isabelle Stenger's notion of *Gaia* as an 'intrusionary' force eroding the foundations of the "tales and refrains of modern history," states:

> This intrusion threatens not life on Earth itself -microbes will adapt, to put it mildly- but threatens the livability of Earth for vast kinds, species, assemblages, and individuals in an "event" already under way called the Sixth Great Extinction [...] *Gaia* does not and could not care about human or other biological beings' intentions or desires or needs, but *Gaia* puts into question our very existence, we who have provoked its brutal mutation that threatens both human and nonhuman livable presents and futures. *Gaia* is not about a list of questions waiting for rational policies.[80]

In the first two decades of the twentieth century, the American military has come to conclude that global climate-change poses a significant strategic challenge to the United States and classifies it as a national security threat. It is thought that between 2020 and 2050 global warming will raise the need for U.S. and other nations' military interventions to deal with humanitarian crises due to violent storms, drought, mass migration and pandemics. In the U.S., the Pentagon, intelligence agencies, security analysts and war game modelers speculate that such crises have the potential to depose governments, nourish terrorist organizations, destabilize world regions—particularly in sub-Saharan Africa, the Middle East and South and Southeast Asia—due to catastrophic flooding and political and economic conflict over water, food and energy resources.[81] One of the impacts of the U.S. military's perspective on the threats of global warming in the American political sphere, is that conservative individuals and groups are more likely to express concern about "climate change", when it is framed as a national security risk and communicated by the armed services, than by scientists or members of the perceived 'biased' professorate.[82]

78 De Pascale and Roger (2020).

79 Mohammed (2014).

80 Haraway (2016).

81 Barnett, Security and climate change, *Global Environmental Change*, 13.1., 2003: 7–17, Broder (2009), Melton (2019).

82 Motta et al. (2021).

5 Conclusion

Relations between environment and war stem from an ancient lineage, an association symbolically featured as William Bowen and Robert Gleeson observe in the urban landscape of ancient Mesopotamian metropolises: "At the center of the city were two large sanctuary complexes, one for Inanna, the goddess of love and war, and one for An, the sky god."[83] In our current age, Putin's invasion in the name of resurrecting a Neo-Czarist Russian Empire, like Gilgamesh and Enkidu's savaging of the Cedars of Lebanon, is underscored by the Russia's desire to possess Ukraine's oil and wheat reserves and warm water Black Sea ports, resources that can be further weaponized against N.A.T.O. and Western Europe. Reporting on benchmarks of human progress United Nations (U.N.) Secretary General António Guterres observes: "seventy-five years ago, the world emerged from a series of cataclysmic events: two successive world wars, genocide, a devastating influenza pandemic and a worldwide economic depression. Our founders gathered in San Francisco promising to save succeeding generations from the scourge of war."[84] In 2016 a U.N. Resolution on the "New Urban Agenda" called for special attention to "be given to countries in situations of conflict, as well as countries and territories under foreign occupation, post -conflict countries and countries affected by natural and human-made disasters."[85] The particular physiographic and climatic features of such countries and the ability to weaponize them must also be taken into account, if we are to implement the *Sendai Framework for Disaster Risk Reduction 2015–2030* report's recommendation to strengthen "baselines and periodically assess disaster risks, vulnerability, capacity, exposure, hazard characteristics and their possible sequential effects at the relevant social and spatial scale on ecosystems, in line with national circumstances."[86] Furthermore, the report *Transforming Our World: The 2030 Agenda for Sustainable Development* recognizes that "sustainable development cannot be realized without peace and security; and peace and security will be at risk without sustainable development."[87] Therefore, holistic approaches to studying and mitigating warfare, global warming, and promoting peace and sustainability could benefit studying relations between Gaia and war. In doing so, we can better understand how planetary systems are weaponizing against the toxic impact our species is having upon the lands, oceans and atmosphere of this planet. To conclude, as Sun Tzu mused in *The Art of War* (5 BCE): "We are not fit to lead an army on the march unless we are familiar with the face of the country—its mountains and forests, its pitfalls and precipices, its marshes and swamps."[88] Therefore we need to better reconnaissance the weaponized and weaponizing environments of our planet. To reiterate, if as Anthropocene discourses speculate that *homo-sapiens* has indeed become 'a

[83] Bowen and Gleeson (2018), 129.

[84] United Nations (2021).

[85] Unite Nations (2016).

[86] United Nations (2015).

[87] United Nations (2015).

[88] Tzu (2020).

geological force'—then the classical definition of a geohazard should be modified from its traditional Holocene precepts, to include the human induced phenomena that spawned the industrial and nuclear ages of the nineteenth and twentieth century. The confluence of these cultural and natural forces has paradoxically weaponized our environments, raising the specter of a counter-insurgency by *Gaia*-the earth goddess, to manifest as the true 'destroyer' and extinguisher of our global, human civilizations.

References

Abrecht RR (2020) Places of possibility: shape shifting in the roman and Chinese borderlands. In: Tamai LAYW, Dineen-Wimberly I, Spickard P (eds) Shape shifters: journeys across terrains of race and identity. University of Nebraska Press, Lincoln, pp 57–82

Anthropocene Working Group (2019) Results of bidding vote by AWG. http://quaternary.stratigraphy.org/working-groups/anthropocene/

Aronowsky L (2021) Gas guzzling Gaia, or: a prehistory of climate change denialism. Crit Inq 47(2):306–327

Bagno VE (2013) The war of 1812 and Russian culture. Herald Russ Acad Sci 83(3):235–239

Banco LM (2012) The biographies of J. Robert Oppenheimer: Desert Saint or destroyer of worlds. Biography 492–515

Bowen WM, Gleeson RE (2018) The evolution of human settlements: from Pleistocene origins to Anthropocene prospects. Springer

Brady LM (2005) The wilderness of war: nature and strategy in the American civil war. Environ Hist 10(3):421–447

Broder JM (2009) Climate change as threat to U.S. security. The New York Times. https://www.shoreline.edu/gac/documents/great-discussions-2013/nyt-climate-change-as-threat.pdf

Browning J, Silver T (2020) An environmental history of the civil war. University of North Carolina, Press, Chapel Hill

Byock JL (2012) The saga of the Volsungs: the Norse epic of Sigurd the dragon slayer. Univ of California Press

Clausewitz CV (1873) On war. Trans. J.J. Graham. N Trübner & Co., London

Cleaves FW (ed) (1982) Trans. The secret history of the Mongols. Harvard University Press, Cambridge, pp 100, 173, 189

Crutzen PJ (2002) Geology of mankind. Nature 415:23

Dickinson E (1983) The poems of Emily Dickinson. In: Johnson TH (ed) Belknap Press of Harvard University Press, Cambridge

Dutch SI (2009) The largest act of environmental warfare in history. Environ Eng Geosci 15(4):287–297

Freneau PM (1809) Poems written and published during the American revolutionary war, and now republished from the original manuscripts: interspersed with translations from the ancients, and other pieces not heretofore in print, vol 1. From the Press of Lydia R. Bailey, no 10, North-Alley, p 416

Geddes AM (2006) The history of smallpox. Clin Dermatol 24(3):152–157

Glenn W (1996) Harold. Integrating teaching about the little ice age with history, art, and literature. J Geosci Educ 44(4):361–365. https://doi.org/10.5408/1089-9995-44.4.361

Goulding Jr VJ (2000) Back to the future with asymmetric warfare. US Army War Coll Q Parameters 30(4):7

Haning T (2009) Geography of war: the significance of physical and human geography principles. Focus Geogr 52(1):32–36

Haraway D (2016) Tentacular thinking: Anthropocene, Capitalocene, Chthulucene, e-Flux J #75. https://www.e-flux.com/journal/75/67125/tentacular-thinking-anthropocene-capitalocene-chthulucene/

Hess EJ (2018) Fighting for Atlanta: tactics, terrain, and trenches in the civil war. Chapel Hill, University of North Carolina Press

Hoffman TB (1994) Emily Dickinson and the limit of war. Emily Dickinson J 3:1–18

Homer (1884) The Iliad of homer with a verse translation. Trans. William Charles Green. Arkose Press, UK, p 3

Hughes JD (2013) Warfare and environment in the ancient world. In: Campbell B, Trittle LA (eds) The oxford handbook of warfare in the classical world. Oxford University Press, Oxford, p 129

Humphreys M (2016) This place of death: environment as weapon in the American civil war. South Quart 53(3):12–36

Hupy JP(2008) The environmental footprint of war. Environ Hist 405–421

Kreike E (2021) Scorched earth: environmental warfare as a crime against humanity and nature. Princeton University Press, p 25

Latour B (2017) Facing Gaia: eight lectures on the new climatic regime. Polity, Cambridge, p 107

Lermontov MY (1979) Collection of works in 4 Vols, vol 1. Leningrad (in Russian), p 265

Lovelock J (1972) Gaia as seen through the atmosphere. Atmos Environ 6:579–580

Lovelock JE (1989) Mill, geophysiology, the science of Gaia. Rev Geophys 17:215–222

Lovelock J (2019) Novacene: The coming age of hyperintelligence. Penguin, Random House, UK, p 33

Longfellow HW (ed) (1876–79). Poems of places: an anthology in 31 volumes. America: vols. XXV–XXIX

Ludlum DM (1976) The weather of independence-5: Trenton and Princeton. Weatherwise 29(2):74–83, 76

Madsen JM (2005) Bio warfare and terrorism: toxins and other mid-spectrum agents. Army Med Res Inst Chem Defense Aberdeen Proving Ground MD

Marrs C (2017) Dickinson in the Anthropocene. ESQ: J Nineteenth-Century Am Lit Cult 63(2):201–225

Marsh GP (2003) Man and nature. University of Washington Press, p 40

Marzec RP (2015) Militarizing the environment: climate change and the security state. University of Minnesota Press

Mayor A (2008) Greek fire, poison arrows, and scorpion bombs: biological & chemical warfare in the ancient world. Penguin

McNeill JR (2004) Woods and warfare in world history. Environ Hist 9(3):388–410, 391

McPherson JM (1988) Battle cry of freedom: the civil war era. Oxford University Press, Oxford

Melton M (2019) Climate change and national security, part II: how big a threat is the climate? Lawfare (blog). https://www.lawfareblog.com/climate-change-and-national-security-part-ii-how-big-threat-climate

Michelakis P (2020) Routes of the Plague in Homer's Iliad, Sophocles' Oedipus the King and Thucydides' History of the Peloponnesian War. Soc Classical Stud. New York University, New York. https://classicalstudies.org/annual-meeting/150/abstract/routes-plague-homer%E2%80%99s-iliad-sophocles%E2%80%99-oedipus-king-and-thucydides%E2%80%99

Mohammed A (2014) Kerry calls climate change 'weapon of mass destruction. Sci Am. Springer Nature America

Motta M, Ralston R, Spindel J (2021) A call to arms for climate change? how military service member concern about climate change can inform effective climate communication. Environ Commun 15(1):85–98. https://doi.org/10.1080/17524032.2020.1799836

Neufeld E (1980) Insects as warfare agents in the ancient near east (Ex. 23:28; Deut. 7:20; Josh. 24:12; Isa. 7:18–20). Orientalia, NOVA SERIES 49(1):30–57

O'Sullivan PO, Miller JW (2015) The geography of warfare. Routledge, London, p 14

De Pascale F, Roger JR (2020) Coronavirus: an anthropocene's hybrid? the need for a geoethic perspective for the future of the earth. AIMS Geosci 6.1:131–134, 132

Pippin T (2018) Fire and fury: standing with John at the end of the world. Perspect Relig Stud 45(2):167–182

Pontin J (2007) Oppenheimer's ghost can we control the evolution and uses of technology? MIT Technol Rev. https://www.technologyreview.com/2007/10/15/223531/oppenheimers-ghost-3/#:~:text=Robert%20Oppenheimer%2C%20who%20had%20been,cried%2C%20most%20people%20were%20silent

Tacitus (1993) Germania 9: Trans. Mattingly, referenced in 64, H.D. Davidson. The lost beliefs of Northern Europe. Routledge, p 64

Tacitus (1999) Publius Cornelius, Agricola and Germany. Oxford University Press, UK

Tierney JJ (2006) Chasing ghosts: Unconventional warfare in American history. Potomac Books, Inc

Travis C, Ludlow F, Matthews A, Lougheed K, Rankin K, Allaire B, Legg R et al (2020) Inventing the grand banks: a deep chart. Geo 7(1)

Travis C (2001–2002) Heart of darkness redux. J Postgrad Res, Trinity College Dublin 1–15

Tzu S (2020) The art of war: the oldest military treatise in the world, Trans., Lionel Giles, Strelbytskyy Multimedia, p 56

United Nations (2015) Sendai framework for disaster risk reduction 2015–2030:14

United Nations (2015) Transforming our world: The 2030 agenda for sustainable development. https://sustainabledevelopment.un.org/post2015/transformingourworld/publication

Unite Nations (2016) Resolution adopted by the general assembly on 23 December 2016. 71/256. New Urban Agenda, p 6

United Nations (2021) Our common agenda, report of the secretary general, p 12. https://www.un.org/en/content/common-agenda-report/assets/pdf/Common_Agenda_Report_English.pdf

Westing AH (1980) Warfare in a Fragile world: the military impact on the human environment. Routledge

Winters HA (2021) Battling the elements: weather and terrain in the conduct of war. JHU Press

Winters HA, Galloway GE, Reynolds WJ, Rhyne DW (1988) Battling the elements: weather and terrain in the conduct of war. John Hopkins, Baltimore

Wright JL (2008) Warfare and wanton destruction: a re-examination of Deuteronomy 20: 19–20 in relation to ancient siegecraft. J Biblic Lit 127(3):423–458

Zalasiewicz J, Williams M, Fortey R, Smith A, Barry TL, Coe AL, Bown PR et al (2011) Stratigraphy of the Anthropocene. Philos Trans Roy Soc A: Math Phys Eng Sci 369(1938):1036–1055

Living on Mount Etna Between Risk, Beauty and Need: A Field Survey on Villages Struck by 2018 Earthquake

Leonardo Mercatanti

Abstract Sicily's uneven geological structure is positioned within a system of tectonic plates. Its (very) active volcano, Mount Etna, has always made it prone to disasters. Nevertheless on the flanks of Etna there are many urbanized areas and the population living on the slopes of Etna has increased a lot in the last century, because there is an ancient and intimate relationship between the land and the people who live and work on the land. In this chapter I will try to investigate possible changes in the perception of the risks associated with the volcano by the resident population by conducting survey in the villages of Fleri, Pisano, Pennisi and Santa Maria la Stella. This study is part of a growing body of research on changes in the perception of risk in the Etna area.

Keywords Etna · Sicily · Resilience

1 Introduction

Sicily, an island characterized by a varied geological structure, is located in the centre of the Mediterranean and is the largest of all the islands in this large basin. It is also at the centre of a complex system of faults that was the cause of the formation of volcanoes, as well as the origin of the earthquakes that have recently marked its history.

Many smaller islands of Sicily are of volcanic origin: Pantelleria, Ustica, Linosa and the seven Aeolian islands. Stromboli, one of these, is one of the most active volcanoes in the world. The activity of the volcano is so particular that it has become a known type of eruptive activity, referred to in the literature as the Strombolian eruption. This is defined by powerful expulsions of glowing cinders and volcanic bombs for several kilometres into the sky. This gives rise to real natural firework displays which are much appreciated by tourists who fill social media with night-time photographs.

L. Mercatanti (✉)
University of Palermo, Palermo, Italy
e-mail: leonardo.mercatanti@unipa.it

B. Dahiya et al. (eds.), *Disaster Resilience and Human Settlements*,
Advances in 21st Century Human Settlements,
https://doi.org/10.1007/978-981-99-2248-2_9

The Tyrrhenian Sea, north of the Sicilian coast, is also home to underwater volcanoes. Some are considered as extinct, such as the Magnaghi or the Vavilov. The Marsili, on the other hand, located between Palermo and Naples, is active and is the largest submerged volcano in Europe, occupying an area of about 2,100 km^2 (D'Alessandro et al. 2009; Ventura et al. 2013). To the south of the southern coast of Sicily there are at least twelve underwater volcanoes, including Empedocle, Terribile, Nerita and the Ferdinandea island, which make up a volcanic region called Campi Flegrei del Mar di Sicilia—Phlegraean Fields of the Sicily Sea (Global Volcanism Program 2002; Lodolo et al. 2012).

Sicily is therefore truly a land of volcanoes. Of all the Sicilian volcanoes, however, the most important is Etna, located in the eastern part of the island, of which it represents the dominant landscape element. It has a diameter of about 40 km, a perimeter of over 200 km, occupies an area of about 1500 km^2 and is the largest active volcano in Europe. The volcanic structure, born about 500,000 years ago at the continental collision line between the African plate to the south and the Euro-Asian plate to the north, borders with the Peloritani Mountains to the north-east, with the Nebrodi Mountains to the north and north-west, with the Catania plain to the south and with the Ionian Sea to the east. Its perimeter is bordered to the north by the course of the Alcantara river and to the west by that of the Simeto river, both of which originate in the Nebrodi mountains. Etna is classified as a stratovolcano (Chester et al. 1985). In 1987 the Etna Regional Park was established, the first in Sicily and in 2013 Etna was included in the UNESCO World Heritage List.

Over the centuries, the disasters caused by the volcano have affected human settlements, the structure and location of inhabited centres and the related work activities. In fact, it is useful to remember that, if on the one hand over the centuries Etna has been able to cause the merciless destruction of large territories, on the other it has provided a diversified landscape, unique of its kind, appreciated over the centuries by visitors, writers, poets and artists from all over the world. It has also made it possible to exploit the materials produced by the volcano in economic activities (Mercatanti and Privitera 2020). Thinking about this, Etna's activity has always been associated with that of the concept of rebirth and fertility. The ash of the volcano, which contains precious minerals, is in fact an excellent natural fertilizer that enriches the agricultural lands around the volcano and is also used today for the production of excellent quality mortars and concrete. The plain of Catania, south of the volcano, with an area of over 400 km^2, is one of the largest plains in southern Italy. It owes its well-known fertility to the constant contribution of this natural product (ash deposits) over the centuries (Fiantis et al. 2019). An excellent description of these advantages was given by Sadao Shoji and Tadashi Takahashi "The periodic additions of volcanic ash generally improve the soil's physical and chemical properties and renew the soil productivity. Volcanic ash soils provide a comfortable living environment and create a favourable landscape for recreation and human health. They also accumulate a large amount of organic carbon and nitrogen as important components of soil organic matter that are the main sources of nitrogen for plants, and various nutrients and energy for soil organisms" (Shoji and Takahashi 2002: 113).

The possibility of cultivating in a healthy environment, in the specific case of the Catania plain, is also due to the profound landscape transformations following incisive territorial reorganization actions for the realization of an integral reclamation implemented since the 1930s (Formica 1970; Sorbello 2020). Before those years, the Plain of Catania had inspired several writers because it was associated with malaria, a disease transmitted by the Anopheles mosquito which for several months a year was found in that area due to the poor care with which the water-course and the drains were regulated (Franchetti and Sonnino 1877).

Etna erupts quite frequently and also in a spectacular way. Many of these eruptions are of the Strombolian type and result in ash emissions that fall on the surface of the volcano, but also hundreds of kilometres away.

On the flanks of the volcano there are many urbanized areas. The population living on the slopes of Etna has increased a lot in the last century, albeit with different intensities on the various sides. This leads to some reflections, the subject of this contribution, on liveability, on the quality of life and on the perception of risk in various Etna areas, also in light of the new scientific investigations on the volcano and the recent changes recorded in its behaviour. The objective of this chapter is precisely to investigate possible changes in the perception of the risks associated with the volcano by the resident population, with specific reference to a part of the eastern side. In addition, a reflection will be made on the connection between risk perception and sustainability and how risk perception has influenced the changes in human settlements in the area studied, remembering that in the first phase of the Anthropocene the social, cultural and structural adjustments due to global changes were more coherent and logical (Bowen and Gleeson 2019). To this end, it was decided to conduct the survey in the villages of Fleri, Pisano, Pennisi and Santa Maria la Stella, analysing the results of a survey on the ground conducted by meeting some residents. The physical presence of the author also intended to record, for complementary research, the state of the reconstruction phase. This study is in fact part of a broader research on changes in the perception of risk in the Etna area.

2 Demographics of the Etna Area

The characteristics of the population of the Etna area, made up of over 40 municipalities (Fig. 1), have been conditioned by the natural and economic environment and above all by the influence of the volcano (Cirelli 1979).

During the twentieth century and, in particular, after the Second World War, an unprecedented population growth was recorded in the Etna area, in line with what was happening throughout Italy. The phenomenon has been accentuated in recent decades especially for the municipalities on the southern side and is mainly duc to a long phase of counter-urbanization relating to the city of Catania, the second largest city in Sicily and among the top ten in Italy by the number of inhabitants. In fact, from the unification of Italy (1861) until the 1971 census, the population of the city grew from 70,000 to 400,000 inhabitants (Source: Istat). These were the years in

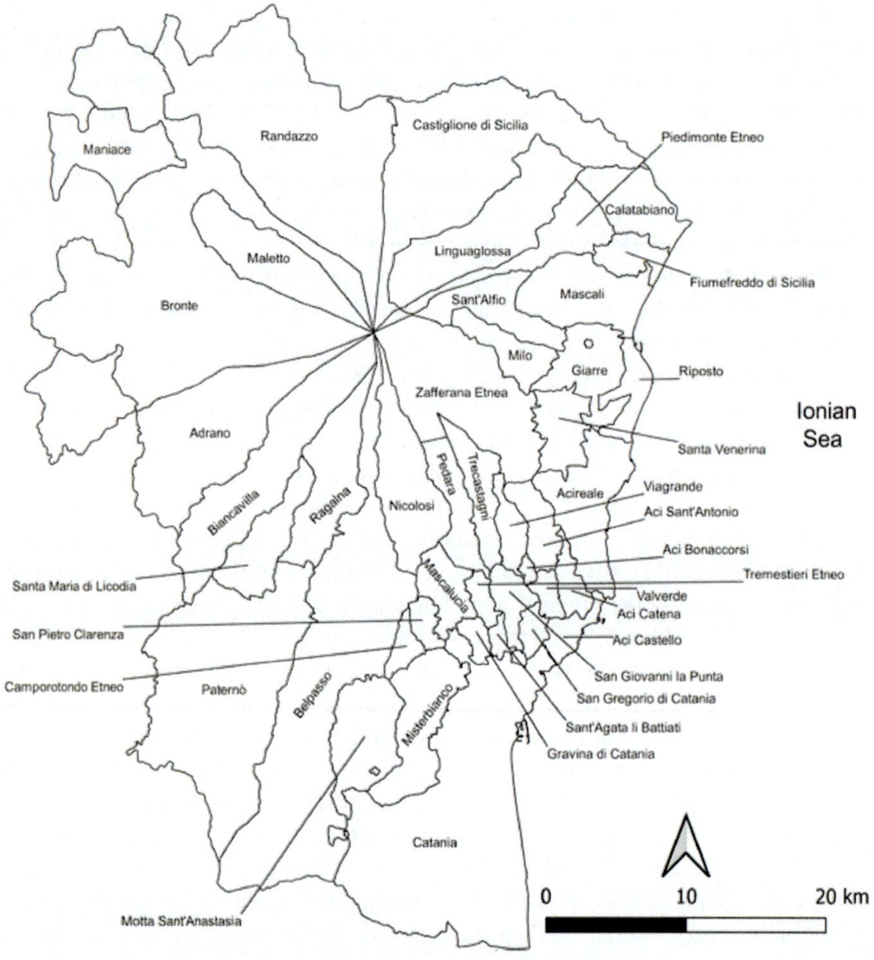

Fig. 1 The Etna municipalities. *Source* Created on QGIS by the author

which the Etna city became modern, attractive, but also more chaotic and, in some respects, unliveable (Di Blasi 1967; Cirelli 1997).

Following the increase in the population of the Etna area, which has reached about one million inhabitants, and the expansion of infrastructure, the communities living on the slopes of Etna are becoming increasingly vulnerable to the consequences of volcanic activity. In fact, Etna, during its existence, has caused profound upheavals from an environmental and territorial point of view, with important consequences on human lives and settlements. One of the most devastating was certainly the eruption of 1669 when a crater opened in the municipality of Nicolosi (south side) whose lava destroyed the town and other villages further south, reaching the city of Catania in a few days, devastating the western part. More recently, in 1928, the municipality of

Mascali, on the eastern side of the volcano, was destroyed by a lava eruption. The related images and video of the disaster are readily available on the web.

The analysis of demographic data provided by the Italian National Institute of Statistics—ISTAT, through the censuses that have been carried out in Italy at the beginning of each decade since 1861, shows a peculiar demographic behaviour in various Etna municipalities. In general, until 1971 the municipalities north of Catania had a regular and physiological increase in population. Due to the strong urbanization process of Catania, which began in the 1960s, and the consequent increase in rental and property prices, there has been a real demographic explosion of these municipalities of the first crown (Cirelli 1994). A fundamental role is also played by the search for a better quality of life. Isidoro Mazza defined the Catania of that period, referring to its growth, as "a metropolis without method" (Mazza 1988). This counter-urbanization process has led to a decrease in the residents of the municipality of Catania which went from 400,000 inhabitants in 1971 to just under 300,000 inhabitants in 2021. The proximity to Catania, a metropolitan city with superior services (international airport, port, train station, metro, university, large hospitals and an important motorway junction), explains the demographic evolution of these specific municipalities on the southern side of Etna, whose residents continue to have a privileged relationship with the capital city for reasons of work and study. Since the demographic movements from Catania towards the outside have not been followed by a development of economic and social infrastructures or by a greater functional autonomy of these municipalities, they are to be understood simply as a mere residential choice with a high rate of dependence on the capital. Significant commuting processes have thus been triggered towards Catania (Finocchiaro 1999; Picone 2006).

Some of these municipalities, such as Gravina di Catania, Mascalucia or Sant'Agata li Battiati, have reached a very high population density in recent decades, losing on the one hand all the advantages of marginality compared to the large urban centre of Catania, and therefore that benefit in terms of quality of life mentioned above, favouring a further choice of location of the population towards municipalities located at a higher altitude. Here, municipalities further north, such as Nicolosi, Pedara or Trecastagni, located at an altitude between 600 and 700 m a.s.l., have seen their population, which in 1971 was, for each of them, of about 4000 inhabitants (Fig. 2), significantly increase in a few decades.

This obviously requires reflections on the perception of volcanic risk, a cognitive process subject to various subjective, economic and cultural factors. Here it is very difficult to rationalize and make a complex phenomenon such as the perception of risk objective. It is difficult to propose a general theory on this topic (Crescimbene et al. 2014: 69). A field survey, possibly repeated after a few years and conducted through questionnaires or interviews, remains the most appropriate form of knowledge acquisition.

A survey published in Geographical Review and conducted between 2011 and 2012 in the territory of the municipality of Nicolosi clearly highlighted that there is no negative perception of risk by the population of any age and gender, and this is demonstrated precisely by the growth demographic and the new location choice of

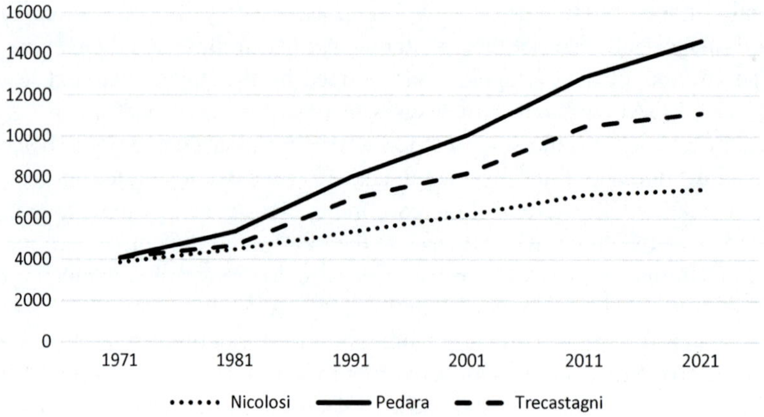

Fig. 2 Demographic growth in Nicolosi, Pedara and Trecastagni (1971–2021). *Source* Author's elaboration on ISTAT data

young families (Mercatanti 2013). But we must be careful: this is a perception that can suddenly be changed in the case of signs of anomalous behaviour of the volcano or even on the basis of scientific studies with a wide media coverage that warn of an imminent danger. This is exactly what happened in 1980 in the United States with the collapse of part of Mount Saint Helens which resulted in 57 deaths despite the evacuation order. In this case, the sudden collapse of the side of the volcano lowered its altitude by about 400 m and devastated an area of about 15 km in diameter.

More recently (September 2021) the eruption of the Cumbre Vieja volcano on the island of Las Palmas (Canary Islands), led to the destruction of over 500 buildings and the evacuation of thousands of residents and tourists who until then certainly had a very low perception of risk.

The municipalities of the western side, such as Adrano, Bronte, Maletto and Randazzo (north-western side), far more distant from the Etna capital and for this historically more independent from it from a socio-cultural and economic point of view, have had fewer occasions of devastation due to the activity of the volcano in recent centuries and this is evident from the analysis of the geological map of Etna which reconstructs the eruptive history of the volcano from its origins to today (Branca et al. 2011). These municipalities are characterized, in the twentieth century and up to the present day, by minimal, zero or even negative demographic growth (Fig. 3).

We can say that these are rather closed territories that do not have a large city to which to refer constantly and that have maintained an agricultural vocation which, in some cases, has given excellent results, as for the famous Bronte pistachio.

On the eastern side, on the other hand, there are municipalities characterized by a certain socio-cultural dynamism, such as Acireale, Giarre, Riposto. This also has consequences on the level of demographic growth, which has been constant since the second half of the twentieth century and therefore less problematic from the point of view of urban planning, services and quality of life.

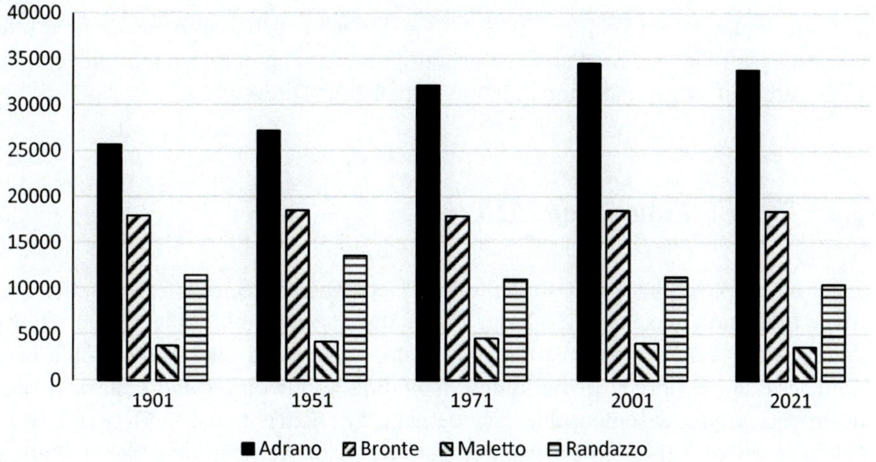

Fig. 3 Demographic trend in Adrano, Bronte, Maletto and Randazzo (1901–2021). *Source* Author's elaboration on ISTAT data

3 Discussion

For the purposes of this work we will consider some territories of the eastern side of the volcano, where, in recent years, due to some events, the perception of volcanic risk has changed. The recent events are a scientific publication of 2018 that demonstrates the slow but continuous detachment of a part of the volcano, a devastating earthquake in December 2018, and an unprecedented paroxysmal activity of 2021.

3.1 The 2018 Publication on the Collapse of the Eastern Flank of Etna

In the first case, reference is made to the 2018 publication entitled "Gravitational collapse of Mount Etna's south-eastern flank", published in the journal Sciences Advances, in which it is shown that the south-eastern flank of the volcano is sliding a few centimetres per year (from two to more than four) into the Ionian Sea. In this scientific article, it is written that "We cannot exclude flank movement to evolve into catastrophic collapse, implying that Etna's flank movement poses a much greater hazard than previously thought" and "our shoreline-crossing deformation analysis implies a greater hazard for flank collapse than previously assumed, as deep-seated gravitational sliding can potentially lead to catastrophic collapse" (Urlaub et al. 2018: 1 and 5). What is relevant for the purposes of our study is that the media have reported the results of this study very strongly for some time, causing, on a psychological level, a certain unease among the residents. For a few weeks nothing else was talked about in eastern Sicily. We know how important the role of the media is in these cases.

Just one example: on 25 June 2018 ANSA, the leading Italian press agency, titled an article as follows: "Svelato il lato oscuro dell'Etna, pericoloso e imprevedibile" (The dark dangerous and unpredictable side of Etna is revealed).

3.2 *The Earthquake of 2018*

After an eruptive phase that started on 23 December 2018, on 26 December a 4.9 magnitude earthquake struck a large area in the eastern flank of the volcano. Fortunately, there were no victims, but only some injured and over 1000 people made homeless and hundreds of buildings, including architectural and historical ones, destroyed or otherwise unusable. The damage was concentrated in Fleri and Pisano Etneo (parts of Zafferana Etnea), Pennisi (a part of Acireale) and Santa Maria la Stella (a part of Aci Sant'Antonio). The earthquake caused the rupture for a length of approximately 8 km along the trace of the NNW-trending active Fiandaca Fault. The areas most affected are located right near the fault. Faults are pejorative elements of seismic forces since they can cause areas of particular fragility and a concentration of seismic energy.

Already in 1984, a similar earthquake had produced the same devastation in Fleri and Pisano, as well as in the municipality of Zafferana Etnea (Fig. 4).

The reconstructions took place in the same places, as always. This is precisely one of the points on which to reflect: the historical seismicity of the Fiandaca fault has been known since the end of the nineteenth century and manifests itself with

LA SICILIA

QUOTIDIANO DEL MATTINO

Il governo supera al Senato lo scoglio del «caso Cirillo»

Ieri notte nella zona di Zafferana un forte terremoto (ottavo grado) avvertito anche a Catania

Fleri - mille abitanti - semidistrutta

Fig. 4 Front page of the newspaper *La Sicilia*—26 October 1984. *Source* Digital archive of the newspaper *La Sicilia*

earthquakes that can have a ten-year frequency (Baratta 1894; Patanè 1989). Why rebuild in the same places positioned on an active fault which has been shown to be dangerous? Perhaps it is cheaper to repair than to rebuild elsewhere?

A recent work has highlighted, precisely with reference to the effects of this earthquake, how the perception of risk can be changed due to the choice of the type of narrative proposed and which has implications on the behaviour of an entire community: newspaper headlines like "Etna, Frightful Night. Earthquake of magnitude 4.8 in Catania. Wrecked buildings and injured people" or telling the story of a surviving family who escaped the partial collapse of their home "It's a miracle we're alive. We were in bed, woken with a start while the house walls were falling on us" (Mercatanti and Sabato 2019: 454–455).

3.3 The Paroxysmal Events of 2021

Starting from February 16, 2021, the volcanic activity of Etna has been of considerable intensity and frequency. Up to September of the same year, over 50 paroxysmal episodes were recorded, generated by the south-east crater. The amount of material produced has raised the altitude of the volcano, which has reached 3,357 m a.s.l. The shape of the top of the volcano also changed in a few months. At each paroxysmal event, signalled by loud roars audible up to Catania, the volcanic ash cloud has almost always covered the municipalities of the eastern side, due to the winds that generally blow from west to east, causing enormous inconvenience concentrated on a part of the territory (Fig. 5).

In September 2021, the mayors of various municipalities on the eastern side asked the state for help to deal with the emergency due to the large amount of ash that had fallen on the ground, with very serious consequences for the mobility of citizens and on the possibility of providing the various public services correctly. The cost to remove many tons of volcanic ash is very high and even citizens are forced to spend a lot of money to have the roofs of their homes cleaned several times. The ash also clogged the rainwater drainage systems and caused damage to agriculture, contributing to the financial distress of these municipalities. In fact, sometimes there was a rainfall of small stones that even damaged cars.

4 Survey

In order to investigate more than two years after the earthquake that devastated the inhabited centres of the villages of Fleri, Pennisi, Pisano and Santa Maria la Stella, some 30 oral interviews with residents and key informants were conducted between June and July 2021, lasting about 30 min each. In particular, for each of the four villages, twelve key witnesses were interviewed: eight residents (with gender equality and belonging to an age group from 35 to 60 years old), two commercial

Fig. 5 Beginning of the paroxysmal event on 21 September 2021. Photo taken in Nicolosi. *Source* Photo by the author

traders, a traffic warden, a municipal administrator. The residents are also owners of properties damaged to varying degrees by the earthquake.

As already mentioned, in 1984 two strong earthquakes struck Zafferana Etnea and the villages of Fleri and Pisano, causing serious damage to the building structures, 70% of which were subsequently declared uninhabitable. During the following 25 years, which passed relatively quietly due to the absence of direct threats from the volcano and devastating earthquakes, the perception of risk repositioned itself to a low level. The time factor is therefore decisive. This is demonstrated by the fact that the resident population of the entire municipality of Zafferana Etnea increased from 5796 to 9449 residents from 1981 to 2021. The Pennisi area (belonging to the municipality of Acireale) in the same period increased from 373 to over 1100 inhabitants and Santa Maria la Stella (municipality of Aci Sant'Antonio) instead had a demographic increase from 462 to about 4500 residents. No reliable data are available for the other fractions. Yet the problematic localization of these zones above an active and dangerous fault is known to all both because the documents on the seismicity of the eastern side of Etna are public and because the framework of knowledge on Etna seismicity in recent decades has considerably expanded, thanks also to the adoption of cutting-edge tools and advanced technologies that make it possible to carry out scientific studies and research of the highest profile.

The survey conducted in Fleri in February and March 2019, a few months after the earthquake, highlighted "the will of younger residents to find a new place to live in, while the elderly, whose houses were not irreversibly damaged by the earthquake, seem to want to continue living in the same area" (Mercatanti and Sabato 2019: 452). Today things have changed.

The question on which we focused a lot in 2021 was: "what are the resilience actions implemented by the populations residing in the volcanic area at risk?". The interviews in general revealed that in the vast majority of cases no risk reduction strategies have ever been initiated. Almost no one has an emergency lighting system that is activated in the event of a blackout during an earthquake or an emergency kit for immediate evacuation of home. Bookcases or wardrobes could very simply be fixed to the wall with dowels, but nobody does this practice. When earthquakes occur at night, most of the injuries to people derive from the absence of these simple precautions. Most of the houses, built before 2009, when the obligation to build new houses by adopting specific anti-seismic technical standards had not yet entered into force, have never been reinforced and adapted to anti-seismic criteria. Not to mention the rural buildings, often with neither foundations nor reinforced concrete pillars (Fig. 6).

Fig. 6 Rural house destroyed by the 2018 earthquake (Fleri). *Source* Photo by Iolanda A. Mercatanti (June 2021)

The interviews highlight that, even if aware of the structural weakness of their homes and the consequent risk, the residents were unlikely to be able to implement the necessary adjustments since these would be very expensive interventions, if not facilitated or financed by public bodies.

All of this is in stark contrast to the clear and convincing recommendations of the United Nations General Assembly (UNGA 2015a, b, 2016). The *Sendai Framework for Disaster Risk Reduction 2015–2030*, applied to the risk of small-scale and large-scale, represents a cornerstone of the UN system for disaster reduction. It outlines seven targets to be achieved by 2030, with a strong emphasis on disaster risk management, reducing existing risk and strengthening resilience. It states that "Effective disaster risk management contributes to sustainable development" and "Policies and practices for disaster risk management should be based on an understanding of disaster risk in all its dimensions of vulnerability, capacity, exposure of persons and assets, hazard characteristics and the environment" (UNGA 2015a: 9 and 14). The New Urban Agenda was adopted in 2016 at the United Nations Conference on Housing and Sustainable Urban Development (Habitat III) in Quito. It offers a shared vision for a more sustainable future and involves new urban rules (UNGA 2016). It implements Sustainable Development Goal 11 titled "sustainable cities and communities" focused to make cities and human settlements more inclusive, resilient and sustainable.

Some interviewees from the various villages actively participated in the initiatives (torchlight processions, demonstrations, assemblies of the earthquake victims) to push the policy to adopt urgent measures for the restoration of the territory and for the return to a normal situation.

The news of the sliding of the eastern flank of the volcano, widely reported in 2018 by the media for about a month and before the earthquake, affected the perception of risk, but only for a certain period.

The interviews clearly highlight that, especially for younger residents up to a few months after the 2018 earthquake event, the idea of moving their residence to a safer place, i.e. a second home or go to rent elsewhere, was very strong. After almost three years, with the beginning of the reconstructions and with the resumption of a certain social life within the territories devastated by the earthquake, this idea has already changed in favour of the certainty of remaining anchored to one's own territorial roots. It is evident that the attachment to one's roots, to one's affections, to one's habits, but also to one's landscapes and also the fear of finding an inhospitable environment in other contexts, are all elements that, after a few years, push residents to resettle in the same place again (Figs. 7 and 8).

The interviews reveal in many the awareness that after the media recall of the 2018 seismic event and after the continuous fallout of volcanic ash in recent times, the value of properties in the areas has been greatly reduced. This, combined with the economic effects of the recession that the entire nation has been in for years, also considering the further crisis generated by the pandemic period, makes it absolutely inconvenient and inadvisable to sell properties and land for at least several years. The choice to stay is therefore almost an obligatory choice due to psychological and economic factors. The interviewees are therefore perfectly aware that they live in

Fig. 7 Semi-destroyed residential building awaiting renovation (Fleri). *Source* Photo by Iolanda A. Mercatanti (June 2021)

a territory that is periodically the subject of devastation, but due to various factors they are almost forced not to change their choice of location. As a great expert on Etna, David K. Chester of the University of Liverpool, rightly wrote, "the key [...] is to recognize that there are aspects of risk which are independent of a region being volcanic and relate to dynamic changes in population, history, culture, and politics" (Chester 2005: 426).

5 Conclusions

In the case studied in this paper, it has been seen that the perception of risk and the relative behaviour of people in the face of the prospect of natural hazards is constrained not only by the concrete manifestation of a disastrous event, but also by social, economic and political forces that are beyond their control. Although almost three years after the earthquake of 2018, the reconstruction of the villages under investigation has not been completed, but is rather slow due to very slow procedures, the lack of quality design and the absence of adequate funds. The idea that emerges from the interviews carried out is that everyone expects a complete rebirth of the places so that they can return to live exactly as in the years preceding the seismic

Fig. 8 Church of Santa Maria del Rosario (Fleri). *Source* Photo by Iolanda A. Mercatanti (June 2021)

event. This is in partial contrast with the results of previous research, conducted in Fleri in the same way, in 2019. It is therefore shown that with the passage of time the factors related to the contingent situation, to the economy, to one's roots and even to faith become more evident and stronger in perception of risk and the consequent location choice. We could say that these are variable geometry risk perceptions.

Not only in the volcanic eruptions occurring as we write this it has been seen that, although the risks (volcanic, seismic) due to the dangerous locations of the area in which they reside are very clear to residents, still no simple resilience strategy has been implemented to prepare for the eventuality of this new disastrous event, which, fortunately, has not caused any victims and which therefore continues to allow Etna be perceived, all in all, as a peaceful volcano. Up to now, it seems that UN recommendations have not been implemented in any way at the local level. This most likely derives from serious gaps in the political chain and in policy responsibilities from the national to the local level. It is necessary to concretely transpose the UNGA documents into actions. This is a core challenge for marginal areas subjected to adverse environmental conditions.

References

Baratta M (1894) Intorno ai recenti fenomeni endogeni avvenuti nella regione etnea (Agosto 1894). Bollettino della Società Geografica Italiana 3(7):740–760

Di Blasi A (1967) La dinamica demografica della provincia di Catania dal 1861 al 1961. Archivio Storico della Sicilia Orientale I-III, pp 189–249

Bowen WM, Gleeson RE (2019) The evolution of human settlements: from Pleistocene origins to Anthropocene prospects. Palgrave Macmillan, London

Branca S, Coltelli M, Groppelli G, Lentini F (2011) Geological map of Etna volcano, 1:50,000 scale. Italian J Geosci 130(3):265–291

Chester DK (2005) Volcanoes, society, and culture. In: Marti J, Ernst G (eds) Volcanoes and the environment. Cambridge University Press, Cambridge, pp 404–439

Chester DK, Duncan A, Guest JE, Kilburn C (1985) Mount Etna, the anatomy of a volcano. Chapman and Hall, London

Cirelli C (1979) Struttura morfologica e sviluppo demografico della città di Catania nell'ultimo trentennio. Annali del Mezzogiorno 19:179–204

Cirelli C (1994) Variazioni demografiche e ridistribuzione della popolazione nella provincia di Catania. In: Citarella F (ed) Studi geografici in onore di Domenico Ruocco. Loffredo, Napoli, pp 455–469

Cirelli C (1997) Mobilità, Politiche del traffico e qualità della vita a Catania. In: Cori B (ed) La città invivibile. Nuove ricerche sul traffico urbano, Pàtron, Bologna, pp 63–86

Crescimbene M, La Longa F, Camassi R, Pino NA, Peruzza L (2014) What's the seismic risk perception in Italy. Eng Geol Soc Territ 7:69–75

D'Alessandro A, D'Anna G, Luzio D, Mangano G (2009) The INGV's new OBS/H: analysis of the signals recorded at the Marsili submarine volcano. J Volcanol Geoth Res 183:17–29

Fiantis D, Ginting FI, Gusnidar NM, Minasny B (2019) Volcanic ash, insecurity for the people but securing fertile soil for the future. Sustainability 11:1–19

Finocchiaro E (1999) Città in trasformazione: le logiche di sviluppo della metropoli contemporanea. Franco Angeli, Milan

Formica C (1970) La Piana di Catania, Istituto di Geografia Economica, Naples

Franchetti L, Sonnino S (1877) La Sicilia nel 1876, Barbera, Florence

Global Volcanism Program (2002) Report on Campi Flegrei Mar Sicilia (Italy). In: Mayberry G (ed) Weekly volcanic activity report, 27 November–3 December 2002. Smithsonian Institution and US Geological Survey

Lodolo E, Civile D, Zanolla C, Geletti R (2012) Magnetic signature of the Sicily Channel volcanism. Mar Geophys Res 33:33–44

Mazza I (1988) Catania: aspetti di una metropoli senza metodo. In: Rapporto Cer-Censis. Il governo dell'economia, Edizioni Il Sole 24 Ore, Milan, pp 347–365

Mercatanti L (2013) Etna and the perception of volcanic risk. Geogr Rev 103(4):486–497

Mercatanti L, Privitera S (2020) I Geositi del monte Etna: un caso studio del patrimonio geologico. In: Cannizzaro S (ed) Ecomuseo dell'Etna tra natura, mito e cultura. Pàtron, Bologna, pp 54–65

Mercatanti L, Sabato G (2019) Volcanic risk and the role of the media. A case study in the Etna area. AIMS Geosciences 5(3):448–460

Patanè A (1989) Fleri tra storia e cronaca (1329–1989), Tringale, Catania

Picone M (2006) Il ciclo di vita urbano in Sicilia. Rivista Geografica Italiana 113(1):129–146

Shoji S, Takahashi T (2002) Environmental and agricultural significance of volcanic ash soils. Glob Environ Res 6:113–135

Sorbello M (2020) L'irrigazione e la bonifica della piana di Catania. Studio geostorico sulla trasformazione del paesaggio e il riordino ambientale, Aracne, Rome

UNGA-United Nations General Assembly (2015a) Sendai framework for disaster risk reduction 2015a–2030, UN, New York. www.undrr.org/publication/sendai-framework-disaster-risk-reduction-2015a-2030

UNGA-United Nations General Assembly (2015b) Transforming our world: the 2030 agenda for sustainable development, UN, New York. https://sdgs.un.org/sites/default/files/publications/21252030%20Agenda%20for%20Sustainable%20Development%20web.pdf

UNGA-United Nations General Assembly (2016) New urban agenda, UN, New York. http://habitat3.org/the-new-urban-agenda/

Urlaub M, Petersen F, Gross F, Bonforte A, Puglisi G, Guglielmino F, Krastel S, Lange D, Kopp H (2018) Gravitational collapse of Mount Etna's southeastern flank. Sci Adv 4(10):1–7

Ventura G, Milano G, Passaro S, Sprovieri M (2013) The Marsili Ridge (Southern Tyrrhenian Sea, Italy): an island-arc volcanic complex emplaced on a 'relict' back-arc basin. Earth Sci Rev 116:85–94

Post-disaster Management and Recovery: Urban Landscape, Pandemic and Community Resilience

Narratives of Urban Resilience and Sustainability in Southern Italy: The Case Studies of Matera (Basilicata) and Filadelfia (Calabria)

Francesco De Pascale

Abstract This paper focuses on the concept of urban resilience and on some thematic declinations of the term, by comparing two case studies in Southern Italy: Matera (Basilicata) and Filadelfia (Calabria). The historical, social, and cultural processes manifesting in these cities can be viewed as symbols of sustainability and urban resilience. The latter is defined as the ability of a city's systems, businesses, institutions, communities, and individuals to survive, adapt, and grow, regardless of the chronic stress and acute shocks they experience. In these two specific cases, different experiences, and similarities are narrated focusing attention on the contrasting paths to urban resilience and sustainability after such shocks. The hypothesis considered in this paper concerns the concept of urban resilience linked to socio-cultural structures. For example, Matera, transformed from the definition of a "national shame" to the European Capital of Culture in 2019, due to synergy between policymakers, stakeholders, and citizens. In comparison, Filadelfia, destroyed by an earthquake in 1783, was rebuilt on a solid, bottom-up approach, serving as an example of an innovative and resilient urban scheme.

Keywords Filadelfia (Southern Italy) · Matera (Southern Italy) · Shock · Sustainability · Urban resilience

1 Introduction

City resilience describes the ability of cities to function, so that people living and working in cities, especially the poor and vulnerable, survive and thrive regardless of the stress or shock they encounter (The Rockefeller Foundation & Arup 2014). According to OECD (Organization for Economic Co-operation and Development), "resilient cities are cities that have the ability to absorb, recover and prepare for future shocks (economic, environmental, social & institutional). Resilient cities

F. De Pascale (✉)
Department of Culture and Society, University of Palermo, Palermo, Italy
e-mail: francesco.depascale@unipa.it

B. Dahiya et al. (eds.), *Disaster Resilience and Human Settlements*,
Advances in 21st Century Human Settlements,
https://doi.org/10.1007/978-981-99-2248-2_10

promote sustainable development, well-being and inclusive growth" (OECD 2022). Musacchio and Wu (2002), an urban planner and an ecologist coined the "city of resilience", a metaphor useful in urban planning, but also in the technical discourse of ecology, that creates important connections between ecology and planning (Pickett et al. 2004). Resilience in cities, the main concern of most urban ecologists (Pickett et al. 2004; Alberti and Marzluff 2004; Andersson 2006; Colding 2007) is closely related to urban form and land-use patterns, on the other hand, and to local and spatial ecological processes on the other (Ernstson et al. 2010).

Several elements of the theory of resilience are fundamental and applicable in the context of cities (Batty 2008). Given the ecological origins of the concept, most resilience scholars have carried out empirical analyses of non-urban areas (Berkes and Folke 1998; Gunderson and Holling 2002; Berkes et al. 2003) focusing less attention on the cultural and social elements of human-dominated systems, such as cities (Ernstson et al. 2010). Resilience in urban areas is concerned with supporting local–regional ecosystem services. The use of resilience often deals with the ability to restore the equilibrium of a system affected by severe critical events, such as environmental or anthropogenic disasters (Lombardo and Salvucci 2014). The term resilience is derived from the science of materials and indicates the properties that some materials must retain in their structures to revert to their original shape after being subjected to crushing or deformation (Burns 1996). This means adopted by the social sciences, not only regarding disasters triggered by natural hazards but more generally a territory's ability to resist and recover from the losses suffered (Tecco 2011). Resilience is, therefore, the ability of communities to absorb disturbances, undergo changes, and reorganize themselves in a new dynamic equilibrium (Marincioni et al. 2014). Similarly, the construct of psychological resilience refers to the ability of individuals to show appropriate levels of competence following exposure to significant adversity and difficulties. Within the psychological sciences, the construct of resilience was introduced in the 1970s (Masten and Garmezy 1985; Werner and Smith 1982), although from the first half of the twentieth century, the ability of individuals to resist negative events has been of interest to researchers. Disasters related to natural hazards, terrorist acts, and technological disasters represent a category of adverse events that concern authors who research on people's resilience. Numerous studies conducted in this area aim to analyze the variables and psychosocial processes that moderate the impact of natural extreme events. They are concerned with the adaptation and psychological well-being of people to develop heuristic models that can also provide guides for interventions to facilitate recovery and support (Masten and Osofsky 2010). Holling (1973) presents the concept of resilience in the field of ecology "as a measure of the persistence of systems and their measure of the ability to absorb change and disturbance and still maintain the same relationships between populations or state variables" (Holling 1973, p. 14). Resilience is built as an interaction between the individual, society, and the environment, and therefore has different expressions in different cultures and territorial contexts (Tecco 2011). Several definitions and conceptual frameworks for resilience have been developed and applied in the field of disaster management (Buckle et al. 2000; Burby et al. 2000; Comfort et al. 1999; Cutter et al. 2008; Godschalk et al. 1999; Indirli 2019; Klein et al. 2003;

Maguire and Hagan 2007; Manyena 2006; Mileti 1999; Paton et al. 2000; Paton and Johnston 2006; Pelling 2003; Tierney and Bruneau 2007; Timmerman 1981; Twigg 2007; UNISDR 2005; Walter 2004; Wildavsky 1991; Yoon et al. 2016).

The concept of resilience is applicable to cities because they are complex systems that constantly adapt to changing circumstances. The notion of a resilient city becomes conceptually relevant when chronic stresses or sudden shocks threaten a widespread disruption or collapse of physical or social systems. Urban resilience is understood as the ability of an urban system to maintain or rapidly restore the required or desired functions of the system in the face of disturbances, to adapt to change, or to redesign systems that limit current or future adaptive capacity accordingly (Meerow et al. 2016, p. 45). According to Pickett et al. (2004), a previous view considered resilience to be the ability of a system to return to a state of equilibrium after the disorder. However, this definition of resilience is not applicable to cities and urban regions, which are subject to historical change (Pickett et al. 2004). Indeed, narratives of destruction and reconstruction have dominated the literature on cities following any disaster related to natural or human-made hazards.

According to Chelleri (2012, p. 296), "all of this expresses a specific perspective of resilience over time. From Plato to Thomas Man, the city has always been (and recognized) as a living cultural and social artifact". As Lewis Mumford argued, in front of the metropolis "the city, the village, the cave and the cairn there was an essential disposition to social life. It (the city) begins as a meeting place" (Mumford 1961, p. 5).

Therefore, this paper's perspective and hypothesis focus attention on the concept of urban resilience as linked to socio-cultural structures. For example, in cases of cities rebuilt on their original sites or on other sites after an earthquake or catastrophic event—such as a volcanic eruption—, although time and the effects of disaster have dissolved built structures, social structures continue to remain durable (as in the "lost city" of Pompeii, a place still alive in memory, and tourism) (Mumford 1961). Indeed, it is the social properties of human existence in which cities express, through the tenacity of their urban lives, resilience over time. Campanella, for example, after the devastation of New Orleans caused by Hurricane Katrina argues that urban resilience is a function of resilient and "enterprising" citizens; the author highlights how city reconstruction plans must not only be implemented in physical-material terms but also through the reconstruction of the social fabric and community networks (Campanella 2006). To achieve this, T.J. Campanella affirms that only through a bottom-up approach of citizens can the reconstruction of New Orleans, or any city, be foreseen in a solid and inclusive way (Campanella 2006).

Italy is characterized by a variety of small, medium, and large cities. Most cities have developed historical centers within them, which, with their buildings and architectural monuments, cultural and natural heritage, shopping streets, squares and public areas, residential areas, and tourist destinations, shape the urban landscape and influence the identity of their inhabitants. The "Inner Spaces" of Italian cities have therefore always been places of urban life, characterized by exchanges, housing, work, culture, tourism, historical appreciation, meeting places, and an aggregator of people (BMI 2021; DST 2021; Sparks 2021; Hangebruch and Othengrafen 2022).

In this paper, two case studies, concerning two cities of different sizes, present similar tendencies of resiliency to recover from shocks and meet significant goals. The first case, Matera (Basilicata region, Southern Italy) had been defined as a "national shame", due to the degraded conditions in which Sassi inhabitants lived. However, the city experienced a resilient rebirth and was named to the UNESCO World Heritage list and subsequently the European Capital of Culture in 2019. The City of Sassi introduced a new culture of planning and cultural investments based on reuse—rather than on new construction—by adhering to key principles of environmental sustainability and resilience (Bernardo and De Pascale 2016a, b). The Sassi of Matera comprises two districts of Sasso Caveoso and Sasso Barisano (Fig. 1), composed of buildings, houses, and dwellings carved from the region's geology and inhabited from prehistoric times.

Filadelfia, linked to the American city (Philadelphia) of the same name, possesses a certain affinity in structure, geometric uniformity, and homogeneity of elements compared to its larger "sister" overseas. According to Barone (2019), it is the plan of Filadelfia itself that confirms a close link to the city in America. Two main arteries intersect to form the skeleton of the Italian city, each 17 m wide, and oriented according to the points of the compass. A plan perfectly corresponding to the rationalistic canons of Masonic inspiration, identical to that of the American Philadelphia. Due to its particular urban layout and archeology that survived the 1783 earthquake, Filadelfia remains the subject of research by scholars from all over the world and an important Calabrian tourist destination.

2 Matera: From "National Shame" to European Capital of Culture 2019

Until the end of the eighteenth century, the city of Matera, characterized by the Sassi, had maintained a system of sustainability, thanks to early modern innovative principles, that have been maintained to the present time: water conservation, waste storage, and the reuse of spaces. The Sassi of Matera, according to the architect Pietro Laureano, "are cliffs, stones, rocks, or rather a housing system created in the geological material itself, in a limestone rock, locally called tuff, along the slopes of a deep valley with unique and natural characteristics, the Gravina (Ravine)" (Laureano 2012, p. 22). The ravines are erosive incisions very similar to canyons, carved by water in the limestone rock, providing an example of karst alluvial erosion (Laureano 2012). The economy of the city of Matera is based on transhumance and the agricultural use of the Sassi Valley provided protection and shelter from the climate in the past, for humans and animals, in addition to serving as a site of storage, agro-pastoral production, and water collection. Water flows from the upper plateau and the clay hills are trapped, and channeled, eroding caves and terraces in the karst landscape. Flows are organized by a system of terraces that breaks the momentum of the water and favors the formation of fertile soil. The structure plot matrix of the urban complex

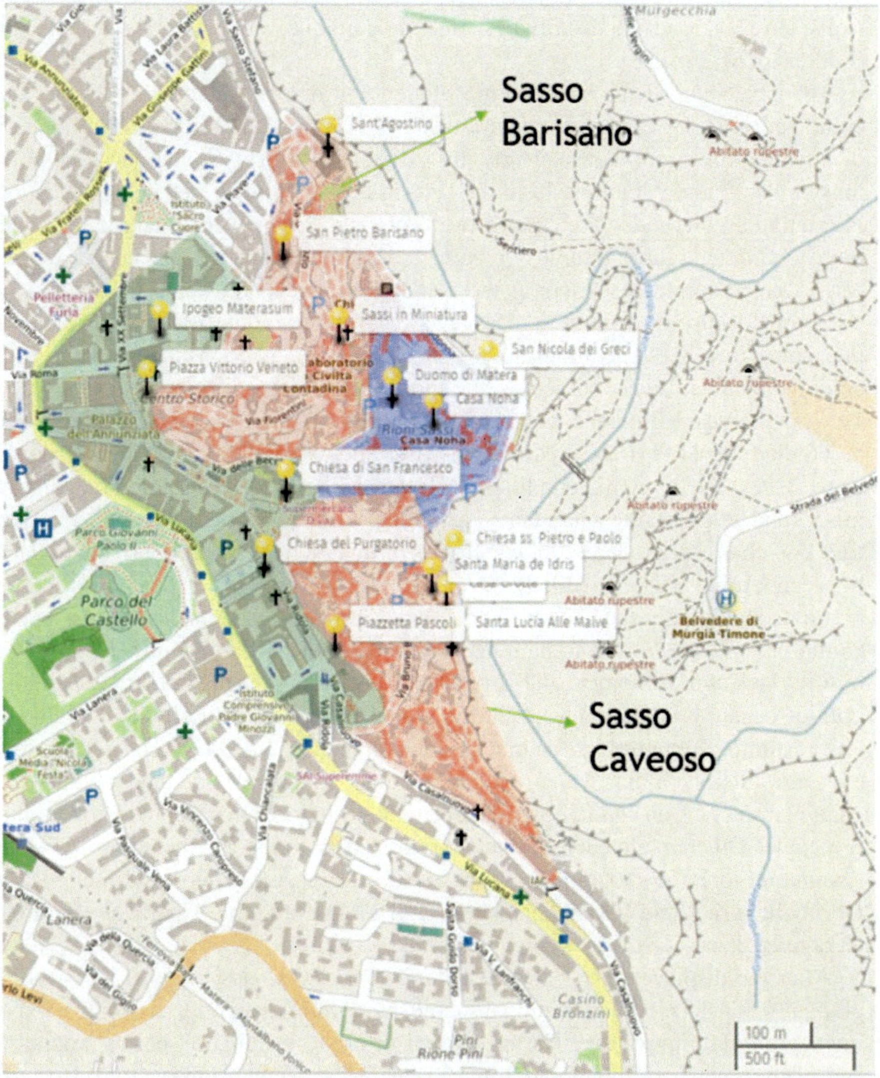

Fig. 1 The map of Matera. The two districts under study are highlighted: Sasso Barisano to the north and Sasso Caveoso to the south. Further east, from the Belvedere di Murgia Timone, it is possible to admire the cultural landscape of the Sassi of Matera. Additionally, a potential tourist can visit the nearby rock churches or the entrenched Neolithic village of Murgia Timone. *Source* https://umap.openstreetmap.fr/

is composed of basic elements: caves, tuff buildings, hanging gardens, canals and basins, paths, and associated neighborhoods (Bernardo and De Pascale 2016a, b). Over time the slopes of the Gravina of Matera were dug, perforated, and sculpted to create tunnels, cisterns, elaborate underground architectural complexes, dry stone walls, terraces, roads, and stairways, creating a unique, and geological urban fabric (Laureano 2012, p. 23).

The expansion of the city and the sudden and dizzying demographic increase, combined with a crisis in pastoralism crisis, were factors that led to a gradual upheaval of urban equilibrium. Consequently, the city's perimeter was enlarged with buildings on the "Piano", ranging from the current Piazza Vittorio Veneto to via Ridola, Piazzetta Pascoli, and Vico Case Nuove. The Sassi caves were enlarged to accommodate more and more families. In some cases, further excavations deeper into the earth were required, in others cisterns and churches carved from the rocks were transformed into rooms and homes. As a result, the amount of water available to the approximately twenty thousand people who lived massed in small and unhealthy environments, and were subject to the most varied diseases, rapidly decreased. In 1937, in the Sassi, illiteracy was widespread, and out of a total of 15,250 newborns, 6,760 died, equal to 44.32%, due to diseases such as malaria and tuberculosis (Di Caro 2019). The hygienic conditions of the cave houses were intolerable, and the misery and fatigue of the inhabitants—peasants, laborers, workers, artisans, and small traders—characterize daily life, which was mitigated only by the sense of solidarity between families (Di Caro 2019).

Immediately after the Second World War, Carlo Levi's denunciation, the squalid conditions of Matera rose to national prominence and became a prime example of how the lack of development and poverty had dug deep roots in Southern Italy. Italian politics began to take an interest in the issue: the leader of the Italian Communist Party Palmiro Togliatti was the first to arrive in the city in 1948 to see with his own eyes how the inhabitants were forced to live with animals. He defined the Sassi as a "national shame", an evil to be eradicated with brute force in order to restore dignity to people. Other intellectuals became interested in the story, including the writer Tommaso Fiore, the geographer Francesco Compagna, the economist Manlio Rossi-Doria, the writer and filmmaker Pier Paolo Pasolini, and the American sociologist George Peck.

After the displacement of the 1950s and 1960s, the Sassi (Figs. 1, 2, 3 and 4), gradually transformed from national shame, by becoming an important site of experimentation and innovation. The process of urban re-revaluation began around the end of the 1980s and culminated in 1993 with the recognition by UNESCO as a World Heritage Site. Matera's candidacy as a World Heritage Site had an important symbolic meaning in Southern Italy and represented a significant comeback for the region. The case of Matera overturned the usual UNESCO evaluation system focusing on art centers and made it possible for a criteria for new cultural values to emerge; Matera, in fact, represents a precious example for all other cities in the southern hemisphere (Bernardo and De Pascale 2016a, b). It signaled a shift in the evaluation of a city's artistic-monumental heritage in which the presence of single large monuments is less critical than its urban fabric and overall environmental space. Citizens, stakeholders, and local policymakers were commended for their efforts and genius in employing resources sparingly and efficiently. In Matera, there is a continuum between landscape, architecture, and human relations that revives a "sense of place" representing the period of the ancient historical settlements. The overall integration

Fig. 2 The landscape of the Sassi of Matera

Fig. 3 Church of St. Pietro Caveoso is a Catholic place of worship in Matera, originally from the end of the thirteenth century

Fig. 4 A suggestive glimpse of the Sassi in Matera

of the natural environment, the immense excavation work, and the buildings overlooking the Sassi make Matera an exceptional example of symbiosis between human beings and nature. The Sassi of Matera has the same name as Petra, in Jordan, due to the common rock features, but they do not have the same importance as this archeological site. However, the Sassi does serve as an example of a historically sustained ability to create architectures and cities with the appropriate use of minimal resources (Bernardo and De Pascale 2016a, b; Laureano 2012).

Matera offers an important example of such resilience and sustainability. To achieve this goal, values such as audacity, dedication, sobriety, care and attention to the territory, hospitality, and symbiosis between nature and culture become a model that can be exported to Europe, thanks to the visibility that Matera has obtained as the European Capital of Culture in 2019. Matera, in fact, was the first city in Southern Italy to receive this recognition, obtained after entering a shortlist that included the candidacies of five other Italian cities (Cagliari, Lecce, Perugia, Ravenna, and Siena).

This new vision has permeated the Sassi with meaning; thirty years after a complete depopulation in the 1960s, it witnessed the return of inhabitants in the 1990s. Resolving a conflict between tradition and modernity, Matera represents a unique example of a sustainable and resilient city in symbiosis with its geological landscape.

3 The Urban Resilience of Matera: The Underground Architecture Model

The urban resilience of Sassi of Matera is characterized by the need to make appropriate and common use of scarce resources, the economy of the soil and water, and the control of the energies of the sun and wind, all mediated by knowledge of the laws of mechanics and fluids. Actions carried out by human beings do not simply administer the environment but are comprised of a stratification of interventions based on harmonious and symbiotic management with space. The terraces and shelters of transhuman agropastoralism, archaic methods to protect oneself from heat and cold, preserve products, collect the water and convey it to gardens in front of the cave's cistern, and the fulcrum of the local organization, constitute the ancient, but resilient matrix on which Matera's urban fabric grows (Bernardo and De Pascale 2016a, b).

To be aware of the excellent skills of Matera's citizens to implement resilient behaviors appropriate to address the difficulties of the inhabited area, it is necessary to refer to the construction of water collection works, visible in the subsoil of Piazza Vittorio Veneto, where the large cistern installed at the origin of the grabiglione of Sasso Barisano is accessible. The grabiglione are drainage channels that flow into the Gravina stream, near Matera, and channel both rainwater and wastewater. In operation until the nineteenth century, the channels, part of the system of underground structures of Matera, an example of effective planning and management that has made it possible over the centuries to enhance urban resilience (Bernardo and De Pascale 2016a; De Pascale et al. 2014, 2015). In the case of Matera, underground channels in the city can be explained by the fact that being located in an arid territory, the knowledge of the risks of drought, has allowed inhabitants to activate a protection strategy. This is accomplished by the distribution of water to citizens, which is intercepted and collected at the edge of the steep sides of canyons. Captured in cisterns the water is filtered and directed in a controlled manner into the Sassi, thanks to the vertical development of the city which allows a gravity flow network. It is an ingenious and harmonious housing and water system, which remained intact until the eighteenth century, as is clear from the fresco of the city painted in 1750 on the vault of the Archbishop's Palace (Fig. 5). In the painting, it is possible to follow the road system and the organization of the Sassi according to the drainage lines up to the current Piazza Vittorio Veneto. In this accurate representation of the urban system of Matera is the missing part of the Sasso Caveoso riverbed, protected by the Idris cliff, the system of caves of the plateau in front of the Murge, the pits, and cisterns that surround the Sassi from the plain. All the water flow lines are mapped, providing landscape observers with a complete view of the complex ecosystem of the Sassi of Matera (Laureano 2000).

Resilience features of a city include knowing the awareness of risk and knowing how to implement behaviors appropriate to the circumstances, in order to improve not only the relationship between humans and nature but also to represent "the relations of territoriality, through which society, transforming the Earth, transforms itself" (Dematteis 2003, p. 950). Matera's geomorphological framework is the important

Fig. 5 Fresco of the Archbishop's Palace, Salone degli Stemmi, showing Matera in the eighteenth century

key to the city's resilience over time, in addition to the urban design and construction of this resilient ecosystem.

The experience of Matera can be generalized to the cities of the Southern Mediterranean and serves as a concrete example of a sustainable city of which the Sassi is cited as a successful case on an international level (Bernardo and De Pascale 2016a, b). The creation of new skills and the rehabilitation of ancient crafts and marginalized identities have been possible, thanks to a careful restoration, favoring economic well-being, urban sustainability, and human progress: the conservation of physical and material signs becomes the protection of intangible ones, including the recovery of identity and cultural and spiritual values. The Sassi, where, until the mid-nineties, not only caves but also important buildings were sold at ridiculously low prices, is now in great demand, with more and more inhabits that constantly increases its value (Bernardo and De Pascale 2016a, b). Matera has transitioned from a place with a complete lack of hotels to a great variety some built in the caves and in the spectacular subterranean hypogea, which has become a rising destination for tourism. Recovery is no longer just the will of intellectuals but a bottom-up process, stimulated by the citizens themselves and supported by private investments and an economic return (Bernardo and De Pascale 2016a, b). Matera is a resilient and sustainable city; able to recover after the social and cultural shock suffered in the middle of the last century, for which the city was defined as a "national shame", it has now become one of the most premium assets in the world, a European model of sustainability and resilience. As reported in the next paragraph, these two concepts are closely related because urban resilience is linked to the broader challenges of global sustainability and sustainable mobility.

3.1 Matera: Example of Urban Sustainable Mobility

The Matera municipal administration recently adopted the urban sustainable mobility plan (PUMS) (Municipality of Matera 2022a). The PUMS offers an accurate snapshot of the choices for the coming years and up to 2031, in terms of mobility and urban viability. This is what was resolved with DGC (Resolution of the Municipal Council) no. 797-2022 of 02/22/2022 (Municipality of Matera 2022b).

In this context, the Sustainable Development Goals (UNDP 2021), made up of 17 goals, are an action plan to achieve a sustainable future for cities all around the world by 2030. Among them, Sustainable Development Goals 3 and 11 are relevant to urban sustainable mobility. The third sustainable development goal emphasizes ensuring a healthy life and promoting well-being for all, including the elderly (Patil et al. 2022). In this context, safe and affordable mobility infrastructures are a fundamental lifeline for older people to reduce transport-related social exclusion and improve access to essential services needed to maintain a better quality of life (Patil et al. 2022; Reardon and Abdallah 2013; Lucas et al. 2016; Rambaldini-Gooding et al. 2021). Research works also highlight the links between improved transport infrastructure and older people's welfare outcomes (UNDP 2021). For example, interventions such as improving transport infrastructure and urban regeneration have shown improved mental health and quality of life outcomes (UNDP 2021). According to Tobin, good urban resilience makes city sustainability operational: by keeping urban functions at an acceptable level of functioning, the resilience capacity of territorial systems contributes to the achievement of the economic, social, and environmental aspirations of a sustainable city (Tobin 1999).

The vision of Matera's municipal administration is that of a sustainable mobility that progressively aims to discourage the use of cars—at least in the historic center—to encourage greater and gradual pedestrian use of the historic center with important cycle and walking paths that will allow the connection between the new North–Center–South urban locations.

The sustainability objectives of the Plan (Municipality of Matera 2022a) include:

- To guarantee all citizens ways of travel that allow them to access key destinations and services;
- To achieve a balanced shift to environmentally friendly modes of transport for a sustainable transport and mobility system;
- To increase the sustainable mobility of people and goods, guaranteeing everyone, by 2030, access to a safe, convenient, accessible, and sustainable transport system, especially by enhancing public transport;
- To improve public passenger transport services to encourage greater efficiency and better performance;

- A reduction of road congestion;
- In terms of air quality and climate change, minimizing emissions and reducing pollutant concentrations in the atmosphere;
- Reduce energy consumption, reduce greenhouse gas emissions, reduce noise pollution, increase the safety, health, and quality of the urban environment in terms of regenerated cities, ensuring accessibility and ensuring the sustainability of connections (Municipality of Matera 2022a).

According to the Italian Ministerial Decree n. 396 (Ministero delle Infrastrutture e della Mobilità Sostenibili 2019), the PUMS (Sustainable Urban Mobility Plan) is "a strategic planning tool that, over a medium-long term time horizon (10 years), develops a system vision of urban mobility proposing the achievement of environmental sustainability objectives, social and economic through the definition of actions aimed at improving the effectiveness and efficiency of the mobility system and its integration with the urban and territorial structure and developments".

The elements that characterize the Urban Sustainable Mobility Plan (Municipality of Matera 2022a) are:

- A participatory approach that involves the local community (citizens and stakeholders) from the phase of sharing the knowledge framework up to the definition of the guidelines of the Plan on which the Plan choices will depend;
- A concrete commitment of the city and its decision-makers for sustainability in the mobility sector in terms: economic, social equity, and environmental quality;
- Integrated planning approach to dialogue with the planning tools promoted by the various sectors (territory, environment, etc.);
- A clear vision of the objectives of the PUMS and their measurability;
- A clear representation of the costs of transport and its benefits, taking into account the different components including environmental and social ones (Municipality of Matera 2022a).

However, the PUMS is the tool to implement the principles of sustainable development in the field of mobility according to socially, economically, and environmentally more sustainable models, obtained by declining objectives and general intervention strategies on the basis of the peculiar characteristics of each territorial area (Municipality of Matera 2022a).

Making cities more sustainable by investing resources in public transport through urban planning and management is part of the Sustainable Development Goal 11 (UNDP 2021; Patil et al. 2022).

Decarbonisation goals promoted by the EU must have as their premise a transition to sustainable urban mobility (Tsavachidis and Le Petit 2022). The mid-century carbon neutrality goals set out in the European Green Deal (European Commission 2019) and the strengthening of the EU's climate ambition to achieve a 55% reduction in greenhouse gas (GHG) emissions by 2030 compared to levels the 1990s require significant social and cultural adaptation by the population (Tsavachidis and Le Petit 2022). The objectives indicated are crucial for cities to implement, since 70% of these emissions are produced by cities, which generate about 80% of all economic growth (European Commission 2021).

To define a future of urban mobility that will improve the quality of life in cities, mitigate climate change, reduce the impact of natural hazards, and create new jobs to strengthen the European mobility sector, following the example of Matera, a new form of cooperation between private, stakeholders, and public organizations is needed, beyond regional borders and involving citizens at the local level using a bottom-up approach (Tsavachidis and Le Petit 2022).

4 The Innovative and Resilient Urban Scheme of Filadelfia (Calabria, Southern Italy) After the Earthquakes of 1783

In 1783, an earthquake devastated southern Calabria and created tens of thousands of victims. Two hundred towns were wholly or partially destroyed. The soil and the hydrogeological system were upset. The news of the earthquake took nine days to arrive to the capital Naples. However, the huge disaster forced the Bourbon government (a European dynasty that conquered the Crowns of Naples and Sicily and restored a united and independent kingdom) to take measures to reform the economic and housing systems of Calabria. Under the influence of new Enlightenment Era ideas, the reconstruction of entire cities and towns was conceived and carried out by military engineers according to revolutionary new rules and urban plans. This model was progressively applied during the nineteenth century. This is how new "ideal cities" were established, with references to urban utopias and the concreteness of geometric plans.

Among the new cities founded after 1783, alongside Palmi, Bagnara, and Mileto, there is also Filadelfia (Figs. 6 and 8), in the province of Vibo Valentia (Calabria, Southern Italy). The earthquake completely destroyed the old town of Castelmonardo. It was decided to rebuild in a more favorable place, a plateau not far from the hill where only the ruins remained. The convergence between the Enlightenment ideas of the baronial nobility, the participatory democracy of the citizens, and the contribution of local clergy designed Filadelfia, a city inspired by liberal principles, with the name and a plan reminiscent of Philadelphia in the US state of Pennsylvania.

4.1 The History Related to the City Foundation

The inhabited center of Castelmonardo (Fig. 7)—according to the writer Nazzareno Salvatore Carioti (2012) underwent a series of seismic events of varying intensity, in 1184 and then three times before 1500. These were followed by the earthquake of 1638, resulting in much damage and many victims, and that of 1659 whose destructive

Fig. 6 Top view of Filadelfia (Calabria, Southern Italy). *Source* Vito Rondinelli

effects were even more powerful. All these preceded the infamous earthquake of 28 Friday, March 1783, recognized for centuries as "the scourge" and the total ruin of Castelmonardo. Filippo Serrao recalls that "of the 390 hamlets, villages and cities that made up the Further Calabria, 181 were completely razed to the ground and [...] Castelmonardo, we would like to emphasize, was anything but a wild village" (Serrao 1983). A detailed description of the days that preceded and followed the terrible event is offered to us by Elia Serrao in "Dei Tremuoti di Castelmonardo e della nuova Filadelfia in Calabria" (1974).

The material damage was immense and irreparable, but considering the devastating scale of the event, all in all, the human losses in Castelmonardo were limited by the willingness of many to flee to the countryside after signs of tremor were inevitable (Serrao 1974). However, the physical damage was incalculable. After the first powerful phase, the surviving inhabitants had to make a painful decision, to choose whether to accept the opportunity, proposed by the mayor of the time, Tommaso Serrao, of an escape to the nearby Piano della Gorna. After the initial uncertainties, due to the emotional bond with the Vaglio of Castelmonardo hill (the place where Castelmonardo, originally, was built), the citizens, abandoned possessions and hopes under the rubble, and wisely decided to follow the Mayor's advice and move to the Piano della Gorna (Serrao 1974). Serrao speaks of it as an excellent flat site not far from the previous one, two miles long and one wide, with an extended horizon over the mountains and on the sea. Furthermore, if the conditions necessary to make a site pleasant and habitable are air, land, and water, this location possessed all three to an excellent extent (Serrao 1974). Serrao, taking his cue

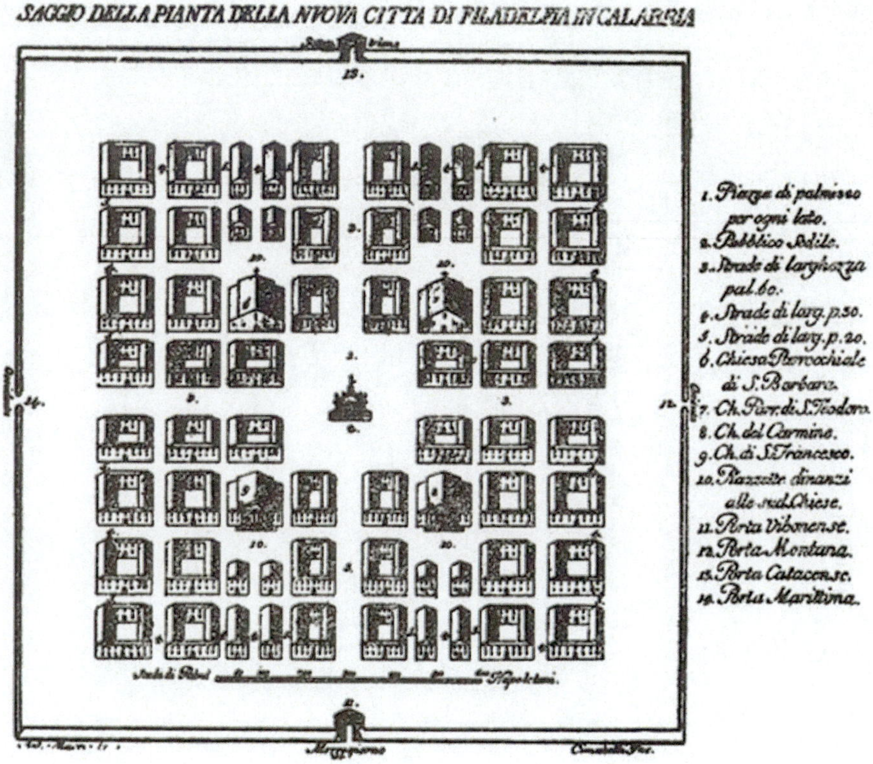

Fig. 7 General outline of the "project" of the city of Filadelfia reported in the first edition (1785) of Elia Serrao's work: "Dei Tremuoti di Castelmonardo e della nuova Filadelfia in Calabria"

from the general dismay, pronounced what appears to be one of the most dramatic invitations to abandon the place: "People of Castelmonardo, brothers, let's flee! We seek escape from this miserable life, which only today advances us. It is no longer to be unresolved. The dispute of staying and leaving no longer takes place" (Serrao 1974, p. 38). The exhortation is supported by an incisive speech, which clarifies the concrete and sentimental reasons for the rise of the places and their abandonment. The town is sacred because there are churches, places of worship, and the graves of the ancestors. Yet the citizens must submit to the laws of necessity. And as he is delivering the speech with conviction, affirming the need to find a new city elsewhere, the Earth resumes shaking and the invitation to flee takes on more dramatic tones: "In the future we will be pointed out as founders of a new city. But, Oh God, while I speak the earth does not cease to tremble! Run away, here I am, and all my parents will give you an example" (Serrao 1974, p. 41). Hearing the words of Serrao, the people abandon the city in tears and screams, and even one of the parish priests, Vincenzo Amorosi, invites everyone to flee. And so all "abandoned everything under the ruins, and given the last farewell to the beloved places, in a truly miserable aspect,

Fig. 8 The current urban map of Filadelfia (Calabria, Southern Italy). *Source* https://umap.openstreetmap.fr/

and almost in procession, set out on their way" (Serrao 1974, p. 43) and headed to the plain where Filadelfia, the new country, would later arise (Teti 2004).

At this point, it is necessary to recreate previous ownership relationships and the social stability required to find the continuity that was interrupted by the earthquake. Thus, the chosen space was divided into a Greek cross, whose two lines of formation went to represent the great main streets, now known as Corso Italia and Corso Castelmonardo (Fig. 8). From these arose the four large square districts, which gave the new city a rational and organized urban structure; these areas, according to their location, took the name of the Montana Zone, the Southern Zone, the Eastern Zone, and the Western Zone (Carioti 2012). The plan of the new city included a large central square, with the Public Seat in the center, near the corners of the aforementioned square, the two parish churches of St. Teodoro and St. Barbara and the two churches of St. Francis and del Carmine, one for each neighborhood and each with its own little piazza. Furthermore, there were walls and four access gates, in line with the two main roads, of which the one toward the sea was called Marittima, the one toward Monteleone (the current Vibo Valentia, the provincial capital) called Vibonese, the one toward the mountains to the north called Montana, and finally, the one called Caticense, because it faces Catanzaro (the current capital of the Calabria region) (Carioti 2012). However, among these works, the walls, doors, and seats in the center of the square were never built.

4.2 The Urban Planning Idea

The scholar Elia Serrao states that the urban planning idea (Fig. 7) was conceived by the architect Francesco Serrao (Carioti 2012), However, Barone (1978) asserts that not only during this period there was no official title of architect, but that the project was entirely desired and conceived by the officials and experts of the Kingdom of Naples and that Serrao and the others operated as a consequence of such provisions. The most conciliatory hypothesis, and perhaps probable, is offered by Filippo Serrao, who argues, pointing to a notarial deed of 1784, that the city's orthogonal plan was the idea of the citizens' assembly and speculates that the inspiration was not from Enlightenment models, but rather the scheme of the Roman camp, and its classical culture that the nobles of Castelmonardo would have been well acquainted with (Serrao 1983). However, the name Filadelfia is of Enlightenment origin and was consciously chosen to identify the new city. Giovanni Andrea Serrao, born in Castelmonardo and trained at the school of Dominican monks before furthering his education in Naples, where he distinguished himself for his excellent applications of humanistic culture, recommended the name. Later the Bishop of Potenza (Basilicata, Southern Italy) was shocked by the July 1783 destruction of his beloved Castelmonardo. He immediately urged a transfer to Piano della Gorna (a neighboring site) and suggested the name of Filadelfia, meaning "brotherly love" for the city. With Greek origins, the name reminds the new inhabitants of their ancestors and the need for fraternal feelings toward all human beings. It seems that Andrea Serrao joined one of the many intellectual associations of the Enlightenment in Naples, a bearer of republican and constitutional values with branches throughout Europe, whose affiliates were identified as "Filadelfi". Although with some confusion due to the habit of identifying the new name, with the old, Filadelfia was quickly accepted by its citizens. Of course, the new city was not built in three days, in fact, construction activities progressed very slowly; indeed, the makeshift shacks erected in 1783 persisted through the mid-nineteenth century. Despite the name of the new city, ruling families reverted to type, and grabbed the best parts of the subdivision, which comprised an entire block at the corners of the main square and adjacent to the churches. According to the scholar Giovan Domenico Barone (1978), all the buildings were limited to the ground floor and the first floor, all utilizing the same construction technique of affixing chestnut wood pillars to the ground, connected to each other by iron chains. Between each pillar local stones and lime mixture formed the walls of houses for middle-class workers, which had a maximum of four rooms with a height of no more than two meters and eighty centimeters.

The scholar Ilario Principe (2001) states that the central square and the four peripheral piazzas (Fig. 9) dominated most of the districts of the new Castelmonardo and was influenced by William Penn's 1682 plan for Philadelphia. Beyond the classical

Fig. 9 Filadelfia with the square in the center. Not all the programmatic lines of the plan have been realized: if some lots have been transformed into homogeneous blocks and built along the perimeter, others, on the other hand, especially in the peripheral areas to the north and south-west, have been divided into two, three, and also four parts; moreover, not all the lots facing the main square have been occupied by as many buildings of a monumental nature, indeed, one of these remained under construction. The central square, which represents the heart of the city and which should have been the first urban structure to be completed, remained unfinished. *Source* Vito Rondinelli

references of the name, it is no coincidence that this new Calabrian city was named after the American city, probably spurred by contacts between Benjamin Franklin and the Italian jurist Gaetano Filangieri, and Masonic circles in Naples and Calabria (Principe 2001).

According to the anthropologist Vito Teti, Filadelfia was born, therefore, with an innovative, rigidly square urban scheme, the result of an ideal operation revealing the contradictions of a class in search of an autonomous affirmation of power (Teti 2004). Filadelfia, and in part the other new cities that arose after the 1783 earthquake, broke with traditional urban schemes. The city is studied contemporarily and considered a prime example of the modern and innovative work of its period. With the earthquake, paradoxically, Calabria, on the one hand, revealed that it already had elites capable of communicating with the outside world, and on the other, it became part of a deeply innovative urban planning and architectural circuit (Teti 2004).

5 Discussion and Conclusions

The two cases of Matera and Filadelfia, albeit with different histories, paths, and geographical locations, have in common a cycle of destruction, abandonment, and innovative reconstruction. Their rebirths are distinguished not only by a rebirth of the material fabric of the cities but by their cultural and social rejuvenation, which makes it possible to define them as resilient cities, having recovered after shocks of a different nature. In the case of Matera, a city of medium size, the shock experienced was cultural; in fact, the mass exodus from the Sassi was stimulated by the pressure of a violent social shock, the triumph of the paradigm of shame, caused by the revelations of the impossibility of civilized living in a traditional settlement. In the Italian post-war period, the state of the Sassi was in comparison to the modern structural transformations occurring throughout the country viewed as economically and culturally neglectful (https://laureano.it). The collective representation of the Sassi after it was transformed with innovative methods, replaced its earlier misery and shame by joining the UNESCO World Heritage List in December 1993,—an event of enormous significance—reported on the front pages of the main national newspapers. Matera was the first city included in the UNESCO list of Southern Italy and was the only one based on the recognition of popular values relating to the reconstruction of a cultural landscape. According to the Commission, the whole of the Sassi and the Archaeological and Natural Park of the Rupestrian Churches of Matera constitute unique testimonies of innovative human activity and allowed the city to become a tourist destination known for its resilience all over the world. In addition, the city's profile has been cinematically elevated by the film *James Bond: No Time to Die* (2021) and by serving as Jerusalem in director Mel Gibson's *The Passion of the Christ* (2004).

The Sassi, has repopulated and become a fertile ground for museums, "cave houses", restaurants, Bed and Breakfast, luxury hotels, spas, and many artisan shops that find inspiration in Matera. In the same caves where only a few decades ago people died of malaria and hardship, visitors from all over the world now come to admire the rock churches, the numerous exhibitions, to taste a typical dish, or to buy a tuff or terracotta souvenir.

In the case of Filadelfia, a smaller city, its shock was caused by a disaster created by an extreme natural event—the earthquakes of 1783. While for Matera the reconstruction and enhancement of the entire cultural and natural ecosystem of the Sassi took place on the same site, in the case of Castelmonardo/Filadelfia, its survivors preferred to build the new city on a new site. Filadelfia, located in central Calabria, in a predominantly hilly area called Piano della Gorna, a few kilometers from the Tyrrhenian Sea, is surrounded on the east side by mountains, rich in vegetation and water sources. Lake Angitola, an oasis renowned and recognized by the WWF (World Wide Fund for Nature), is located near the town. The history of Filadelfia is also one of resilience and resistance on the part of the inhabitants and has intrigued scholars from all over the world for the particular urban layout and its archeological ruins (Fig. 10). From 1970 to 1973, surveys and excavations on the hill of Castelmonardo were carried out by the Archaeological Groups of Italy. As for the artistic-monumental heritage, the central point of the city is the large square, intended for public assemblies and the market, with the war memorial. The two large main streets meet on the square (Corso Italia and Corso Castelmonardo) (Fig. 8) and overlook the noble palace Serrao (Fig. 11), the Auditorium, the Outpatient Clinic, and the public gardens. To characterize the urban districts, four monumental avenues stretch toward the four local churches (St. Francis of Paola, the Madonna of Carmine, St. Barbara, and St. Teodoro). Filadelfia represents a unique model of urban resilience. It is an example of rebirth after the terrible scourge of the earthquake that razed the old Castelmonardo to the ground. The history of Filadelfia is an example of cultural resistance, brotherhood, and freedom—even despite the Mafia power rooted in the area.

The stories of these two southern Italian cities, different yet similar at certain levels, illustrate that the transmission of historical memory, together with the knowledge of diversity in a multiscalar and multidisciplinary sense (Tecco 2011), can become useful means to enhance the resilience of the territories, treasuring already existing and tested experiences,—some more famous (Matera), others more hidden (as in the case of Filadelfia). Urban resilience, therefore, is a fundamental concept in geographic education. Improving preparedness, awareness of risks, and telling the

Fig. 10 The ruins of the old site of Castelmonardo were destroyed by the earthquakes of 1783. *Source* Vito Rondinelli

Fig. 11 The "Serrao del Compasso" noble palace is a valid example of a shack-up house still existing in Filadelfia. The building, which is currently in a state of neglect, would deserve an immediate restoration for its architectural and historical value. *Source* Vito Rondinelli

stories of our small, medium, and large cities allows us to increase urban resilience and ensure more effective planning. Urban resilience requires cities to look holistically at their capabilities and risks (https://resilientcitiesnetwork.org/what-is-resilience), enhanced by meaningful engagement with the most vulnerable members of a community.

Finally, the aim of this work is to demonstrate through the power of storytelling how the organization of social and cultural structures is indispensable for the achievement of urban resilience. Both case studies have this aspect in common. The Matera dream was made possible thanks to the synergy between policymakers, stakeholders, and citizens who transformed the crumbling cave houses, where the inhabitants slept together with animals, into a resilient system, which was able to make the best use of rare natural resources, managing water, soil, and energy in a harmonious and sustainable way.

Filadelfia, although a different case than Matera because its inhabitants were forced to create a new city in a neighboring site, retained and preserved its social and cultural structures over time. The population of Filadelfia was able to carry out readjustments and tweaks, representing their past, and ancient traditions, through rebuilding a new city from a "physical" point of view. However, the city has remained "old" over the centuries thanks to the transmission of historical and anthropological memory to subsequent generations. And it is at this point that the processes of resilience become visible. Certainly, this topic needs to be explored in future studies, as proposed by Teti (2004), on how old social hierarchies and forms of socialization reproduced themselves in a new urban fabric. The participatory democracy of its citizens in reconstruction planning after the earthquake, inspiration from liberal values, its innovative urban scheme, the historical–cultural memory and the recovery of ancient traditions and archeological ruins , and the ideal of "brotherhood", derived from the etymology of its chosen name, all conspired to make Filadelfia a resilient city. Many designers, architects, geographers, philosophers, and economists agree that the city itself represents the manifestation of the maximum social energy of a territory,—a place where time and human experience become visible through the processes of power and the melding of regional and international cultural symbols (Mumford 1961).

Acknowledgements The author thanks Prof. Vito Rondinelli, a careful and rigorous scholar, for having allowed the publication of his images concerning Filadelfia. Furthermore, the author thanks Prof. Bruno Bartucca for his kindness and availability. Special thanks also go to Prof. Charles Travis for the language review.

Conflict of Interest The author declares no conflicts of interest in this paper.

References

Alberti M, Marzluff JM (2004) Ecological resilience in urban ecosystems: linking urban patterns to human and ecological functions. Urban Ecosyst 7:241–265. https://doi.org/10.1023/B:UECO.0000044038.90173.c6

Andersson E (2006) Urban landscapes and sustainable cities. Ecol Soc 11(1):34

Barone GD (1978) Castel Mainardi e Filadelfia nel Regno del Sud. Dall'VIII secolo d. C. al 1860, Frama Sud, Chiaravalle Centrale

Barone M (2019) Massoneria, istituzioni ed elite politiche nella storia di Filadelfia di Calabria (1783–1920), Nicolò Edizioni, Messina

Batty M (2008) The size, scale, and shape of cities. Science 319:769–771. https://doi.org/10.1126/science.1151419

Berkes F, Folke C (eds) (1998) Linking social and ecological systems. Cambridge University Press, Cambridge, 476 pp

Berkes F, Colding J, Folke C (eds) (2003) Navigating social-ecological systems: Building resilience for complexity and change. Cambridge University Press, Cambridge, 416 pp

Bernardo M, De Pascale F (2016a) Matera (Basilicata, Southern Italy): a European model of reuse, sustainability and resilience. Adv Econ Bus 4(1):26–36. http://www.hrpub.org. https://doi.org/10.13189/aeb.2016.040104

Bernardo M, De Pascale F (2016b). Matera, Capitale Europea della Cultura 2019: simbolo di riuso, resilienza ed ospitalità sostenibile. Annali del Turismo, vol 5, pp 61–84

BMI (2021) (Bundesministerium des Innern, für Bau und Heimat). Innenstrategie des Beirats Innenstadt beim BMI: Die Innenstadt von Morgen—Multifunktional, Resilient, Kooperativ; BMI, Berlin, Germany

Buckle P, Marsh G, Smale S (2000) New approach to assessing vulnerability and resilience. Aust J Emerg Manage 15:8–14

Burby RJ, Deyle RE, Godschalk DR, Olshansky RB (2000) Creating hazard resilient communities through land-use planning. Nat Hazard Rev 1(2):96–106

Burns ET (1996) From risk to resilience: a journey with heart for our children, our future, Dallas, Marco Polo, pp 320–333

Campanella TJ (2006) Urban resilience and the recovery of New Orleans. J Am Plann Assoc 72(2):141–146

Carioti NS (2012) Castel Monardo. Archeologia e storia di un insediamento medievale, Adhoc Edizioni, Vibo Valentia

Chelleri L (2012) From the "resilient city" to urban resilience. A review essay on understanding and integrating the resilience perspective for urban systems. Documents d'Anàlisi Geogràfica 58(2):287–306. https://raco.cat/index.php/DocumentsAnalisi/article/view/252964

Colding J (2007) 'Ecological land-use complementation' for building resilience in urban ecosystems. Landsc Urban Plan 81(1–2):46–55. https://doi.org/10.1016/j.landurbplan.2006.10.016

Comfort LK, Wisner B, Cutter S, Pulwarty R, Hewitt K, Oliver-Smith A, Wiener J, Fordham M, Peacock W, Krimgold F (1999) Reframing disaster policy: the global evolution of vulnerable communities. Environ Hazards 1(1):39–44

Cutter SL, Barnes L, Berry M, Burton C, Evans E, Tate E, Webb J (2008) A place-based model for understanding community resilience to natural disasters. Glob Environ Chang 18:598–606

De Pascale F, Bernardo M, Muto F (2014) Hazardscape, territorial and individual resilience in an interdisciplinary study: the case of Pollino, Calabria, Southern Italy. In: Lollino G et al (eds) Engineering geology for society and territory, vol 7, education, professional ethics and public recognition of engineering geology. Springer International Publishing Switzerland, pp 109–113

De Pascale F, Bernardo M, Muto F, Zumbo R, Galea P, Agius M (2015) Neogeography and seismic risk perception. A comparison between two case-studies: Calabria (Southern Italy), Malta. Eur J Geogr 6(1):64–83

Dematteis G (2003) La metafora geografica è postmoderna? Bollettino della Società Geografica Italiana 12(8):947–954

Di Caro E (2019) Matera 2019, da «vergogna nazionale» a simbolo di sviluppo. Ecco come nasce questo traguardo, Il Sole 24 ore

DST (2021) (Deutscher Städtetag). Zukunft der Innenstadt. In Positionspapier des Deutschen Städtetages; DST, Berlin, Germany

Ernstson H, van der Leeuw SE, Redman CL, Meffert DJ, Davis G, Alfsen C, Elmqvist T (2010) Urban transitions: on urban resilience and human-dominated ecosystems. Ambio 39(8):531–545. https://doi.org/10.1007/s13280-010-0081-9. PMID: 21141773; PMCID: PMC3357675

European Commission (2019) The European green deal. https://ec.europa.eu/info/sites/info/files/european-green-deal-communication_en.pdf. Last access 10 Aug 2022

European Commission (2021) Road safety: European Commission sets out next steps towards \215Vision Zero\216 including key performance indicators. https://transport.ec.europa.eu/news/road-safety-european-commission-sets-out-next-steps-towards-vision-zero-including key-performance-2019-06-19_en

Godschalk DR, Beatley T, Berke P, Brower DJ, Kaiser EJ (1999) Natural hazard mitigation: recasting disaster policy and planning. Island Press, Washington, DC

Gunderson LH, Holling CS (eds) (2002) Panarchy: understanding transformations in human and natural systems, 450 pp. Island Press, Washington, DC

Hangebruch N, Othengrafen F (2022) Resilient inner cities: conditions and examples for the transformation of former department stores in Germany. Sustainability 14:8303. https://doi.org/10.3390/su14148303

Holling CS (1973) Resilience and stability of ecological systems. Annu Rev Ecol Syst 4:2–23

Indirli M (2019) An historical flight and some open questions towards a pluralistic but holistic view of resilience. In: Farabollini P, Lugeri FR, Mugnano S (eds) Earthquake risk perception, communication and mitigation strategies across Europe. Geographies of the anthropocene book series. Il Sileno Edizioni, Lago

Klein RJT, Nicholls RJ, Thomalla F (2003) Resilience to natural hazards: how useful is this concept? Environ Hazards 5:35–45

Laureano P (2000) Giardini di pietra: i Sassi di Matera. In: Pizzetti I, Alberti F (eds) Architetture nel paesaggio, Alinea, Firenze

Laureano P (2012) Giardini di Pietra. I Sassi di Matera e la civiltà mediterranea. New Edition, Bollati Borlinghieri, Torino

Lombardo G, Salvucci G (2014) "Roma città quotidianamente resiliente, proposta metodologica per un'analisi territoriale dei tempi di reazione per l'intervento nell'emergenza quotidiana. Memorie Geografiche 12:55–58

Lucas K, Mattioli G, Verlinghieri E, Guzman A (2016) Transport poverty and its adverse social consequences. Proc Inst Civ Eng Transp 169(6):353–365

Maguire B, Hagan P (2007) Disasters and communities: understanding social resilience. Aust J Emerg Manage 22(2):16–20

Manyena SB (2006) The concept of resilience revisited. Disasters 30(4):433–450

Marincioni F, Appiotti F, Pusceddu A (2014) Resistenza e resilienza ai disastri: parallelismi fra sistemi ecologici e so-cioeconomici, Memorie Geografiche, pp 59–63

Masten AS, Garmezy YN (1985) Risk, vulnerability, and protective factors in developmental psychopathology. In: Lahey B, Kazdin A (a cura di) Advances in clinical child psychology, vol 8. New York, Plenum, pp 1–52

Masten AS, Osofsky J (2010) Disasters and their impact on child development: introduction to the special section. Child Dev 81:1029–1039

Meerow S, Newell JP, Stults M (2016) Defining urban resilience: a review. Landsc Urban Plan 147:38–49

Mileti DS (1999) Disaster by design: a reassessment of natural hazards in the United States. Joseph Henry Press, Washington, DC

Ministero delle Infrastrutture e della Mobilità Sostenibili, Decreto ministeriale n. 396 del 28/08/2019. https://www.mit.gov.it/normativa/decreto-ministeriale-n-396-del-28082019. Last access 18 Aug 2022

Mumford L (1961) The city in history. Secker & Warburg, London

Municipality of Matera (2022a) Delibera di Giunta n°37/2022 del 22/02/2022—Adozione PUMS. https://www.comune.matera.it/images/utilita/PUMS/delibera_di_giunta_del_PUMS.pdf. Last access 17 Aug 2022

Municipality of Matera (2022b) Piano Urbano della Mobilità Sostenibile (PUMS) - Quadro Conoscitivo, https://www.comune.matera.it/piano-urbano-della-mobilita-sostenibile-pums-quadro-conoscitivo. Last access 17 Aug 2022

Musacchio L, Wu J (2002) Cities of resilience: four themes of the symposium. Understanding and restoring ecosystems: a convocation. Ecological Society of America, Washington, DC, p 38 (Abstracts)

OECD (2022) Resilient cities. https://www.oecd.org/cfe/regionaldevelopment/resilient-cities.htm. Last access 15 Aug 2022

Patil DS, Bailey A, Yadav UN et al (2022) Contextual factors influencing the urban mobility infrastructure interventions and policies for older adults in low- and middle-income countries: a realist review. BMC Public Health 22:1489. https://doi.org/10.1186/s12889-022-13875-6

Paton D, Johnston D (2006) Disaster resilience: an integrated approach. Charles C. Thomas, Springfield, IL

Paton D, Smith L, Violanti J (2000) Disasters response: risk, vulnerabilities and resilience. Disaster Prevent Manage 9(3):173–179

Pelling M (2003) The vulnerability of cities: natural disasters and social resilience. Earthscan, London

Pickett STA, Cadenasso ML, Grove JM (2004) Resilient cities: meaning, models, and metaphor for integrating the ecological, socio-economic, and planning realms. Landsc Urban Plan 69(4):369–384. https://doi.org/10.1016/j.landurbplan.2003.10.035

Principe I (2001) Città nuove di Calabria nel tardo Settecento (1976). Allegato immagini dell'Accademia delle Scienze di Napoli dei luoghi della Calabria devastati dai sismi del 1783, prefazione di V. Franchetti Pardo, Appendice di P. Mascilli Migliorini, Gangemi, Roma

Rambaldini-Gooding D, Molloy L, Parrish AM, Strahilevitz M, Clarke R, Dubrau JM, Perez P (2021) Exploring the impact of public transport including free and subsidised on the physical, mental and social well-being of older adults: a literature review. Transp Rev 41(5):600–616

Reardon L, Abdallah S (2013) Well-being and transport: taking stock and looking forward. Transp Rev 33(6):634–657

Serrao E (1974) del Vescovo Patrizio di Filadelfia Avvocato e Giureconsulto di Napoli, Dei tremuoti di Castelmonardo e della nuova Filadelfia in Calabria, Edizioni Frama's, Chiaravalle Centrale

Serrao F (1983) Dall'apocalisse all'esodo: contributo per la storia di Filadelfia nel bicentenario della sua fondazione. In: "Brutium", Anno LXII, n. 1, Reggio Calabria, gennaio-marzo, p 13

Sparks L (2021) Towns, high streets and resilience in Scotland: a question for policy? Sustainability 13:5631

Tecco N (2011) Educazione geografica, resilienza e catastrofi naturali. In: Giorda C., Puttilli M. (a cura di), Educare al territorio, educare il territorio. Geografia per la formazione, Carocci, Roma, pp 308–320

Teti V (2004) Il senso dei luoghi. Paesi abbandonati di Calabria, Donzelli editore, Roma

The Rockefeller Foundation & Arup (2014) City resilience framework April 2014 (Updated December 2015), Arup, London

Tierney K, Bruneau M (2007) Conceptualizing and measuring resilience: a key to disaster loss reduction. TR News 250:14–17

Timmerman P (1981) Vulnerability, resilience and the collapse of society: a review of models and possible climatic applications. Canada University of Toronto, Toronto

Tobin GA (1999) Sustainability and community resilience: the Holy Grail of hazards planning? Environ Hazards 1:13–25

Tsavachidis M, Le Petit Y (2022) Re-shaping urban mobility—key to Europe's green transition. J Urban Mobility 2:100014. https://doi.org/10.1016/j.urbmob.2022.100014

Twigg J (2007) Characteristics of a disaster-resilient community: a guidance note. Benfield UCL Hazard Research Centre, London

UNDP (2021) Sustainable Development Goals 2021. https://www.undp.org/sustainable-development-goals. Last access 11 Aug 2022

United Nations International Strategy for Disaster Reduction (2005) Hyogo framework for action 2005–2015: building the resilience of nations and communities to disasters. www.unisdr.org/wcdr/intergover/official-docs/Hyogo-framework-action-english.pdf

Walter J (ed) (2004) World disasters report 2004: focus on community resilience. Kumarian, Bloomfield, CT

Werner EE, Smith RS (1982) Vulnerable but invincible: a study of resilient children. McGraw-Hill, New York

Wildavsky A (1991) Searching for safety. Transaction, New Brunswick, NJ

Yoon DK, Kang JE, Brody SD (2016) A measurement of community disaster resilience in Korea. J Environ Planning Manage 59(3):436–460. https://doi.org/10.1080/09640568.2015.1016142

https://laureano.it. Last access 1 Aug 2022

https://resilientcitiesnetwork.org/what-is-resilience/. Last access 3 Aug 2022

Disaster Resilience Assessment for Drainage Network and Urban Landscape After Heavy Meteorological Events: Examples from the Middle Adriatic Coastal Area (Abruzzo Region, Central Italy)

Giorgio Paglia, Massimiliano Fazzini, Gianluca Esposito, Vania Mancinelli, Vincenzo Marsala, and Enrico Miccadei

Abstract Central Italy is sadly acknowledged as highly exposed to natural hazards and affected by ever-present disasters. The middle Adriatic coastal area of the Abruzzo Region has been severely affected by heavy rainfall and flood events in recent times. It is characterized by moderate to low annual precipitation and, occasionally, by serious rainfall events (up to 400–500 mm/day), which determine landscape modifications and geomorphological processes. Floods occurring in the study area revealed the state of vulnerability towards natural hazards posing a threat to human settlements, livelihoods, and properties, despite several mitigation activities carried out in the last decades. In the context of climate change, these events' frequency has increased and continues to intensify, also in addition to the ever-increasing of anthropic activities. Only a high knowledge of geomorphological dynamics, connected to drainage network evolution, geological-geomorphological features, and meteorological events, can help develop practical mitigation activities and bridge the gap between disaster resilience and urban sustainability, advocated as one of the core elements of new urban agendas. This work could represent a scientific basis, readily available to interested stakeholders, for implementing sustainable territorial planning for disaster resilience assessment and management in urban and coastal landscapes.

Keywords Disaster resilience assessment · Heavy rainfalls · Meteoclimatic analysis · Landscape management · Urban coastal areas · Abruzzo Region

G. Paglia · M. Fazzini · G. Esposito · V. Mancinelli · E. Miccadei (✉)
Laboratory of Environmental Geomorphology and GIS, Department of Engineering and Geology, Università degli Studi "G. d'Annunzio" Chieti-Pescara, Via dei Vestini 31, 66100 Chieti Scalo, CH, Italy
e-mail: enrico.miccadei@unich.it

V. Marsala
SGI Studio Galli Ingeneria S.r.l., Via della Provvidenza 15, 35030 Sarmeola di Rubano, PD, Italy

B. Dahiya et al. (eds.), *Disaster Resilience and Human Settlements*,
Advances in 21st Century Human Settlements,
https://doi.org/10.1007/978-981-99-2248-2_11

1 Introduction

Disasters can result from natural processes which may be aided and intensified by human actions and inactions. Some develop slowly, while others may happen suddenly and unexpectedly, imposing huge impacts on social and environmental systems with fatalities and economic damages. The reduction of losses in lives and the number of affected human settlements during natural disasters, including geo-hydrological ones, is an expected outcome of the Sendai Framework for Disaster Risk Reduction 2015–2030 (UNISDR 2015). Floods can be classified among the quick and sudden disaster types, but few in this category could be well predicted, anticipated, and controlled to a great extent. Moreover, floods should not be labeled as disasters a priori since they become "real" disasters when they cause damages or undesirable effects on human settlements, livelihoods, and properties (Glago 2021). These impacts are expected to increase because of environmental degradation, climate change, rapid population growth, and improper land-use (Salvati et al. 2014; Napolitano et al. 2018). Floods, together with landslides, are among the most frequent natural processes which can induce disasters across the globe, especially in urban areas with high levels of vulnerability. Heavy and persistent rainfalls are one of the primary triggers of these events, together with other geological and anthropogenic factors, particularly in the framework of recent climate changes (Bathrellos et al. 2016; Xu et al. 2021). However, establishing a relationship between recent climate change and its potential effects on the occurrence of floods or rainfall-induced landslides remains an open issue (Blöschl et al. 2019; Kundzewicz et al. 2018; Sharma et al. 2018).

Central Italy is sadly acknowledged as an area highly exposed to natural hazards. Abruzzo Region (Fig. 1), particularly the middle Adriatic coastal area, is widely affected by geo-hydrological hazards, reflecting the different physiographic contexts and producing significant geomorphological changes. In this framework, the study area is characterized by moderate to low annual precipitation and, occasionally, by heavy rainfall events (up to 400–500 mm/day). Small-to-medium size catchments and coastal urban areas in the northeastern sectors have been widely affected by flood and landslide events (Miccadei et al. 2012; Piacentini et al. 2018; Calista et al. 2019). As a result, the landscape suffered severe changes and modifications highlighted by several geomorphological processes, such as floods, soil erosion (rill, gullies), and landslides (earth flows) with distribution and types varying from event to event, and in many cases led to an official natural disaster declaration at the regional government level (Regione Abruzzo—https://www.regione.abruzzo.it/content/calamit%C3%A0-naturali).

Recently, heavy rainfall and flood events have severely affected the study area. Almost three of them have had daily rainfall >100 mm and >200 mm in a few days. In this work, we analyzed heavy meteorological events that occurred on several dates: 6–7 October 2007 (in a small part of the hilly and coastal Teramo area); 1–2 March 2011 (in the hilly and coastal sectors of Teramo and Pescara); 13–14 March 2012 and 12 January 2018 (in the hilly and coastal Pineto area); 10 July 2019 (extreme hailstorm

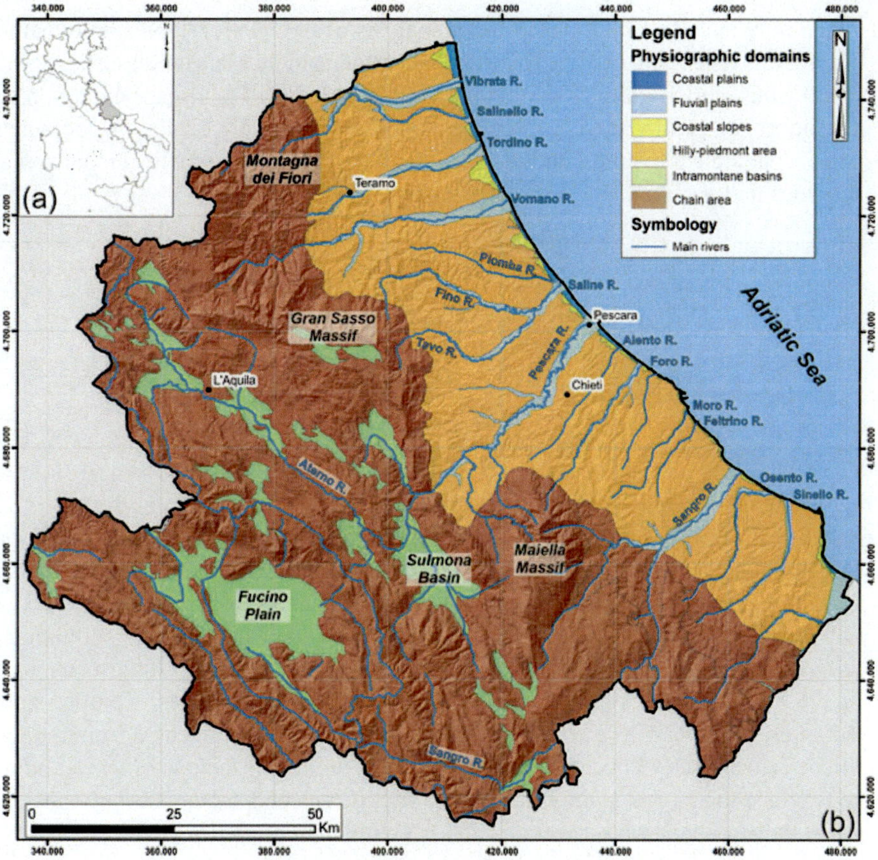

Fig. 1 **a** Location map of Abruzzo Region in Central Italy; **b** main physiographic domains

in the Pescara area). All the events were analyzed with regard to their meteorological aspects, highlighting distinct geomorphological effects on the landscape and human settlements.

The meteorological features were analyzed by processing data belonging to 3 thermometric series and 8 pluviometric series. Geomorphological effects were investigated through a multidisciplinary approach involving field surveys, aerial photo analysis, inventories analysis, technical reports, and photo documentation of the spatial distribution of landslides, soil erosion, and flooding. In detail, heavy flooding mainly affected alluvial plains and coastal plains due to both tributary and main rivers, with the formation of widespread crevasse splay, sheet-rill, and gully erosion. The main aim of the work is to illustrate detailed analyses of geo-hydrological hazards valid to increase knowledge in preventing hazards and related risks. In so doing, it is possible to measure community disaster awareness actions and competence in natural hazards' knowledge. Since these phenomena are challenging to understand, to mitigate the effects of these disasters in the short term, an adaptation approach is fundamental, which may include a multidisciplinary analysis of the relationships among natural hazards and risks, with specific focuses on disaster resilience and urban sustainability. Finally, this work highlights that only a high knowledge of landscape evolution and a geomorphological approach, connected to drainage network evolution, geological-geomorphological features, and meteorological events, allow practical management activities and sustainable territorial planning.

2 Study Area

2.1 Geological and Geomorphological Setting

The study area is located along the hilly piedmont area of the Abruzzo Region, between the Apennine chain and the coastal zone (Fig. 1). It includes the lower part of the main fluvial valleys and the small tributary catchments of the main rivers. The Apennine chain area shows a mountainous landscape (with reliefs up to 2900 m a.s.l. high) interrupted by longitudinal valleys and broad intermontane basins (i.e., Fucino Plain, Sulmona Basin). It is made up of lithological sequences pertaining to different palaeogeographical domains. Carbonate shelf and slope limestones, basin limestone, and marls represent the backbone of the main ridges. Compressional tectonics, due to NW–SE- to N–S-oriented thrusts, affected the chain from the Late Miocene to the Early Pliocene. It was followed by strike-slip tectonics that was poorly constrained in age and mainly masked by later extensional tectonic events since the Early Pleistocene (Miccadei et al. 2021). The piedmont area is characterized by a hilly relief morphology with altitudes ranging from ~100 to 800 m a.s.l. It shows a cuesta, mesa, and plateau landscape on Neogene sandy-pelitic turbidites and Plio-Pleistocene marine clayey–sandy and conglomeratic deposits (Ori et al. 1991). The geomorphological evolution has comprised the incision of WSW–ENE major dip river valleys, characterized by fluvial deposits arranged in a sequence of at least four main orders of terraces constrained from Middle Pleistocene to Holocene (Bigi et al. 1995; Miccadei et al. 2017). Quaternary deposits are widely present in the alluvial valleys, alluvial plains, and coastal slopes. They can be referred to fluvio-lacustrine, morainic, travertine, sandy shore, and eluvial–colluvial deposits (Fig. 2).

The geomorphological setting is mainly related to fluvial and slope processes. Fluvial processes affect the main rivers, alternating between channel incisions and flooding. The slope processes due to running water affect the clayey and arenaceous-pelitic areas of piedmont and coastal sectors, generating minor landforms such as rills, gullies, and mudflows (Piacentini et al. 2018). The area is extensively affected by different landslides (e.g., mostly rotational–translational slides, earth flows, rockfalls, complex slides), mainly characterizing the hilly piedmont and, locally, the coastal area (Esposito et al. 2021).

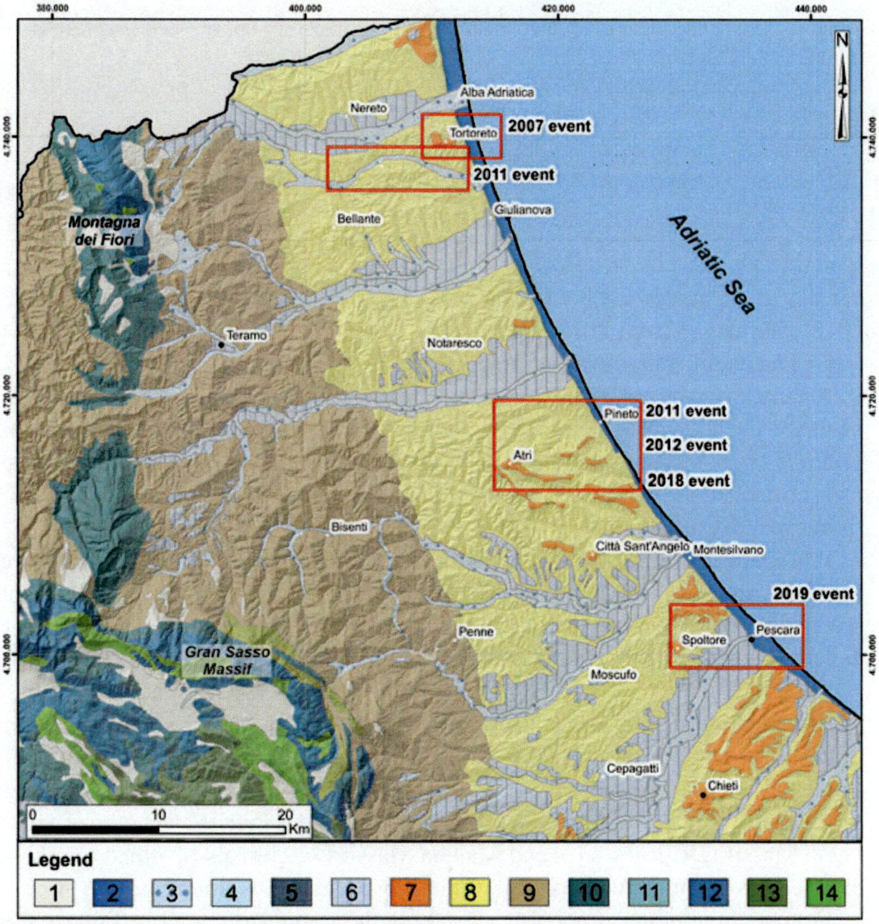

Fig. 2 Lithological scheme of the Abruzzo Region (modified from Accordi and Carbone 1988; Vezzani and Ghisetti 1998) with the location of the areas affected by heavy rainfall events (red boxes). Legend: (1) Eluvial–colluvial deposits; (2) Sandy shore deposits; (3) Recent fluvio-lacustrine deposits; (4) Travertine deposits; (5) Morainic deposits; (6) Old fluvio-lacustrine deposits; (7) Conglomeratic deposits; (8) Clayey–sandy deposits; (9) Sandy-pelitic turbidites; (10) Carbonate ramp limestones; (11) Basin limestones and marls; (12) Slope limestone; (13) Open carbonate shelf-edge limestones; (14) Carbonate shelf limestones and dolomites

2.2 *Climatic Features*

The study area is characterized by a maritime Mediterranean climate, especially in the eastern sector (Di Lena et al. 2012). It is affected by the physiographic setting, changing from a Mediterranean type with maritime influence along the coasts to a more continental-like in the inner sectors (Peel et al. 2007). The average annual precipitation is 600–800 mm/y, but, in particular, in the last two decades, this area was affected by heavy rainfalls ranging from 60 to 80 mm in a few hours to >200 mm in

Table 1 Geographic features of three main thermo-pluviometric stations, showing complete and recent rainfalls' datasets

Thermo-pluviometric station	Latitude	Longitude	Elevation (m a.s.l.)
Giulianova	42° 44′ 51.10″	13° 57′ 00.00″	64
Atri	42° 34′ 29.17″	13° 57′ 59.48″	421
Pescara airport	42° 26′ 13.11″	14° 11′ 12.51″	11

one day (e.g., January 2003, October 2007, March 2011, September 2012, December 2013, February–March 2015, January 2017) (Di Lena et al. 2021; Miccadei et al. 2012; Piacentini et al. 2018).

Finally, over the last three years, this area has been affected by two other heavy meteoric events—in January 2018 and July 2019. These events have triggered different types of geomorphological instability (i.e., landslides and flooding in the most critical urban areas and human settlements), with distribution and types varying from event to event.

It is necessary to quantitatively define the average climatology of the study area, especially in the recent trends of the studied parameters (i.e., temperature and total precipitation), in order to better analyze the most intense meteoric phenomena for understanding their influence on the hydrological balance of the area.

From the thermal point of view, the coastal area analyzed is characterized by an overall Mediterranean climate. Several factors at the meso and microscale concur to define peculiar features at the local scale in the study area. The presence of a small and shallow sea, the relative proximity of the most important mountain ranges of the entire Central Apennines chain—Montagna dei Fiori (2476 m a.s.l.), Gran Sasso Massif (2912 m a.s.l.), and Maiella Massif (2795 m a.s.l.)—tend to mitigate the Mediterranean features, typical of the geographical position (Fig. 1).

From a climatological point of view, only the stations reported in Table 1 show homogeneous and relatively continuous historical series of at least thirty years. For the other 5 stations, the data relating to the selected heavy rainfall events were analyzed (Fig. 3).

The Annual Average Temperature (AAT) for the last 30 years is comprised of between 14.1 and 15.3 °C (Table 2). Temperatures in January (the coldest month) range between 5.5 and 7.1 °C and those in July (the warmest month) vary between 23.9 and 24.5 °C; therefore, the annual temperature varies between 17 and 18.5 °C, while the daily temperature range is between 8 and 10.5 C. Relatively to the last decade, evidently characterized by a considerable thermal forcing, in the Pescara thermo-pluviometric station (Pescara airport in Table 1), the number of hot days (with Tmax > 30 °C) was 28, while the number of frosty days (with Tmin < 0 °C) stood on 22. The values extrapolated for the sixty years 1961–2020 are respectively on an average of 24 and 25.

Let's examine the trend in the last sixty years relating to the Pescara thermo-pluviometric station (Pescara airport; Table 1)—the only one not to have had any

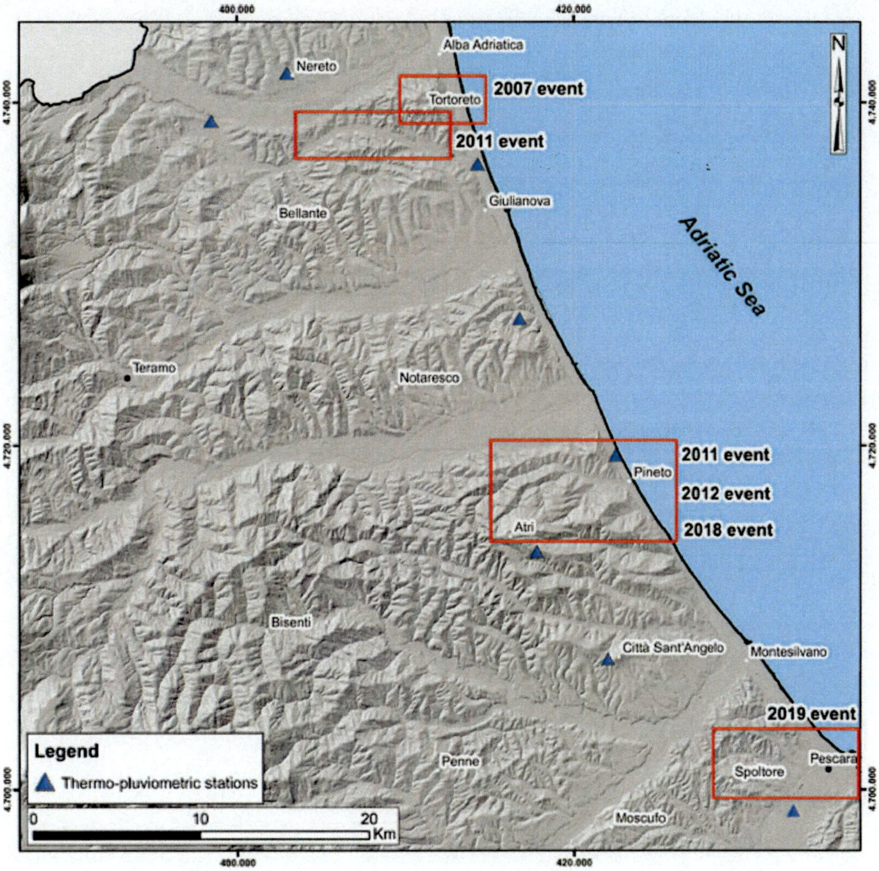

Fig. 3 Spatial distribution of the thermo-pluviometric stations analyzed in the study. The red boxes indicate the areas affected by heavy rainfall events

changes in position. We can observe an increase in the annual average values quantifiable at about 1.4 °C, a not dissimilar value from what was calculated by ISPRA (2020) for the same period. If we observe the thermal trend in Fig. 4, it is evident that the thermal increase just quantified has been observed since the end of the '70s and, therefore, in just over forty years.

Let's go to characterize the thermal climate according to the classification of Pinna (1978). In the last thirty years, it is evident that we have passed, at least on the coastal strip and low hills, from a sublittoral temperate to a sub-Mediterranean temperature thermal type.

To describe a quantitative feature of precipitation, reference was made to the exact thermo-pluviometric stations already used for the thermal study—which are also representative of the entire physical environment analyzed—since, at least for the cumulative daily precipitation values, they hold almost historical series, totally complete and above all homogeneous.

Table 2 Monthly and annual (AAT) average temperature (timespan 1991–2020)

Thermo-pluviometric station	Jan	Feb	Mar	Apr	May	Jun	Jul	Aug	Sep	Oct	Nov	Dec	AAT (°C)
Giulianova	7.1	7.3	10	13.2	17.7	21.7	24.5	24.6	20.6	16.9	11.7	8	15.3
Atri	5.5	6	8.8	12.1	16.5	21.1	23.9	23.6	19.4	15.3	10.5	6.6	14.1
Pescara airport	6.9	7.7	10.1	13.2	17.5	21.5	24	24	20.8	16.7	12.2	8.4	15.3

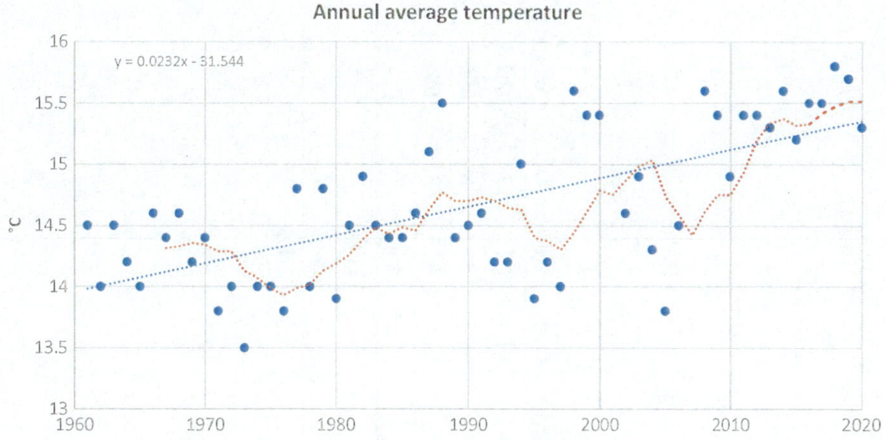

Fig. 4 Trend of annual average temperature at Pescara station (timespan 1961–2020). The red function (dotted red line) represents the 7-year moving average

The daily data of the three thermo-pluviometric stations (Table 1) were analyzed for the period 1951–2020. Therefore, it was possible to derive the monthly distribution of the total rainfall, as graphically reported in Table 3.

The analysis of the annual sum shows values between 650 and 800 mm, therefore poor if compared to the cumulative yearly averages for the entire Italian territory (ISPRA 2020). This meteoric deficit derives mainly from the morphological position of "downwind" controlled by mountain reliefs mentioned above (Montagna dei Fiori, Gran Sasso Massif, and Maiella Massif) compared to the average trajectories of the frontal systems coming from the Atlantic or the northwestern Mediterranean. The most abundant and intense precipitations derive from the advection of unstable air coming from the first and second quadrants, frequently linked to the presence of a dynamic low-pressure on the Middle-Lower Tyrrhenian. Generally, the average meteoric type of the area is of the Adriatic sublittoral (Fazzini and Giuffrida 2005)—slightly bimodal, with a non-dominant monthly and absolute seasonal maximum in autumn, a second slight maximum in mid-spring and two quantitatively not different minimums—in particular along the coast—positioned in the solstice seasons (Fig. 5).

As can be seen in Figs. 3 and 4, rainfall does not seem to show any significant quantitative variations, with slightly contrasting trends even at short distances; it would then seem evident in the last decade a lower interannual variability in values; even on the monthly scale, no significant changes are observed, apart from a moderate increase in September.

Finally, the number of days with rainfall varies between 75 on the coast and 85 in the hilly area; the signal relating to the last three decades, 1991–2020, shows a decrease in events of 5–7% (Fig. 6).

Therefore, the study area shows a temperate sublittoral climate with cool winters, sweltering summers, and moderate rainfall with an evident but not dry summer minimum (Pinna 1978).

Table 3 Average, seasonal, and annual rainfall datasets for the period 1951–2020

Thermo-pluviometric station	Jan	Feb	Mar	Apr	May	Jun	Jul	Aug	Sep	Oct	Nov	Dec	Year
Giulianova	56.3	48	59.4	54.4	42.7	42.4	36.6	39.8	68.3	71.4	72.2	70.9	662.6
Atri	71.5	59	66.4	68	55.8	51.7	40.7	44.6	69.8	84.3	94.7	84.1	790.5
Pescara airport	64.2	53	57.9	51.2	41.2	41.1	33.3	39.1	60.4	77.6	80.3	85.9	685.2

Thermo-pluviometric station	Winter	Spring	Summer	Autumn
Giulianova	175.2	156.5	118.9	212
Atri	214.5	190.2	137	248.8
Pescara airport	203.1	150.3	113.5	218.3

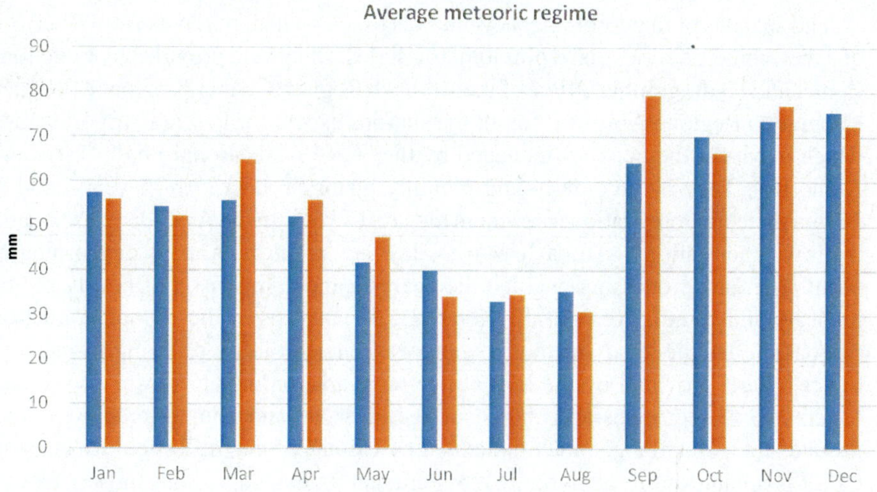

Fig. 5 Average meteoric regime of the study area—timespan 1951–2020 versus 1991–2020

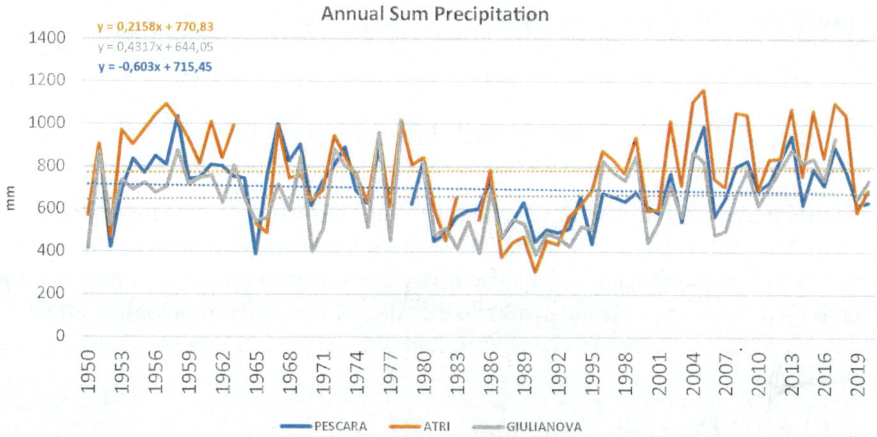

Fig. 6 Annual sum of precipitation at Pescara, Atri, and Giulianova stations for the timespan 1951–2020

3 Methods

The present work was based on the analysis of heavy meteorological events that occurred on several dates: 6–7 October 2007 (in the hilly and coastal Teramo area); 1–2 March 2011 (in the hilly and coastal sectors of Teramo and Pescara); 13–14 March 2012 and 12 January 2018 (in the hilly and coastal Pineto area); and 10 July 2019 (extreme hailstorm in the Pescara area). These analyses focused on the meteoclimatic aspects and the main geomorphological effects on the landscape and human settlements.

The analysis of meteorological events was based on data retrieved from a network of 3 thermometric series and 8 pluviometric series. Data were provided by Functional Center and Hydrographic Office of the Abruzzo Region (Centro Funzionale e Ufficio Idrografico Regione Abruzzo) for all the stations except for Pescara Airport station, which is part of the network managed by the "ENAV" (National Flight Assistance Authority). They include daily and monthly historical data (30–70 years) and 5–15 min meteoric registration (present in the last 15 years in the AWS database—automatic weather stations—) for at least weekdays before the main heavy meteorological event. The data processing enabled the analysis and comparison of hourly rainfall intensity and cumulative rainfall. Moreover, the analysis of the synoptic situations responsible for the "bad weather" events was carried out by referring to meteorological charts provided by the Met Office (available online at: https://www.metoffice.gov.uk/weather/maps-and-charts/surface-pressure) and from the reanalysis at sea ground and a 500 hPa geopotential edited by Climate Forecast System Reanalysis–CFSR (available online at: https://www.wetterzentrale.de/de/reanalysis.php?model=cfsr).

Geomorphological effects were analyzed through a multidisciplinary approach involving site-specific investigations, field surveys, aerial photo analysis, analysis of inventories, and technical reports.

4 Heavy Rainfall Events and Affected Areas

The rainfall distribution analysis and post-event investigations are presented for the years 2007, 2011, 2012, 2018, and 2019 events. For each event, only the most significant rainfall diagrams and the meteorological situation on the synoptic scale are presented, as the main features and distribution of the geomorphological effects.

4.1 2007 Event (6–7 October)

The event affected the hilly and coastal Tortoreto area within the Teramo province, between the Salinello and Vibrata rivers (Fig. 2) for a short time (14–16 h). The rainfall values were >150 mm, and some were >200 mm, up to the maximum recorded value of 291 mm (Nereto station). This event occurred after a relatively dry summer period.

The hourly rainfall reached values ranging from 10 mm/h in the coastal area to 40 mm/h in the hilly area (Nereto station; Fig. 7). The cumulative rainfall was from 60 to 80 mm in the coastal area to >200 mm in the hilly area (Nereto station).

In summary, the 2007 event is probably an extreme event, for this area, with high intensity (10–40 mm/h), a high cumulative rainfall (up to 220 mm), and occurring after two months of low rainfall (Miccadei et al. 2012; Piacentini et al. 2018).

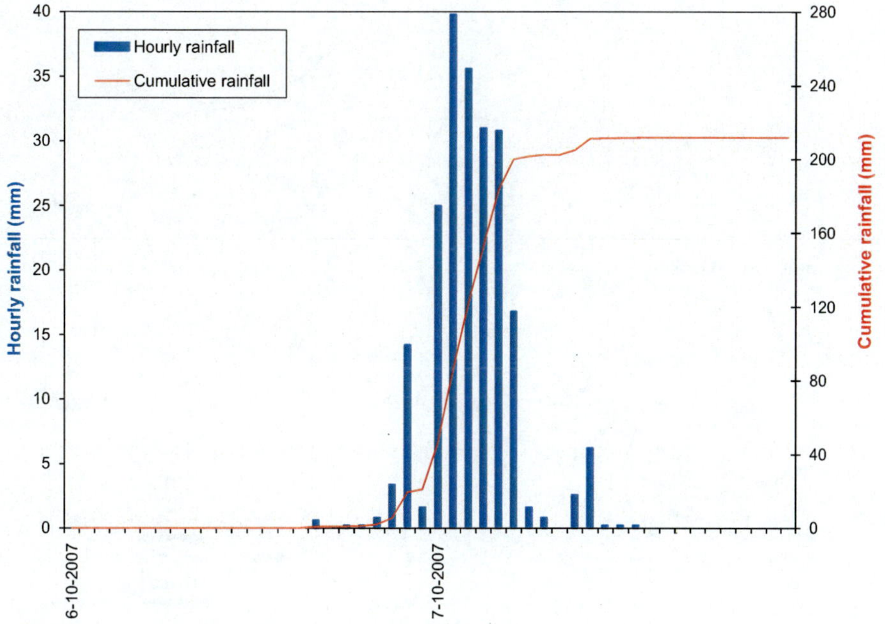

Fig. 7 Heavy rainfall event of 7 October 2007. Hourly and cumulative rainfall at Nereto station (modified from Piacentini et al. 2018)

The combination of a warm front brought about the abundant precipitations, and the following line of instability connected to a deep depression centered on the lower Tyrrhenian and fed on its rear edge by fresh and unstable polar maritime air (Fig. 8).

This event induced heavy soil erosion processes on slopes (sheet, rill, and gully erosion), rapid mudflows at the bases of slopes and minor drainage streams, and flooding within the main river and coastal plains. In the downstream side along the urbanized area and the coastal plain, large volumes of sediment were deposited on roads, at the stream outlet, and along paleovalleys covered by the urban areas (Miccadei et al. 2012; Piacentini et al. 2018). This induced severe problems in road circulation and affected the safety of human settlements (Fig. 9).

4.2 2011 Event (1–2 March)

The event affected the hilly and coastal areas of Tortoreto and Pineto (Fig. 2) for a total duration of 22–26 h. The rainfall values were between 150 and 250 mm. This event occurred after a moderately humid winter. The hourly rainfall was around 15–20 mm/h (coastal area) and up to 35 mm/h (hilly area; Fig. 10). The cumulative rainfall reached 100–130 mm at Pineto station to 210 mm at Nereto station.

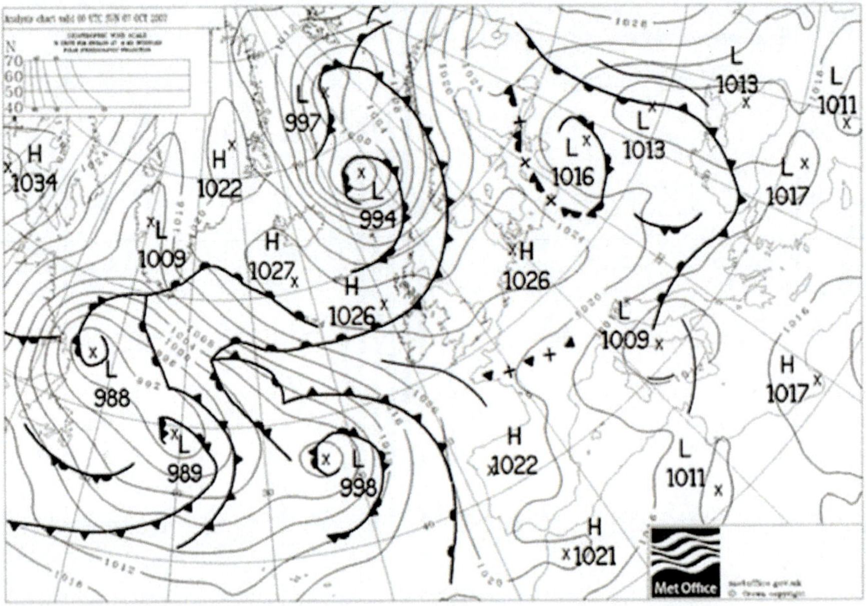

Fig. 8 Meteorological chart showing surface pressure and weather front, at 00Z on 7 October 2007 (*Source* www.metoffice.gov.uk)

In summary, the 2011 event is an extreme one, although intermediate in the previous ones, after ~10 days of moderate antecedent rainfall (Miccadei et al. 2012; Piacentini et al. 2018).

The abundant precipitations were brought about by combining two occluded-cold fronts. The following instability line is connected to a deep depression centered on the south of Sardinia, moving eastward, and fed on its rear edge by fresh and unstable continental polar air (Fig. 11).

The geomorphologic analysis performed after the event outlined soil erosion due to sheet and rill erosion, gully erosion, and sedimentation as mudflows and urban flooding (Fig. 12), similar to the 2007 event but with a different distribution and minor total extent. In the hilly catchment and slopes of the Pineto-Atri area, soil erosion occurred as gully areas. At the bases of the slopes within the main valleys and along the coastal slope, rapid mudflows occurred from 20 to 150 cm thick. Crevasse splays happened along the main rivers and at the outlet to the coastal plain, with a sediment thickness of 10–50 cm. In the Salinello River valley, limited soil erosion occurred on the moderately vegetated slopes, while most of the effects were related to flooding along the main river. Heavy mud and water flooding were the overall effects of this event, affecting the coastal plains and the river plain almost seamlessly. Extensive overbank flooding along the rivers induced the formation of wide and long crevasse splays on the floodplain and flooding phenomena in urban areas (Miccadei et al. 2012; Piacentini et al. 2018).

Fig. 9 Flooding and geomorphological effects induced by the 2007 heavy rainfall event in the urban and hilly Tortoreto area. **a** Urban flooding causing damages to a house gate (*Source* www.valtrigno.org); **b** detail of urban flooding at Tortoreto (*Source* www.valtrigno.org); **c** fluvial erosion scarp affecting the Salinello River valley floor; **d** detail of damages to urban stream network and roads

4.3 2012 Event (5–6 and 13–14 September)

This event results from two rainfall periods separated by a 7-day dry period and must be considered as two isolated events. The first rainfall event occurred from about 17:00 on 5 September to 11:00 on 6 September 2012, for a total duration of some ~18 h. The second event occurred from about 18:00 on 13 September to 18:00 on 14 September 2012, for a total duration of ~24 h. The hourly rainfall was 5 to >15 mm/h for the first event and ~15–45 mm/h for the second one. The cumulative rainfall was ~45 mm for the first event and ~80–190 mm for the second event (Fig. 13). Moreover, these events occurred after more than a month of arid conditions (Piacentini et al. 2018).

In summary, the 2012 event was a double event, 18 and 24 h long, within a 1-week interval. It showed high intensity (15–45 mm/h, up to >190 mm/day) and combined cumulative rainfall of up to 260 mm.

The significant precipitations were brought about by the combination of a warm and cold front and the following line of instability connected to a deep depression centered on the southern Tyrrhenian, moving eastward, and fed on its rear edge by fresh and unstable intermediate polar air (Fig. 14).

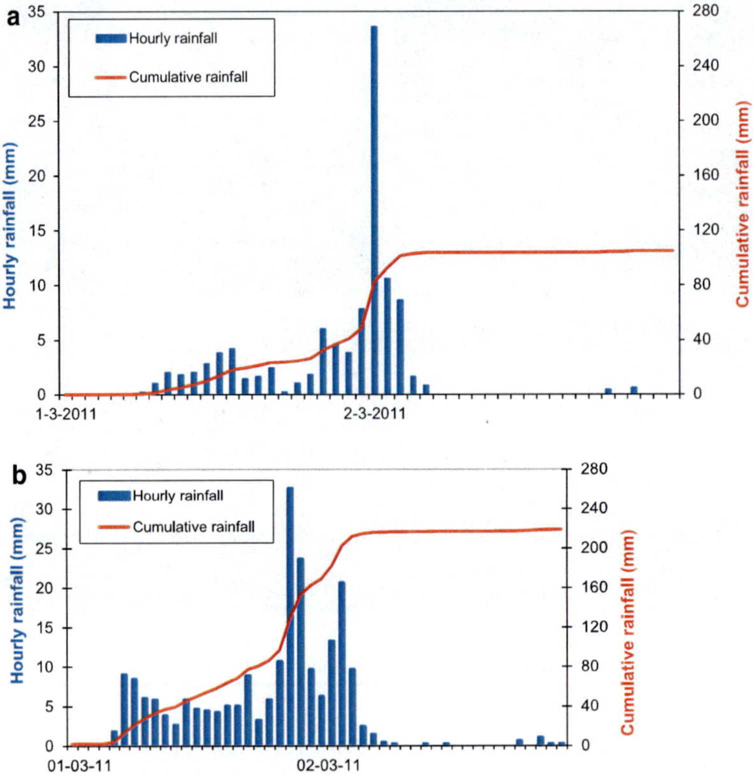

Fig. 10 Heavy rainfall event of 1–2 March 2011. Hourly and cumulative rainfall at **a** Pineto station and **b** Nereto station (modified from Piacentini et al. 2018)

The geomorphologic analysis performed after both events outlined a large distribution of soil erosion features combined with mudflows, flooding areas, and minor crevasse splays (Fig. 15). On September 5–6, minor effects occurred in terms of soil erosion features, such as gully areas with 10–30 cm average erosion and gullies 20–40 cm deep and 50–150 cm; sedimentation features such as mudflows 25–50 cm thick, flooding areas (Fig. 15) with 0–3 cm wide mud and minor crevasse splays. On September 13–14, further heavy soil erosion occurred, primarily due to gully areas with an average 20–40 cm erosion and major gullies 20–70 cm deep and 1–2 m wide. Sheet-rill erosion areas were observed with a 2–5 cm erosion. In the lower part of the slopes and at the junctions of minor tributaries to main streams, mudflows occurred up to 150 cm thick; along the main fluvial plain and the coastal plain, flooding areas occurred with 0–3 cm of accumulated mud as well as some local crevasse splay 10–30 cm thick (Piacentini et al. 2018).

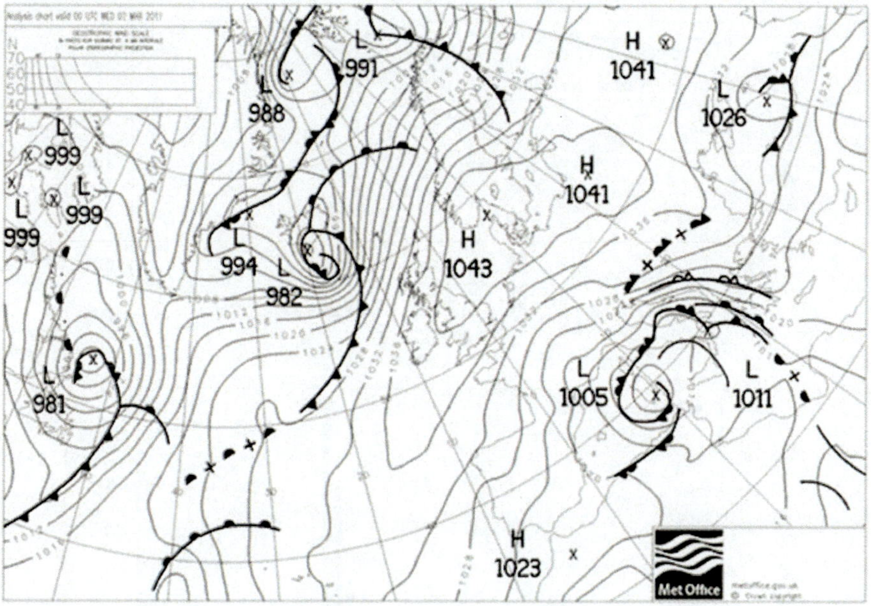

Fig. 11 Meteorological chart showing surface pressure and weather front, at 00Z on 2 March 2011 (*Source* www.metoffice.gov.uk)

Fig. 12 Flooding and geomorphological effects induced by the 2011 heavy rainfall event. **a** Salinello River, damage to a bridge; **b** Salinello River, fluvial erosion scarp affecting the valley road; **c** flooding phenomena at Pineto, **d** gully erosion in Pineto hilly area

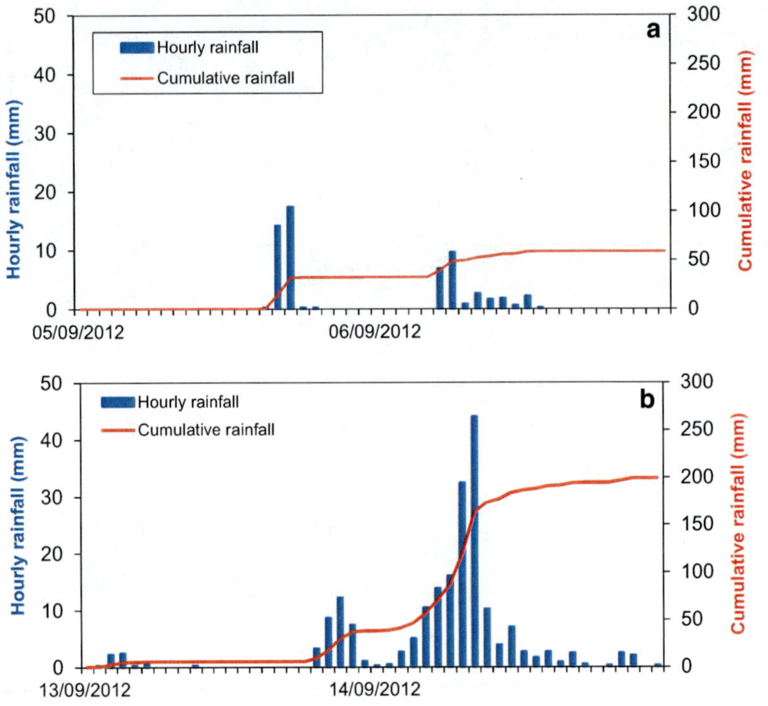

Fig. 13 Double heavy rainfall event of 2012 in the Atri-Pineto area. **a** Hourly and cumulative rainfall (5–6 September 2012; Atri station); **b** hourly and cumulative rainfall (13–14 September 2012; Atri station), modified from Piacentini et al. (2018)

4.4 2018 Event (12 January)

The event affected the hilly and coastal area of Pineto (Fig. 2) and was characterized by decisive spatial punctuality and not exceptionally high cumulative values, neither with hourly steps nor in its totality. Only the Atri thermo-pluviometric station recorded a cumulative of 33 mm distributed in about 7 h—between 5:30 and 12:30—with a maximum half-hourly value of 6.2 mm between 8:30 and 9:00 (Fig. 16).

The synoptic weather map (Fig. 17) shows that the cause of these stretches of instability was an occlusion moving from south-west to north-east, connected to a Mediterranean low-pressure, centered in the east of Sardinia. The trajectory of the aforementioned occlusion did not result in extensive rainfall but spatially irregular and of weak-moderate intensity, given the position of the study area—"downwind" with respect to the Maiella Massif.

The precipitation caused extensive flooding in Pineto and its hamlet Scerne (Fig. 18), impacting, directly and indirectly, human settlements with flooded underpasses, stranded cars, and damages to roads and buildings.

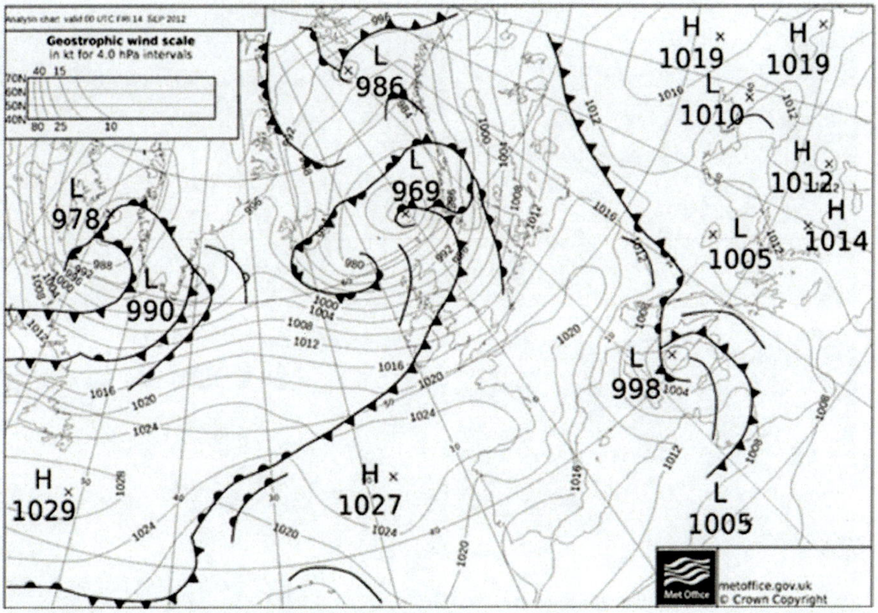

Fig. 14 Meteorological chart showing surface pressure and weather front, at 00Z on 14 September 2012 (*Source* www.metoffice.gov.uk)

4.5 2019 Event (10 July)

This event affected the urban and coastal areas of Pescara municipality.

On Wednesday, 10 July 2019, the central and southern sectors of the area were hit by a heavy meteorological event. This latter was related to the swift passage of a Mesoscale Convective System (MCS) moving from the northwest to the southeast, as graphically shown in Fig. 19.

The thermo-pluviometric station of Pescara airport quantified cumulative meteoric accumulations of 98 mm. At 12:50, the phenomenological characterization is evident as thunder and hail, therefore hailstorm, accompanied by gusts of wind of about 43 mph equal to about 70 km/h. Finally, at 12.15 LT, a 5-min rain rate of 548.60 mm/h was reached in the unofficial rain gauge of Colle Marino, in the northwestern part sector of Pescara City.

The formation of this intensely organized temporal structure comes from the strong thermal contrast on the vertical column of air existing between an occurrence of polar maritime air with the pre-existing damp and humid continental subtropical air (Fig. 20).

It causes "severe bad weather conditions" with intense rainfall, thunderstorms, and temporary hail—with grains of huge size—up to 12 cm, as shown in Fig. 21. The precipitation was accompanied by typical downburst gusts of wind up to over 70 km/h.

Fig. 15 Flooding and geomorphological effects triggered by the 2012 double heavy rainfall event: **a** gully erosion in Atri-Pineto slopes; **b** flooding area in the valley floor, at the base of the hilly slopes; **c** flooding area with mud accumulation in the Pineto coastal plain; **d** sheet-rill and gully erosion areas along Atri hilly slopes

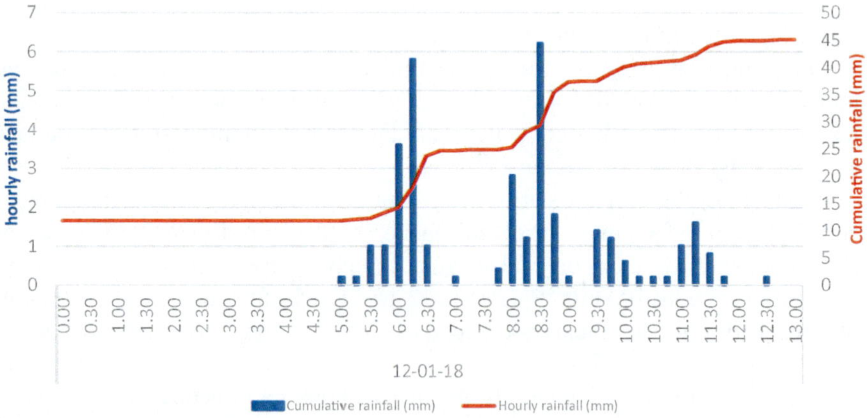

Fig. 16 Heavy rainfall event of 12 January 2018. Hourly and cumulative rainfall at Atri thermo-pluviometric station

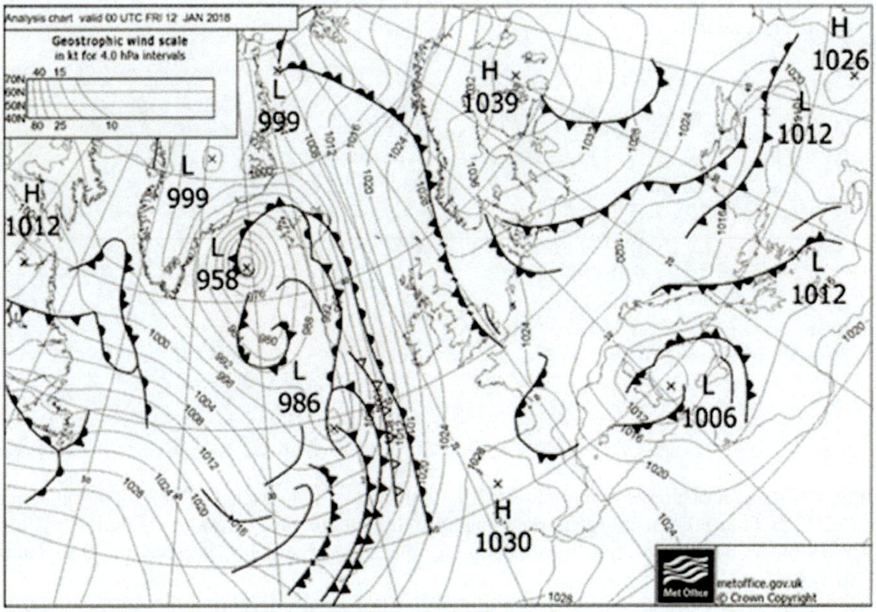

Fig. 17 Meteorological chart showing surface pressure and weather front, at 00Z on 12 January 2018 (*Source* www.metoffice.gov.uk)

Fig. 18 Flooding and geomorphological effects triggered by the 2018 heavy rainfall event: **a** underpass flooded and cars stranded in the Pineto area; **b** detail of damages to roads with rill erosion and mud accumulation (*Source* www.rete8.it)

Severe damages and problems were recorded in multiple locations over the metropolitan area (Fig. 21), highlighted by broad flooded areas, damages to roads and buildings, and several hail-damaged cars.

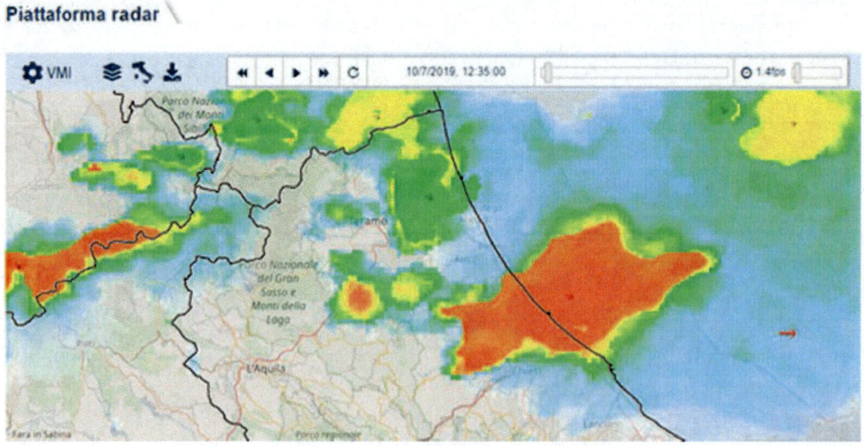

Fig. 19 Radar measurements on 10 July 2019. Vertical Maximum Intensity (VMI) of the reflectivity at 12.35 LT. Orange color indicates exceptionally high values for the Pescara area (Courtesy of Italian Civil Protection Department—https://radar.protezionecivile.it/radar-dpc; accessed on 20 September 2021)

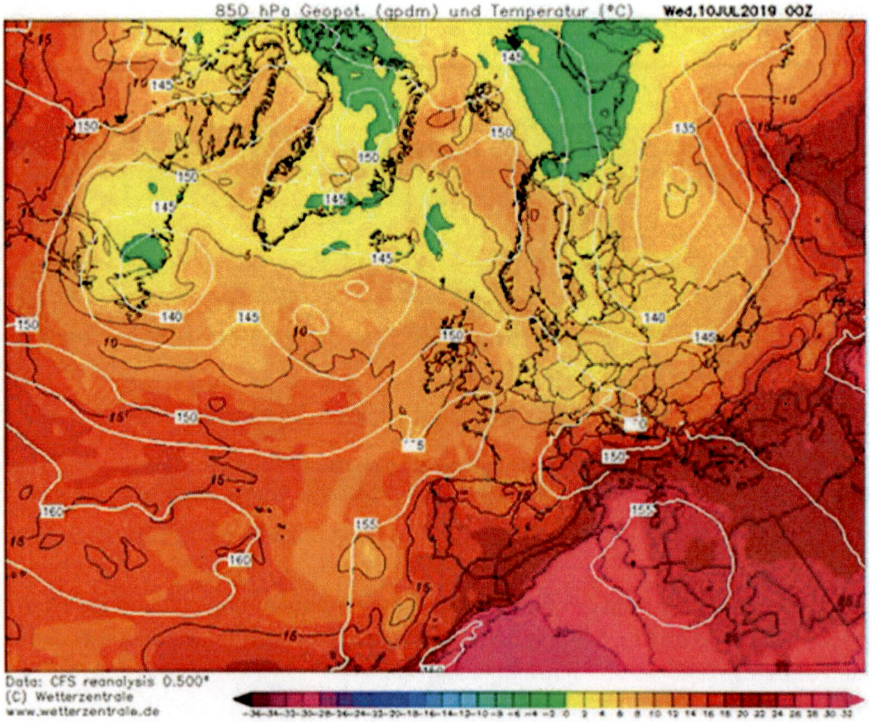

Fig. 20 High-resolution forecast chart showing temperatures (°C; see the colored bar on the lower right) at 850 hPa (about 1500 m) on 10 July 2019 at 00LT (*Source* www.wetterzentrale.de)

Fig. 21 Flooding effects and damages induced by the 2019 heavy rainfall event: flooded roads (**a**) and **b** in Pescara urban area (*Source* www.pescarapost.it) **c** detail of centimetric-sized hail; **d** hail-damaged car (*Source* ANSA)

5 Linking Disaster Resilience and Urban Areas: A Geomorphological Approach

Natural disasters, such as landslides and flood phenomena caused by heavy meteorological events, pose many urban areas at risk. Therefore, improving urban resilience to reduce injuries and accelerate post-disaster recovery is a multifaceted challenge for urban policymakers and interested stakeholders (United Nations 2017). In detail, flooding has been recognized as a significant threat due to the spatial extent and magnitude of damage it poses worldwide and in Italy (Barbaro et al. 2021; Diez-Herrero and Garrote 2020). Moreover, considering the nature and evolution of human settlements, several activities such as expanding urban areas and improper land-use have led to a marked increase in flood risk over the last 50 years (Bowen and Gleeson 2019; Brath et al. 2003; Berndtsson et al. 2019). Since the '90s, the United Nations decided to pay particular attention to fostering international cooperation in the field of natural disaster reduction with the objective to reduce loss of life, property damage, and social and economic disruption caused by natural disasters (International Decade for Natural Disaster Reduction—UN General Assembly 1990). Consequently, projects, activities, and dissemination plans have been developed worldwide to enhance people's awareness and preparedness for natural disasters to reduce their adverse impacts on human settlements. The Italian scientific community provided significant contributions, with a detailed historical report which highlighted that the recognition of past events is a not negligible component of prevention, being able to reuse in a positive sense—as a scientific heritage and social memory—also adverse events, often made so by human improvidence (Catenacci 1992). Combining these literature data with the analysis here presented, geo-hydrological events occurring in the middle Adriatic coastal area revealed the state of vulnerability towards natural hazards posing a threat to human settlements, livelihoods, and properties. In this regard, it would be desirable to use the "Ten Essentials for Making Cities Resilient" as a conceptual roadmap to attempt to help urban leaders and researchers implement DRR policies. In 2015, the Third United Nations World Conference for Disaster Risk Reduction adopted the Sendai Framework for DRR 2015–2030, focusing on urban risk management. Hence, the "Local Urban Indicators" were developed to "*allow cities to assess their resilience to disasters*" based on the updated framework (UNISDR 2017). Additionally, to faithfully consider all the core elements of new urban agendas (United Nations 2017, 2021a, b), there is a clear need to improve the practices towards the 2030 Agenda goals with a landscape approach *s.l.* Geomorphologists have a crucial role in achieving several of the 17 Sustainable Development Goals (SDGs) (United Nations 2021b), aiming to bridge the gap between geosciences and resilience and investigate multidisciplinary ways to "translate" geomorphological-based actions into resilience aspects.

6 Discussions and Conclusions

In recent years, climate-related risks have increased because of changing climate and unplanned urbanization, improper land-use, land-cover changes, and ecosystems' degradation. Combining these factors is now causing a difficult situation to manage, which places us at the forefront of real emergencies. Many human settlements and urban areas face potential threats from various natural hazards (i.e., floods and landslides). However, evaluating the resilience of those areas is not well established in both the academia and practice field (Sim et al. 2018). Developing precise analytical tools, sensible land-use planning, and mitigation activities will be necessary. In any case, reducing climate-related risks and strengthening natural disaster resilience are significant societal challenges demanding a better understanding of complex interactions between societies and natural hazards under current and future climate scenarios (Bodoque et al. 2016). Strategic measures for monitoring and reporting progress made in disaster resilience assessment and associated risk reduction are core elements of disaster risk management and climate change adaptation (Park et al. 2021; Villani et al. 2019).

In the middle Adriatic coastal area, hydrometeorological hazards, including floods and landslides, are considered the most common of all hazards. Floods occurring in the study area revealed our state of vulnerability towards geo-hydrological processes despite many accurate mitigation interventions carried out in the last decades and, above all, the inadequacy of the current urban fabric concerning the hydrographic network, both on the hydraulic and geomorphological level (Marzi et al. 2019; Paliaga et al. 2020; Faccini et al. 2021).

Despite the number of rainy days being negative and the neutral signal in the annual cumulative rainfall, perturbations capable of generating heavy meteorological events are increasingly frequent in the Mediterranean area (Boudou et al. 2016).

In the study area, the pluviometric climatic extremes mainly manifest themselves with phenomena of vigorous intensity that would seem more and more recurrent but not necessarily more intense; rarely of a self-regenerating type but anyway capable of causing rapid changes in the landscape, such as soil erosion, rainfall-induced landslides, and damages to infrastructures. A recurring climatic pattern was identified as responsible for frequent situations of geomorphological instability. The formation of a low-pressure system on the Gulf of Genoa, which migrates towards the southern Tyrrhenian, draws on its eastern edge mild and humid air of Mediterranean origin ("*Scirocco*" winds) and, on the western one, a colder and unstable air of polar origin. It could be acknowledged as a cause of the phenomena of intense precipitation, often abundant and amplified by the presence of the Central Apennines chain (orographic component), which concur with the development of V-shaped solid thunderstorms. The comparison between the events that occurred in limited times (15 years) highlights the need to promote risk mitigation activities beyond the essential structural interventions on the minor hydrographic network (extraordinary maintenance), which involves inevitable expensive and delayed financial programs. It seems necessary to prepare soft measures that include both active (ordinary maintenance of secondary ditches, correct methodologies in agricultural practices on the slopes above urban centers, strengthening of weather-hydrological monitoring) and passive (urban planning regulations with intensive use of sewer networks, sustainability, inventory of historical events, dissemination to enhance community's capacity to respond to extreme events and residual risk). As for the actions to be taken in the short and long term to adapt to the reappearance of the effects of heavy meteorological events, it is necessary to quickly calculate at the local level new climate adaptation activities that involve urban, social, and economic choices.

These activities need to be developed considering classic and innovative methodologies for investigating and mapping landscape sensitivity to geomorphological processes, such as floods, landslides, and soil erosion (Gustavsson et al. 2006; Miccadei and Piacentini 2013; Javidan et al. 2021). As a result, it may represent a practical integrated approach in geomorphological studies at different spatial scales, readily available to interested stakeholders. Finally, this work highlights that only a high knowledge of the landscape, connected to drainage network evolution, geological-geomorphological features, and meteorological events, can help activate sustainable territorial planning and disaster resilience assessment in urban areas in a changing environment.

Acknowledgements The authors wish to thank the Functional Center and Hydrographic Office of the Abruzzo Region (Centro Funzionale e Ufficio Idrografico Regione Abruzzo) for providing pluviometric data of the events and the historical datasets analyzed in this work. This work was supported by the Department of Engineering and Geology, Università degli Studi "G. d'Annunzio" Chieti-Pescara funds (E. Miccadei University fund).

References

Accordi B, Carbone F (1988) Carta delle litofacies del Lazio-Abruzzi ed aree limitrofe. CNR—Progetto Finalizzato "Geodinamica": Rome, Italy. Quad Ric Sci 114:223

Barbaro G, Miguez MG, de Sousa MM, da Cruz Franco ABR, de Magalhães PMC, Foti G, Valadão MR, Occhiuto I (2021) Innovations in best practices: approaches to managing urban areas and reducing flood risk in Reggio Calabria (Italy). Sustainability (Switzerland) 13(6):3463. https://doi.org/10.3390/su13063463

Bathrellos GD, Karymbalis E, Skilodimou HD, Gaki-Papanastassiou K, Baltas EA (2016) Urban flood hazard assessment in the basin of Athens Metropolitan city, Greece. Environ Earth Sci 75(4):319. https://doi.org/10.1007/s12665-015-5157-1

Berndtsson R, Becker P, Persson A, Aspegren H, Haghighatafshar S, Jönsson K, Larsson R, Mobini S, Mottaghi M, Nilsson J, Nordström J, Tussupova K (2019) Drivers of changing urban flood risk: a framework for action. J Environ Manag 240:47–56. https://doi.org/10.1016/j.jenvman.2019.03.094

Bigi S, Cantalamessa G, Centamore E, Didaskalou P, Dramis F, Farabollini P,Gentili B, Invernizzi C, Micarelli A, Nisio S, Pambianchi G, Potetti M (1995) La fascia periadriatica marchigiano-abruzzese dal pliocene medio ai tempi attuale: evoluzione tettonico-sedimentaria e geomorfologica. Studi Geologici Camerti Vol Spec 37–49

Blöschl G, Hall J, Viglione A, Perdigão RAP, Parajka J, Merz B, Lun D, Arheimer B, Aronica GT, Bilibashi A, Bohác M, Živković N (2019) Changing climate both increases and decreases European river floods. Nature 573(7772):108–111. https://doi.org/10.1038/s41586-019-1495-6

Bodoque JM, Amérigo M, Díez-Herrero A, García JA, Cortés B, Ballesteros-Cánovas JA, Olcina J (2016) Improvement of resilience of urban areas by integrating social perception in flash-flood risk management. J Hydrol 541:665–676. https://doi.org/10.1016/j.jhydrol.2016.02.005

Boudou M, Danière B, Lang M (2016) Assessing changes in urban flood vulnerability through mapping land use from historical information. Hydrol Earth Syst Sci 20(1):161–173. https://doi.org/10.5194/hess-20-161-2016

Bowen WM, Gleeson RE (2019) Prospects for human settlements in the anthropocene: challenges to equality, sustainability, and agency. In: Bowen WM, Gleeson RE (eds) The evolution of human settlements. Palgrave Macmillan, Cham, pp 273–300

Brath A, Montanari A, Moretti G (2003) Assessing the effects on flood risk of land-use changes in the last five decades: an Italian case study. IAHS-AISH Publication, no 278, pp 435–441

Calista M, Miccadei E, Piacentini T, Sciarra N (2019) Morphostructural, meteorological and seismic factors controlling landslides in weak rocks: the case studies of Castelnuovo and Ponzano (northeast Abruzzo, central Italy). Geosciences (Switzerland) 9(3):122. https://doi.org/10.3390/geosciences9030122

Catenacci V (1992) Cronistorie abruzzesi. In "Il dissesto geologico e geoambientale in Italia dal dopoguerra al 1990." Mem Descr della Carta Geologica d'Italia 47:156–166

Di Lena B, Antenucci F, Mariani L (2012) Space and time evolution of the Abruzzo precipitation. Italian J Agrometeorol 1:5–20

Di Lena B, Curci G, Vergni L (2021) Analysis of rainfall erosivity trends 1980–2018 in a complex terrain region (Abruzzo, central Italy) from rain gauges and gridded datasets. Atmosphere 12(6):657. https://doi.org/10.3390/atmos12060657

Diez-Herrero A, Garrote J (2020) Flood risk analysis and assessment, applications and uncertainties: a bibliometric review. Water (Switzerland) 12(7):2050. https://doi.org/10.3390/w12072050

Esposito G, Carabella C, Paglia G, Miccadei E (2021) Relationships between morphostructural/geological framework and landslide types: historical landslides in the hilly piedmont area of Abruzzo region (Central Italy). Land 10(3):287. https://doi.org/10.3390/land10030287

Faccini F, Luino F, Paliaga G, Roccati A, Turconi L (2021) Flash flood events along the west Mediterranean coasts: inundations of urbanized areas conditioned by anthropic impacts. Land 10(6):620. https://doi.org/10.3390/land10060620

Fazzini M, Giuffrida A (2005) Une nouvelle proposition quantitative des régimes pluviométriques dans le territoire de Italie: Premiers résultats. In: Climat Urbain, Ville et Architecture—Actes XVIII Colloque Internationale de Climatologie, pp 361–364, 7–11 September 2005, Genova, Italy

Glago FJ (2021) Flood disaster hazards; causes, impacts, and management: a state-of-the-art review. In: Natural hazards—impacts, adjustments, and resilience, p 18. https://doi.org/10.5772/intechopen.95048

Gustavsson M, Kolstrup E, Seijmonsbergen AC (2006) A new symbol-and-GIS based detailed geomorphological mapping system: renewal of a scientific discipline for understanding landscape development. Geomorphology 77(1–2):90–111. https://doi.org/10.1016/j.geomorph.2006.01.026

ISPRA (2020) Gli indicatori del clima in Italia nel 2020 – Anno XVI, 76 pp

Javidan N, Kavian A, Pourghasemi HR, Conoscenti C, Jafarian Z, Rodrigo-Comino J (2021) Evaluation of multi-hazard map produced using MaxEnt machine learning technique. Sci Rep 11(1):6496. https://doi.org/10.1038/s41598-021-85862-7

Kundzewicz ZW, Pin'skwar I, Brakenridge GR (2018) Changes in river flood hazard in Europe: a review. Hydrol Res 49(2):294–302. https://doi.org/10.2166/nh.2017.016

Marzi S, Mysiak J, Essenfelder AH, Amadio M, Giove S, Fekete A (2019) Constructing a comprehensive disaster resilience index: the case of Italy. PLoS ONE 14(9):e0221585. https://doi.org/10.1371/journal.pone.0221585

Miccadei E, Piacentini T (2013) The role of knowledge in the prevention of natural hazards and related risks. Italian J Plan Pract 3(1):46–68

Miccadei E, Piacentini T, Daverio F, Di Michele R (2012) Geomorphological instability triggered by heavy rainfall: examples in the Abruzzi Region (Central Italy). In: Studies on environmental and applied geomorphology, pp 45–62. https://doi.org/10.5772/39121

Miccadei E, Piacentini T, Buccolini M (2017) Long-term geomorphological evolution in the Abruzzo area, Central Italy: twenty years of research. Geol Carpath 68(1):19–28. https://doi.org/10.1515/geoca-2017-0002

Miccadei E, Carabella C, Paglia G (2021) Morphoneotectonics of the Abruzzo Periadriatic Area (Central Italy): morphometric analysis and morphological evidence of tectonics features. Geosciences (Switzerland) 11(9):397. https://doi.org/10.3390/geosciences11090397

Napolitano E, Marchesini I, Salvati P, Donnini M, Bianchi C, Guzzetti F (2018) LAND-deFeND—an innovative database structure for landslides and floods and their consequences. J Environ Manag 203:203–218. https://doi.org/10.1016/j.jenvman.2017.11.022

Ori G, Serafini G, Visentin C, Ricci Lucchi F, Casnedi R, Colalongo ML, Mosna S (1991) The Pliocene-Pleistocene Adriatic foredeep (Marche and Abruzzo, Italy): an integrated approach to surface and subsurface geology. In: 3rd E.A.P.G. conference, Adriatic Foredeep Field Trip, Guide Book, 85, May 26th–30th, 1991, Firenze, Italy

Paliaga G, Luino F, Turconi L, Marincioni F, Faccini F (2020) Exposure to geo-hydrological hazards of the metropolitan area of Genoa, Italy: a multi-temporal analysis of the Bisagno stream. Sustainability (Switzerland) 12(3):1114. https://doi.org/10.3390/su12031114

Park K, Oh H, Won JH (2021) Analysis of disaster resilience of urban planning facilities on urban flooding vulnerability. Environ Eng Res 26(1):1–8. https://doi.org/10.4491/eer.2019.529

Peel MC, Finlayson BL, McMahon TA (2007) Updated world map of the Köppen-Geiger climate classification. Hydrol Earth Syst Sci 11(5):1633–1644. https://doi.org/10.5194/hess-11-1633-2007

Piacentini T, Galli A, Marsala V, Miccadei E (2018) Analysis of soil erosion induced by heavy rainfall: a case study from the NE Abruzzo Hills Area in Central Italy. Water (Switzerland) 10(10):1314. https://doi.org/10.3390/w10101314

Pinna M (1978) L'atmosfera e il clima. UTET, Torino, p 478

Salvati P, Bianchi C, Fiorucci F, Giostrella P, Marchesini I, Guzzetti F (2014) Perception of flood and landslide risk in Italy: a preliminary analysis. Nat Hazard 14(9):2589–2603. https://doi.org/10.5194/nhess-14-2589-2014

Sharma A, Wasko C, Lettenmaier DP (2018) If precipitation extremes are increasing, why aren't floods? Water Resour Res 54(11):8545–8551. https://doi.org/10.1029/2018WR023749

Sim T, Wang D, Han Z (2018) Assessing the disaster resilience of megacities: the case of Hong Kong. Sustainability 10(4):1137. https://doi.org/10.3390/su10041137

UN General Assembly (1990) International decade for natural disaster reduction: resolution/adopted by the General Assembly, 21 December 1990, A/RES/45/185. https://www.refworld.org/docid/3b00efe50.html. Accessed 5 March 2022

UNISDR (2015) Sendai framework for disaster risk reduction 2015–2030. In: UN world conference on disaster risk reduction, 2015 March 14–18, Sendai, Japan. United Nations Office for Disaster Risk Reduction, Geneva. http://www.wcdrr.org/uploads/Sendai_Framework_for_Disaster_Risk_Reduction_2015-2030.pd

UNISDR (2017) How to make cities more resilient a handbook for local government leaders. United Nations Office for Disaster Risk Reduction, Geneva, Switzerland, p 122

United Nations (2017) Adopted draft of the new urban agenda, Quito, 17–20 October. In: United Nations conference on housing and sustainable urban development (Habitat III). https://habitat3.org/the-new-urban-agenda/. Accessed 01 March 2022

United Nations (2021a) Our common agenda—report of the Secretary-General. Published by the United Nations New York, NY 10017, United States of America

United Nations (2021b) Transforming our world: the 2030 agenda for sustainable development. In: UN General Assembly 21 October 2015, A/RES/70/1. https://www.refworld.org/docid/57b6e3e44.html. Accessed 4 March 2022

Vezzani L, Ghisetti F (1998) Carta Geologica Dell'Abruzzo—Scala 1:100,000. SELCA, Firenze, Italy

Villani G, Nanni S, Tomei F, Pasetti S, Mangiaracina R, Agnetti A, Leoni P, Folegani M, Mazzini G, Botarelli L, Castellari S (2019) The rainBO platform for enhancing urban resilience to floods: an efficient tool for planning and emergency phases. Climate 7(12):145. https://doi.org/10.3390/cli7120145

Xu W, Cong J, Proverbs D, Zhang L (2021) An evaluation of urban resilience to flooding. Water (Switzerland) 13(15):2022. https://doi.org/10.3390/w13152022

Revitalizing the Wounded Territory: The "Geo-Hiking's" Potential

Piero Farabollini, Fabrizio Bendia, and Francesca Romana Lugeri

Abstract The postmodern society has an urgent need to develop effective strategies in favor of land management and risk prevention: priority objectives that are among the components of resilience. A fundamental tool to achieve these goals is the shared knowledge of the environment where we live. Modern GIS effectively facilitates the whole society in this sense, thanks to its many functions. Moreover, it is perfectly compatible with the Landscape Ecology approach, which considers all the components that characterize the complexity of nature in an integrated way. Knowledge sharing is based on communication: new approaches privilege unconventional modes of communication based on people's emotional and experiential involvement. In this sense, sports in the natural environment become communicative vectors of the environment. Therefore, some case studies in the Marche region, recently hit by the earthquake, are examined. The project has taken care of the realization of a geo-hiking itinerary that connects various geosites of great landscape importance to offer society scientifically valid information about the territory during the tourist fruition of the same to raise awareness of the enhancement of the territory and risk prevention.

Keywords Society · Resilience · Knowledge · Geosites · Marche Region Italy

P. Farabollini (✉) · F. Bendia · F. R. Lugeri
School of Science and Technology, University of Camerino, Camerino, Italy
e-mail: piero.farabollini@unicam.it

F. R. Lugeri
Istituto Superiore Protezione E Ricerca Ambientale, Rome, Italy

B. Dahiya et al. (eds.), *Disaster Resilience and Human Settlements*,
Advances in 21st Century Human Settlements,
https://doi.org/10.1007/978-981-99-2248-2_12

Part 1

1 Introduction

The development of everyone, of a community, of the whole society, cannot disregard to the knowledge of environment. The living spaces of humanity are subject to dynamics and balances whose knowledge is essential for the survival itself. The Anthropocene (Crutzen, 2006), now complemented by further definitions—Capitalocene, Chthulucene—all based on the mostly harmful effects of anthropogenic actions (Moore 2017; Haraway 2019), lead us to confront us with an urgent need to change course, to reverse the trend (Bauman 1997), to save our planet and allow the survival, possibly the well-being of each human, as an integral part of a dynamic and complex ecosystem.

The real challenge for the Postmodern society is to find new ways, responding to an extraordinary need of security, intended as safety of life, and private/collective properties. An essential step is to provide the society with effective tools, to strengthen resilience, and start renaissance as well.

Pragmatically, referring to hydrogeological and hydraulic risk, the elaboration of a geomorphology-based matrix is essential to define the spatial and temporal variation of resilience to landslide and flood hazard. This approach can be effectively applied in different climatic and geomorphological frameworks to provide information for disaster mitigation works and land planning strategies (Marincioni et al. 2019; Miccadei et al. 2017; UNISDR 2015).

Analogously, referring to volcanic risk, basing on the assessments carried out, the areas and territorial contexts within which it will be necessary to operate for the implementation of structural and non-structural interventions, will be identified for the mitigation of risk conditions.

More complex is the seismic risk, both in terms of social perception an in technical-scientific approach. Seismic hazard is a function of many variables, some of them known or calculable (such as the maximum expected magnitude of a fault, which is a function of its geometry), others unpredictable (i.e., fault directivity, which influence its destructive potential, or the accumulated stress before an activation).

Because of this, among the types of natural risk, seismic risk is the most extreme and difficult to manage. Knowing that a seismic event is probable, while not knowing its precise location in space, and mostly, in time: these are the factors that determine a low awareness of the risk, both in the population and policymakers.

2 Central Italy Earthquake

Starting from 24 August 2016, central Italy has been struck by a seismic crisis with an unprecedent extension in the recent Italian history. Its effects, in facts, regarded four Italian regions: Marche, Umbria, Lazio and Abruzzo, while the shaking produced was perceived along the whole peninsula. The seismic sequence was very intense and with a long earthquake swarm, unnerving for the resident communities. The first event (24 August 2016 event (epicentre in Accumoli, Rieti province, Lazio region) registered a Magnitude moment of 6.0. Seismic sequence continued its spatio-temporal evolution with the earthquakes of 26 October 2016 (epicentres respectively in Castelsantangelo sul Nera, Mw 5.4 and Ussita, Mw 5.9, both in Macerata province, Marche region). When the seismic sequence seemed to have released all its potential, another devasting earthquake occurred in the central portion of the previously mobilized area: Norcia (Perugia province, Umbria region), 30 October 2016, with a Mw 6.5, the main event of the entire seismic sequence. The hypocentral depths, as identified by INGV, were registered from 7.5 to 9.5 km below the Earth's surface. The fault released all its potential starting to break itself in the northern tip on 24 August, in the southern tip (26 October) and in the central portion (30 October).

Maybe the most important, positive, aspect to highline in all this confusing scenario is represented by the fact that, after the first, sudden and unpredictable event of 24 August that caused about 300 deaths along the epicentral area, nobody else died. Probably because most of the damaged buildings were examined by inspection of usability to ensure the habitability; because a very large number of people were transferred to coast or, anyway, in other locations far from the hypocentral Apennine areas and because a lot of them slept in civil protection tents or in their cars. Furthermore, the aspect of fear should not be overlooked: people, informed by mass media, civil protection, and expert geologists, tried to pay attention to frequent only safe buildings (Farabollini et al. 2019).

The sequence continued also in 2017: four events took place on January 18 with epicenters in Montereale (Mw 5.1), Capitignano (Mw 5.5) Pizzoli (5.4) and Cagnano Aileron (5.0) in the province of L'Aquila (Abruzzo region).

In the following monts, aftershocks continued to occur, decreasing in number and intensity over time.

Each seismic event, in addition to affecting the building, cultural and infrastructural heritage, can engrave on the environment, with primary and secondary effects on it. These, respectively, consist of permanent deformations of the topographic surface (displacement due to surface rupture caused by faulting; subsidence/raising of the portion of territory in the hanging wall and footwall of the fault) or essentially linked to the ground shaking (seismic-induced landslides; liquefaction of granular soils; fractures). Earthquakes generally also induce hydrological variations in the source regime and can generate anomalous waves (tsunamis) both at sea and in smaller basins. Following the highest magnitude events in Central Italy in 2016, those of 24 August, 26 and 30 October, numerous effects occurred on the environment, both primary and secondary detected and mapped by a lot of research campaigns. (Farabollini et al. 2019).

3 Knowledge, Consciousness, Communication

The knowledge of the environment is the essential basis for starting a project of change that—now more than ever—must be initiated without hesitation (Lugeri and Cardillo 2009).

From a technical-scientific point of view, research aims to analyse and monitor the condition of the natural environment and to study its dynamics and evolution through the creation of models based on updatable, multi-scalar systems capable of containing a wide range of information concerning the physical, biotic, and anthropic aspects of the territory, considered both individually and in their integration and mutual interaction. A systemic approach, therefore, that considers all the aspects that make up the complexity that nature itself presents (Naveh and Lieberman 1994). The communication is an essential component of developing and enhancing social resilience, especially referring to the effectiveness of the results. In this regard, the use of topics with more appealing for the broad public as vectors of scientific concepts, is a successful strategy when attempting at sharing complex information. An example is to provide "unexpected" information through unconventional channels. It means that the extraordinary geo-environmental and cultural landscape characterizing the Italian country, is itself a powerful medium to communicate territorial issues to the whole society. The extraordinary geo-environmental and cultural landscape that characterises Italy is itself a powerful means of communicating territorial issues to society. When recognised and understood, landscapes become part of everyone's cultural heritage, and the dynamics that characterise their natural history—at the basis of the "risk/resource" binomial—are revealed, providing the observer with the knowledge indispensable for understanding complex environmental realities. It is in this sense that the perception of the landscape is at the basis of a cognitive process that can trigger a virtuous behaviour, revitalising and adding new lymph to the roots that bind man and the environment. By this way, it is possible to encourage a more creative participation of society in a balanced management of the territory and sustainable development, which constitutes an essential resource in times of crisis (UN 2015, 2021).

4 New Proposals: Geosites and Geo-Hiking

Some researchers have identified geosites as indicators of geodiversity understood as geological specificity, but they can also be described, according to the more usual definition, as elements, zones or localities of geological-geomorphological interest relevant to conservation and protection; in this sense, geosites become emerging elements of the condition of geological diversity (or uniformity) existing in a given context.

A Geotope is a basic element of the landscape, like the biotope, which can be identified as the smallest unit of the earth's surface, geographically definable, homogeneous in terms of characteristics and structure (Troll 1950). In this sense, a parallelism between geodiversity and biodiversity is justified because, although on different scales, endogenous and exogenous processes and rocks, as elements of landscape formation, condition the evolution of environments and biological diversity, and are at the basis of the spatio-temporal development of a region. The relationships between the components of the landscape and the environment are very close, and the relationships between causes and effects of phenomena affecting the physical part of the territory and the biological part have a logic of mutual reciprocity, with a view to studying environmental diversity understood as the integration of geodiversity and biodiversity.

According to recent definitions agreed upon in the national and international field, the term "geotopes" refers to those geological manifestations, those physical elements of the territory or singularities of the landscape that manifest a scientific value, as an expression and testimony of the processes that originated and modelled the planet Earth, but which, at the same time, can constitute an appeal for additional aesthetic, naturalistic, cultural, historical, tourist and educational attributes (Brilha 2016).

In parallel, a geosite can be any locality, area, or territory where it is possible to define a geological or geomorphological interest for conservation (Prosser et al. 2018; Piacente 2005). In any case, the desire to guarantee the conservation of places of high geological interest stems from the need to ensure that future generations can enjoy the landscape in all its configurations. The Framework Law on Protected Areas (394/1991) includes, among the aims of the protection and management regime, the conservation of geological singularities, palaeontological formations, scenic and panoramic values, natural processes, and hydraulic and hydrogeological balances.

Importance of geological monuments.

The new approaches address unconventional communication techniques, based on the emotional and experiential involvement of the individual/community/society in a process of behavioural evolution.

Those Outdoor sports that take place in the natural environment, such as cycling, hiking and trekking, orienteering, climbing, skiing, sailing and so on, are excellent vectors for communicating environmental issues.

Hiking is a popular sport, deeply linked to the territory. Moreover, it favours discovering processes, revealing a high potential for disseminating scientific information to the public (Lugeri et al. 2018).

The description of the hiked places, explained in a simple, but not simplified way (this is the secret for an effective scientific communication) thanks to the "Landscape Ecology" approach, it allows to integrate nature and culture and to divulge them while walking into the wild. It is a proper way to highlight the most significant places in our country and analyse their constitution and origin, with particular attention to situations of high landscape value and/or territorial fragility (Lugeri et al. 2018 bis).

In the following pages we are going to explain three experiences of geo-hiking through the geosites of some areas hit by the Central Italy earthquake, realised within a project aimed at restarting the social-economical progress.

5 Tools

Through the modern Geographic Information Systems (GIS), it is possible to represent geographic informations with values related to each object, collected in a database that allows to reorganize and catalogue cartographic and bibliographic material, translating it into a well-aggregated set of information, easily accessible and spatially defined. Integration of information, although complex, offers many advantages: the data are easy to recover and update; in addition, the use of the information can be controlled according to the thematic address and differentiated according to the user needs. Spatial databases are managed using computer technologies (GIS, WebGIS) that allow a multi-scale and multi-thematic analysis and representation of all of the georeferenced information. WebGIS technology allows to visualize and query online, georeferenced information and associated maps, not needing additional software, providing a valuable tool for territorial planning, and risk prevention: in these perspective GIS can be considered the ideal instruments to identify risks and resources, to implement risk prevention and sustainable development, to analyse the related problems, and solve them. They are considered the best modality to represent analysis and synthesis maps.

In few words, GIS are the most proper tools both in assessing social-ecological resilience, in improving its functionality, and in promoting resilience of desirable balances by mitigating those stressors affecting these systems.

In parallel, PPGIS (public participation geographic information systems) are a relevant evolution of GIS. Such kind of tools confirm the role of GIS technology in involving a wider participation of individual/community/society in a common action for environmental protection and sustainable development. The high potential of PPGIS is not yet developed at the best level, despite the technological progress, especially referring to shared knowledge, induction of social awareness, involvement of marginalized populations.

There are many fields of application: considering the network of links between the environment and society, the strategic and revolutionary importance of the PPGIS is evident: from the monitoring of landslides and other forms of hydro-geological hazards to the function of observatories on the territory. Public participation in GIS data collection, implies knowledge of the tool and analytical skills, therefore it presumes a social involvement in educational paths that are integrated into the action itself of participation in the GIS, through inductive and deductive processes.

Part 2

6 Case Study: The Geosites of the Fiastrone River Valley

In 2018, some small municipalities of Macerata province in Marche region (including Camporotondo di Fiastrone, Fiastra and Cessapalombo) subscribed an intermunicipal reciprocal agreement with the aim to participate to the "Piano di Sviluppo Rurale"

(Rural Development Plan), a call through which it would have been possible to receive a financing to spend on environmental, social, and cultural interventions.

Submitting the project, administrators hope to contribute to restart after the seismic events of 2016, under a social and economic point of view. This territory, in facts, is in the central part of Italy, a few kilometres from the epicentral areas of earthquakes of August and October 2016 and January 2017. Camporotondo di Fiastrone, Fiastra and Cessapalombo are three so called "rural" municipalities, each one with less than 500 inhabitants, in the National Park of Sibillini Mountains (Regione Marche 1991).

The seismic events damaged about 60% of buildings, both public and private, people lost not only their houses but also public meeting places as Churches, schools, public offices, cinemas and, in many cases, the whole historical centres. Damaged houses, the loss of public services and meeting places, the continuous aftershocks going on for more than a year, the impossibility to work (due to unusability of many factories and shops), the fear and, not least, the political strategies during the emergency, induced people to move far from the epicentral areas. A lot of evacuated people were hijacked in camping and hotels along the Adriatic coast, at a safe distance from the tectonic lineaments activated. The damages produced by the earthquake were irreparable, local administrators had to act immediately to counter the depopulation in progress, sometimes without money, means, energies. It starts from here the idea of a mayor geologist to bet on the only inexpensive thing in their property: the geological heritage.

The project intended to raise the local economy (after years of economic crisis subsequently amplified by the seismic events) strengthening attractiveness and tourism competitiveness trying to take advantage of the reel effect with the Sibillini mountains. This project can count on a multiple valence, it aims to counteract depopulation, improve economy (increase in presences of peoples in the area), tourism (increase in supply), sociability (new public places of aggregation), environment (preservation of geodiversity and enhancement) and culture (didactic activities with schools).

7 Geological and Hydrographic Context

Fiastrone river is a long mountain watercourse long approximately 34 kms. It originates in the hearth of the anticlinorium of Sibillini mountains that crosses cutting calcareous lithologies of "Umbria-Marche succession" until it arrives in the foothill area, where it flows through the siliciclastic terms of "Laga Formation" (Cantalamessa et al. 1980). The stratigraphic succession of this area was deposed under the sea level starting from the Jurassic until the Quaternary period. During the Miocene this area was involved in compressive tectonic, which starts to delineate the intra-Apenninic basins each one with its related turbiditic succession (Cantalamessa et al. 1981).

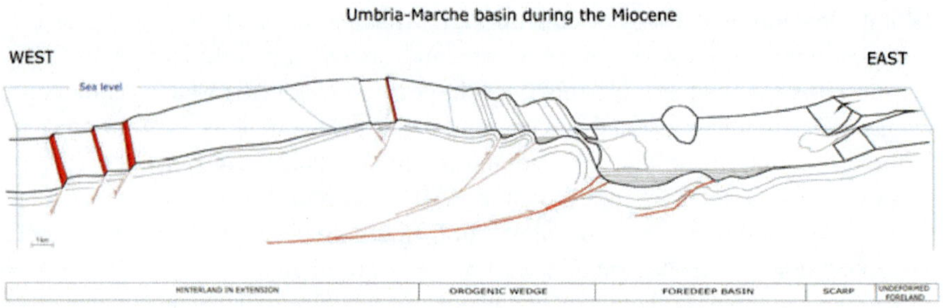

Fig. 1 Paleogeography of the area during the Miocene

During this time, the seabed was divided into different morphostructural units, constituted by alternations of dorsal and depressions elongated following an Apenninic direction (N-NW/S-SE) and characterized by sedimentations differentiated for facies and thickness. One of this depressions, the largest one, was represented by the Laga basin, a depositional environment which was located in the entire territory of Camporotondo di Fiastrone (Fig. 1).

The prosecution of compression led to the definitive emergence in a subaerial environment, occurred in the Plio-Pleistocene, while in the Quaternary period, the area came back into a regime of extensional tectonic.

After the dam of Fiastrone lake, waters of the homonymous river flows channelled along a high and tight rocky gorge—a tourist destination known in Italy under the name of "Gole del Fiastrone"—where it is possible to admire a lot of geological peculiarities (not yet well mapped and valued). Here, it starts the geoturistic itinerary individuated, which subsequently crosses the administrative boundaries of Cessapalombo, San Ginesio and, finally, Camporotondo di Fiastrone, before converging into Chienti river in the town of Belforte del Chienti.

The rocks outcropping in the territory of Camporotondo di Fiastrone have a sedimentary origin. They are constituted by siliciclastic sediments, deposed starting from Miocene sup. under a turbiditic basin (Cantalamessa et al. 1980; Ohneiser et al. 2015). The persistence of compressional tectonic bring to the emersion of the area above the sea level in the Quaternary period (Fig. 2).

8 The Project

The project expects to individuate, mapping and achieve an itinerary able to connect all the geosites individuated along this lowest portion of Fiastrone river (Figs. 3, 4).

The individuated geosites along the Fiastrone gorge can be resumed in the following image, where it is possible to observe the rock formations (here sorted in descending order depending on the age and increasing according to the stratigraphic height). From SW to NE, they are so arranged (ISPRA 2009):

Fig. 2 The territory object of the present study with Fiastrone lake, Fiastrone river in blue, the crossed municipalities, Chienti river in light blue and the edge of Sibillini Mountains National Parc in red

- "Calcare Massiccio Formation" and other Jurassic Formations of "Grotta dei Frati" and "Gole del Fiastrone" (Jurassic period).
- "Bonarelli Level" in the "Scaglia Bianca Formation" (it is a key horizon of Cretaceous epoch).
- "Scaglia Rossa Formation" outcropping on the Sibillini Mountains thrust and in the "Chevron Folds of Coldipietra" (Lower Turoniano-Medium Eocene).
- "Vulcanoclastic Level of Laga Formation" in Camporotondo di Fiastrone (Messinian age).
- "Lame Rosse" (Quaternary period).

During this work we will study in deep the geosite of Camporotondo di Fiastrone, the volcanoclastic level of Laga Formation, because one of the authors contributed to drawing up the project in detail for that municipality.

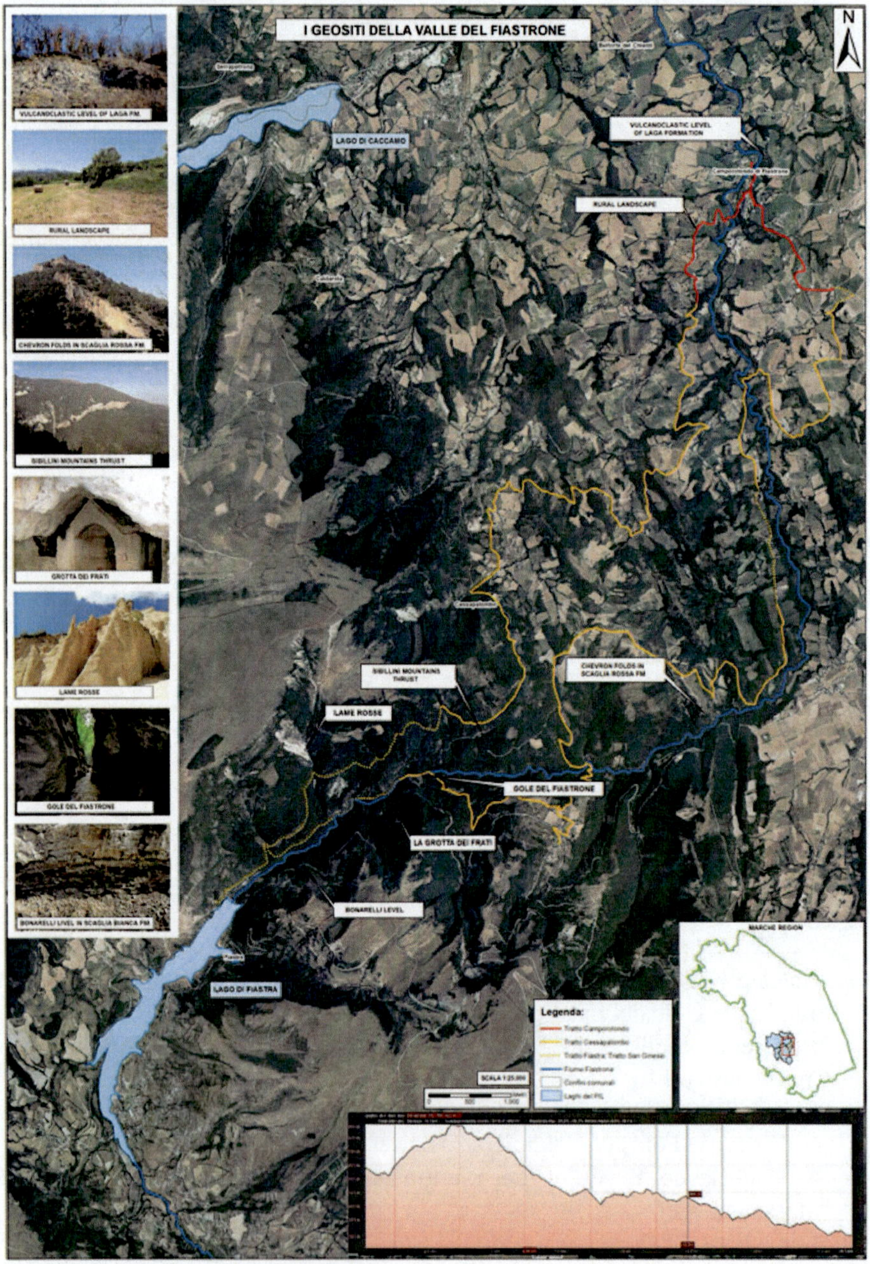

Fig. 3 Geographical distribution of the geosites along the itinerary in project

Fig. 4 Details of each geosite

9 The Geosite of Camporotondo Di Fiastrone

The individuated geosite (not classified by the regional catalogue of geosites ISPRA) consists of a sedimentary succession exposed along a river escarpment of about 40 m high. Along its height the outcropping sediments are referred to the "pelitic-arenaceous association" of Laga Formation, post-evaporitic member, an alternation

Fig. 5 The location where the pitch will be built

of greyish clays with thin arenaceous levels intercalated. In the medium-upper portion of this litofacies, incorporated inside of the clay material, a layer characterized by a more whitish colour of about 2 m thickness is recognizable: the volcanoclastic level of Laga Formation (Guerrera et al. 1986), the real peculiarity of the geosite (Fig. 3). Upwards, the succession passes into the "arenaceous-pelitic association".

The volcanoclastic level outcrops in the medium-highest portion of the scarp, recognizable thanks to its greyish-whitish colouration. It is predominantly composed by volcanic ashes coming from now extinct volcanoes, once probably located in the western part of Italy where the crustal thinning evolved so long to allow the rise of magma from the atmosphere and the subsequential emission of effusive products, whose most volatile portion was able to reach these latitudes. This layer of volcanic ashes covered all the seabed as a drapery, without having significant lateral variations and stratifying itself with continuity along the basin. For this reason, the volcanoclastic level of Laga Formation is considered a key horizon with regional importance (Fig. 5).

From a petrographic point of view, it consists of syneructive volcanoclastic sediments constituted by riodacitic volcanoclasts of graded pyroclasticites, sometimes laminated (microlytic and glassy pumices and glass shards where it is possible to find very small cristals of unaltered biotite, which provided a radiometric age of about 6 million years).

10 Interventions

In the alluvial plan in front of the fluvial scarp will be realized a pitch, equipped with a touch screen to read cartographic and bibliographic material and a "totem" with a monitor and a camera from which it will be possible to zoom towards the outcrop (170 m far from here).

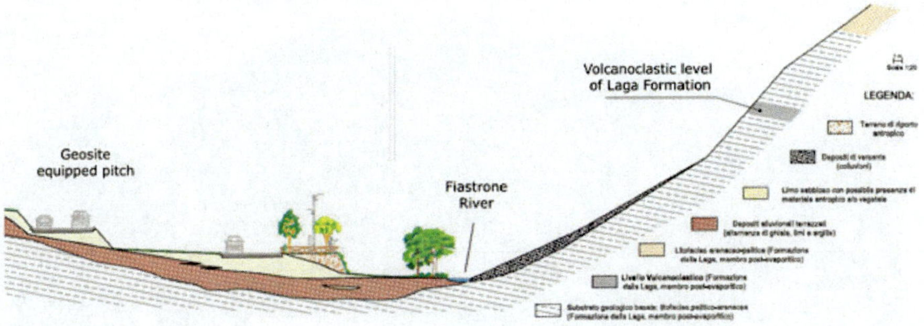

Fig. 6 The project section

In the area tables and benches will be assembled to host school groups, citizens or tourists, then a barbecue, a wifi connection and a charging station for smartphone and e-bikes (Fig. 6).

Part 3

11 Case Study: The Frasassi Caves

Frasassi cave is a complex karst system located near Genga, Ancona province, Marche region. They have been discovered 50 years ago, on the 25 September 1971, when a group of about ten speleologists, with age between 15 and 24 years, found a hole along the rocky slope of the Frasassi gorge with the size of a football from which came out of a souffle of air. Those were the years of the cave explorations, a lot of speleological groups frequented Frasassi gorge and its Calcare Massiccio Formation trying to find new caves or new connections (Figs. 7, 8).

In 1974, after three years from their discovery, Consorzio Frasassi (the consortium which manages the caves) inaugurated the tourist route inside the hypogeum complex, a tour of about one hour and a half, where tourists can admire the magnificence of the karst complex along a walkway.

The tourist attraction immediately revealed itself as a winning bet, with annual presences wandering around 300.000 visitors. During the time, a lot of people coming from all over the world visited Frasassi caves and, with the passing of years, presences slowly start to constantly drop, reaching a lowest of 250.000 around the 2009,[1] year in which available data ends.

Looking at the number of visitors, it appears how each dropping follows a seismic crisis verified around this territory, that discouraged tourists to enter in the cave.

[1] Consorzio Frasassi only allowed data consultation.

Fig. 7 Frasassi gorge view from mount Vallemontagnana

Fig. 8 The famous "Sala delle Candeline" in Frasassi caves

It is possible to note this especially after the earthquakes of 1997 (Colfiorito, province of Perugia) and 2016 (Norcia, province of Perugia), while in the cave collapses or damages were not noticed.

Fortunately, each dropping is followed from a positive peak, usually recorded after one or two years. This is attributable not only to the attenuation of the sense of fear in the visitors, but also to the multiple initiative's boarder up by the local and regional administrations (especially Consorsio Frasassi, but also Municipality of Genga, "Parco naturale regionale della gola della Rossa e di Frasassi", Marche Regional Council).

In 2020, the pandemic crisis was added to the seismic crisis and Consorzio Frasassi had to plan a series of initiatives to react after these catastrophes, pulling out that spirit of resilience that hopefully would also been useful for the society. One of the initiatives undertaken was to accept a proposal of the geology section of the University of Camerino and subscribe a three-way agreement, even together with the Municipality of Genga. The project was planned during the autumn 2019 but the pandemic crisis slowed the signature of the covenant, which therefore was postponed in October 2020.

The aim of this project was to share the geological expertise of the university to deepen the knowledge in the Frasassi area, both outside and outside the caves. In the planned roadmap the organization of scientific conferences in the area, the inauguration of a head office of University of Camerino in Frasassi (as a hub from which organize didactic geo-hiking and, subsequently, a scientific laboratory) and the enhancement of the tourist offer with the introduction of a new tour in the cave was expected, with geologists explaining an exclusively geological tour.

In this way, the geology would be exalted even more, its peculiarities valued and promoted. In return, the University of Camerino could benefit for the conduction of teaching and research activities directly in the caves and it will take advantage from the visibility donated by hundreds of thousands of visitors each year.

The first activity, in December 2020, was to participate together in a call, which foresaw financings for this type of environmental activities. Several interventions were planned, such as the modernization of the speleo-paleontological museum of Frasassi (hosting the Gengasaurus Nicosiai, a Jurassic ichthyosaur fossil of about 3 m long found a few meters from the caves); the predisposition of a geotourist itinerary able to connect many of the geosites present in the area; and training courses for the geological-guides that will accompany people in the caves and along the territory.

The second activity, in September 2021, saw the organization of a national scientific congress for the 50th anniversary of the discovery of the Frasassi caves. They invited the discoverers, the greatest experts of the area (both geologists and speleologists), local administrators and national politicians (Fig. 9).

The third activity, also this written in the convention, was to implement the actual tourist offer allowing the tourist the choice to buy a "geological tour" in addition to the standard tour. On the occasion of the national initiative of "Settimana del Pianeta Terra" (Planet Eart Week) which has been organized a pilot project which foresees a Sunday "In Grotta con il Geologo" (In the Cave with the Geologist), where five university professors and researchers will explain the caves showing all

Fig. 9 The 50th anniversary of Frasassi Caves conference organized by the geology division of University of Camerino

the geological peculiarities not shown during the standard tour. Therefore, it will be initially illustrating the geological succession of Umbria-Marche domain and explained all the processes which involved this portion of territory from the Jurassic period until now, including the explanation of the geomorphology of the gorge and the formation of the hypogea environments. Consequently, in the cave, the group will see stalactites, stalagmites, tepee structures, fossiliferous levels with gastropods, a Jurassic fault view from inside the cave, karst levels, eccentric helictites, gypsum deposits and many other situations (Fig. 10).

All these activities aim to sensibilise tourists and the citizens of Genga about the Earth sciences, transmitting to them the awareness of what they are visiting or what they have under their feet, trying to make them aware of the resources and risks of this territory. At the same time, these initiatives contributed to keep local administrators informed about the geoturistic potential of the territory they manage, developing an

Fig. 10 The initiative "In the cave with a Geologist" organized by the geology division of University of Camerino

awareness of its strengths and its criticalities. All of this must be observed in optics of safeguarding of geodiversity, enhancement of resources and risk reduction.

This is the answer of this territory to the seismic events that occurred a few ten kilometres far from this area (Frasassi caves are located under the administrative boundaries of Fabriano (AN), a town included in the most damaged area most damaged by the earthquakes of 2016 and 2017 (in the list of the 138 municipalities in which the earthquake occurred with more intensity).

12 Cultural Landscape as a Function of Natural Landscape

Geo-hiking experiences and itineraries for the renaissance of the communities affected by a natural catastrophe represent a new model of strategy which is based on the social and economic recovery of itself. Furthermore, it is a sustainable way to move and live in nature, deepening its cultural aspects and contemplating the close relationships between landscape and history. All of this could be possible without impacting the landscape and without foreseeing the use of resources, both economic and environmental (EU 2021).

The cultural landscape has been determined by the natural landscape; human settlements came and are set along the environmental lines, such as gorges, alluvial valleys, watercourses, mountain ridges, karst systems, headlands, etc. Anthropogenic settlements, in fact, were often realized in steep panoramic places, to constitute a natural defence against sieges and, at the same time, guarantee control over the surrounding areas. For this reason, it is possible to find castles, towers, abbeys, monasteries on the top of the mountains, but also hermitages and places of warship very often embedded in the rocks of gorges or even in epigean caves. Therefore, since the dawn of time, nature and culture have been interconnected and the first one conditioned the latter.

During the post-war period, human settlements have been set along fluvial valleys, originating a series of consequences and repercussions along the whole hydrographic basin (Bowen and Gleeson 2019).

After the seismic crisis, local and national administrators discussed a lot if it would have been right to reconstruct "where it was and how it was". This Hamletic question occupied all the technical and political tables of the reconstruction, at all levels and this question arises from the high seismicity of these territories. This can create doubts about how to respect the rebuilding of the cultural landscape by guaranteeing a compatible security level.

In many cases, the cultural landscape is a testimony of wisdom in the use of the territory: the environmental knowledge of our predecessors, in fact, determined a precise and weighted development of the cultural landscape. Almost the whole of cultural sites has remained intact today (obviously, except for structural elements which have deteriorated over time), even if this rule is not always respected in areas with high seismicity.

13 Final Remarks: Territorial Enhancement, and Social Involvement for Resilience and Restart

Sharing knowledge is an ethical objective, an essential process to develop the capacity for recovery and rebirth, in a word, for 'resilience' (literally: the ability to be happy, successful, etc. again after something difficult or bad has happened).

From an examination of the dramatic events that have occurred in Italy, linked to the country's geo-environmental structure, and the effects of its interaction with anthropic pressure, it is an undeniable need to provide the general public with correct and clear information on the complex scenario that characterises our country (Farabollini et al. 2014). Landscape plays a key role in the communication of Earth and environmental sciences: it is the object of human perceptions, the aspect of the environment that people perceive and interact with, and—in this sense—it guides everyone's knowledge processes (Troll 1950). A special reflection is dedicated to people with disabilities: the experiences carried out in the field of disability have demonstrated the function of landscape and its conscious use, to achieve a common and complete harmonisation between the individual and the environment (Lugeri and Cardillo 2009).

According to the "Landscape Ecology" approach (Odum 1961; Amadio et al. 2002; Amadio 2003), landscape is the result of the interaction between physical, biotic and anthropic components, considered with reference to a spatio-temporal scale (Forman and Godron 1986). Similarly, landscape can be defined as the expression of the balances between endogenous and exogenous forces that interact in the dynamics of the planet, melding its surface (Farabollini et al. 2014). Even at different scales, abiotic transformation processes and rocks, as landscape elements, condition the evolution of the environment and form the basis of the spatio-temporal development of a region (Brilha 2002). Shape is synthesis, as Aristotle teaches us: following an Aristotelian approach to the study of nature, it is natural to read the characteristics of the territory as an expression of its geological and environmental dynamics (Panizza and Piacente 2005). A holistic approach to knowledge has much to offer to landscape studies, considering all the components that make up the complexity of nature (Ratcliffe 1976; Badiali and Piacente 2012).

Prevention and resilience are closely linked and both concepts imply awareness, consciousness, and action (UNISDR 2015).

Until knowledge is assimilated and transformed into behaviour, it is essential to work on many fronts. Information is the first step and the role of communication in relation to prevention is vital. It is dramatically evident that current styles of scientific communication are not effective enough. Whether in relation to natural disasters or pandemics, there is no shared awareness of the common effects of natural hazards, as well as public health problems. Despite the global spread of information tools, the contents of 'scientific communication' are too often misunderstood and/or rejected as 'official communication'. And this is modern-day history. The old tradition in 'mainstream' communication styles testifies to the habit of using spectacularization and fatalism; moreover, the highly differentiated availability of many channels causes a fragmentation of the public into thematic groups: users can choose, and this is a right, but the side effect is the lack of plurality in the fruition and the 'sectorialisation' of information. How can these contradictions be resolved? There are no formulae or algorithms: rather, new ways of communicating are needed. At long last, the world of research has begun to consider the importance of unconventional communication, aimed at involving the individual/community/society in an effective dialogue, dedicated to risk prevention as well as to sustainable development.

The communication of those scientific topics that directly involve society must be comprehensible, proposed in simple language, without simplifying fundamental concepts. At the same time, communication must reach the widest and most varied range of audiences; it must be interesting, intriguing, captivating (Peppoloni and Capua 2012). Let us reflect on this: it is the way of advertising. No scandal: after all, we must induce behaviour (not in marketing, but in prevention). Experience teaches us that talking about risks "tout court", provokes a removal effect on the spectator, therefore, it is much more effective to discuss the subject by resorting to an opposite concept, but in a certain sense complementary to the risk: the resource. Two poles of a continuum that connotes the cultural and environmental reality of our country.

From environmental and geological hazards to pandemics, it is always more effective to invite people to behave proactively, rather than to intimidate such that inducing a kind of superstitious disbelief. The presence of information must be constant and continuously offered to the public, and at the same time, it must be so sincere as to instil confidence.

It is then necessary to monitor and analyse the response of the public, but in order to achieve extensive and lasting results, it is necessary to intensify and broaden the implementation of the unconventional communication projects, making their effects such as to induce a new shared consciousness, thus providing the society with effective tools to be able to strengthen resilience and initiate rebirth.

Communication and dissemination are inextricably linked, so that they must be planned together as essential components of any educational project.

The implementation of our proposals is mainly devoted to dissemination and capacity building. Communication itself is designed to be shared, and the main objectives are to involve the widest possible range of audiences, to guide individuals/communities/society to natural hazard awareness and preventive behaviour. An ethical duty.

Communication is both means and message. Start, route and finish. The medium is message and massage (the typographer's error during the printing of Marshall McLuhan's famous treatise, was brilliantly relaunched by the author to mean massage, age of the masses "Massage" and "Mass Age", message, age of chaos: "Message" and "Mess Age". In our case, closer to the concept of massage, the use of topics more appealing to the public as vectors of scientific concepts is a successful strategic solution, especially when trying to share complex, or simply unpopular, concepts, information, or regulations.

It is verified that an effective tactic is to provide 'unexpected' information through unconventional channels and to invite prominent personalities to be actively involved as testimonials. Moreover, the extraordinary geo-environmental and cultural landscape that characterises Italy is itself a powerful means of communicating territorial issues to the whole of society: if recognised and understood, landscapes become part of everyone's cultural heritage, and the dynamics that characterise their natural history—at the basis of the aforementioned 'risk/resource' binomial—are revealed, providing the observer with the knowledge indispensable for understanding complex environmental realities. It is in this sense that the perception of the landscape is the basis of a cognitive process that can trigger a virtuous circle, revitalising the roots that link mankind and the environment, encouraging a more creative participation of society in a sustainable development of the territory.

The new approaches address unconventional communication techniques, based on the emotional and experiential involvement of the individual/community/society in a process of behavioural evolution.

As already mentioned, if scientific communication is to be simple, scientific information must not be simplified: it is a matter of language, which must be decodable (Lugeri et al. 2018).

The proposals outlined in this chapter are based on the previous reflections and have been carried out experimentally, and in a "pioneering" way, however on the basis of careful analysis of the need for integration between the sciences that study the Planet and those that analyse its inhabitants. The integration of the environment and society seems banal, but the first studies on the sociology of the environment are very recent (Catton and Dunlap 2003).

The proposed communication modules combine traditionally separate themes, which instead, in the complexity of reality, are connected by significant links, all of which have yet to be discovered. Sport and Landscapes to reach the widest audience, conveying communication through different themes, made complementary in a new innovative formula, with the aim of increasing the resilience and rebirth of communities and territory.

References

Amadio V (2003) Analisi di Sistemi e Progetti di Paesaggio. Franco Angeli, Milano, Italy, p 23

Amadio V, Amadei M, Bagnaia R, Di Bucci D, Laureti L, Lisi A, Lugeri FR, Lugeri N (2002) The role of geomorphology in landscape ecology: the landscape unit map of Italy, Scale 1: 250,000 ("CartadellaNatura" Project). In: Allison RJ (ed) Applied geomorphology: theory and practice. Wiley, London, UK, pp 265–282

Badiali F, Piacente S (2012) The study of the landscape: From a holistic approach to a social concept of knowledge. Ann Geophys 55:481–485

Bauman Z (1997) Postmodernity and its discontents. Polity, Cambridge

Bowen W, Gleeson R (2019) The evolution of human settlements: from Pleistocene origins to Anthropocene prospects. Springer. https://doi.org/10.1007/978-3-319-95034-1

Brilha J (2002) Geoconservation and protected areas. Environ Conserv 2002(29):273–276

Brilha J (2016) Inventory and quantitative assessment of geosites and geodiversity sites: a review. Geoheritage 2016(8):119–134. https://doi.org/10.1007/s12371-014-0139-3

Cantalamessa G, Centamore E, Chiocchin U, Di Lorito L, Leonelli M, Micarelli A, Pesaresi A, Potetti M, Taddei L, Venanzini D (1980) Analisi tettonico-sedimentaria dei bacini minori torbiditici del Miocene medio-superiore nell'Appennino umbro-marchigiano e laziale-abruzzese: il bacino della Laga tra il F. Fiastrone- T. Fiastrella ed il T. Fluvione. Studi Geol Camerti 6: 81–133

Cantalamessa G, Centamore E, Chiocchini U, Di Lorito L, Giardini G, Marchetti P, Pontoni F, Potetti M (1981) Analisi tettonico-sedimentaria dei bacini minori torbiditici del Miocene medio-superiore nell'Appennino umbro-marchigiano e laziale-abruzzese: 9) Il bacino della Laga tra il F. Fiastrone- T. Fiastrella ed il T. Fluvione. Studi Geol Camerti 7: 17–80

Catton WR, Dunlap RE (2003) Environmental sociology. A new paradigm. Am Sociol 1978, 13:41–49. 32. Comput Geosci 61:83–93

Crutzen PJ (2006) The "Anthropocene." In: Ehlers E, Krafft T (eds) Earth system science in the Anthropocene. Springer, Berlin/Heidelberg, Germany, pp 3–6

EU (2021) Urban agenda for the EU: multi-level governance in action. https://ec.europa.eu/regional_policy/en/information/publications/brochures/2021/updated-version-2021-urban-agenda-for-the-eu-multi-level-governance-in-action Accessed on Mar 2022

Farabollini P, Lugeri F, Amadio V (2014) Geomorpho-landscapes. EGU general assembly, 27 April–2 May 2014—Vienna, Austria, id.4954

Farabollini P, Angelini S, Fazzini M, Lugeri FR (2019) Earthquakes and society: the 2016 central Italy reverse seismic sequence. Geographies of the Anthropocene book series, Il Sileno Edizioni. http://www.ilsileno.it/geographiesoftheanthropocene

Forman RTT, Godron M (1986) Landscape ecology. Wiley, New York, NY, USA, p 620

Guerrera F, Tonelli G, Veneri F (1986) Caratteri lito-sedimentologici e mineralogico-petrografici di vulcanoclastiti mioceniche presenti nella successione umbro-marchigiana. Boll Soc Geol It 105: 307–325, 7 figg., 4 tabb., Roma

ISPRA (2009) Note illustrative della Carta Geologica d'Italia alla scala 1:50000. Foglio 302 Tolentino (MC). Regione Marche- Servizio ambiente e paesaggio

Lugeri FR, Cardillo A (2009) Tutela ambientale e crescita personale. Atti convegno Montagna solidale: i versanti della montagnaterapia. pag. 11–14

Lugeri FR, Farabollini P, Amadio V, Greco R (2018) Unconventional approach for prevention of environmental and related social risks: a geoethic mission. Geosciences 8:54. https://doi.org/10.3390/geosciences8020054

Marincioni F, Gioia E, Zoppi M, Vittadini E (2019) Food management in disasters: the case study of the earthquakes of 24 august 2016 in Central Italy. In: Earthquake risk perception, communication and mitigation strategies across Europe, Geographies of the Anthropocene book series, Il Sileno Edizioni, Lago, Italy

Miccadei E, Piacentini T, Buccolini M (2017) Long-term geomorphological evolution in the Abruzzo area, Central Italy_twenty years of research. In Geologica Carpathica. ISSN: 1335–0552 Editore Bratislava: Veda http://hdl.handle.net/11564/667892

Naveh Z, Lieberman A (1994) Landscape ecology theory and application; series on environmental management. Springer, New York, NY, USA, pp 99–103

Odum EP (1961) Fundamentals of ecology. Cengage Learning, Boston, USA

Ohneiser C, Florindo F, Stocchi P, Roberts AP, De Conto RM, Pollard D (2015) Antarctic glacio-eustatic contributions to late Miocene Mediterranean desiccation and reflooding. Nat Commun 6:8765

Panizza M, Piacente S (2005). Geomorphosites: a bridge between scientific research, cultural integration and artistic suggestion Geomorphological sites and geodiversity. Il Quaternario—Italian J Q Sci 18:3–10

Peppoloni S, Di Capua G (2012) Geoethics and geological culture: awareness, responsibility and challenges. Ann Geophys 2012(55):335–341

Piacente S (2005) Geosites and geodiversity for a cultural approach to Geology. Il Quaternario. Italian J Quat Sci 18(1):11–14

Prosser C, Díaz-Martínez E, Larwood J (2018) The conservation of geosites. In: Reynard E, Brilha J (eds) Geoheritage: assessment, protection, and management Chapter: 11. Elsevier Editors, pp.193–212

Regione Marche, Assessorato urbanistica e ambiente (1991) L'ambiente fisico delle Marche

Troll C (1950) Die GeografischeLandschaft und IhreErforschung; Springer: Heidelberg. Germany 1950:163–181

UN (2015). Transforming our world: the 2030 agenda for sustainable development. https://www.un.org/ohrlls/sites/www.un.org.ohrlls/files/2030_agenda_for_sustainable_development_web.pdf. Accessed Mar 2022

UN (2021) Our common agenda. https://www.un.org/en/un75/common-agenda Accessed Mar 2022

UNISDR (2015) Sendai framework for disaster risk reduction 2015–2030. http://www.wcdrr.org/uploads/Sendai_Framework_for_Disaster_Risk_Reduction_2015-2030.pdf. Accessed Mar 2022

Sense of Belonging and Response to Climate Change: How the Relation with Local Territories Influences Climate Resilience

Cristina Casareale, Noemi Marchetti, and Alessandra Colocci

Abstract Earth's climate has changed throughout history and regional changes in climate have already affected a diverse set of physical and biological systems in many parts of the world. The impacts depend on community's vulnerability and exposure conditions, and mitigation and adaptation actions have been recognized as essential to meet the goals of global agendas (e.g., New Urban Agenda, Sustainable Development Goals and Sendai Framework). Adaptation measures seem to be less common compared to mitigation ones, probably for the limits on ecological, physical, economic, and technological dimensions. The objective of this study, conducted as part of the Interreg RESPONSe Project, is to analyze the resilience of local population in the Veneto Region in relation to the integration in the territory. Hence, the investigation focused on analyzing the locals' perception of climate impacts on their daily life, their sense of place, and how perception and sense of place influence the willingness to take personal and collective actions to counteract climate change. Social and physical characteristics, local climate change scenarios, and the targeted adaptation approach of different communities have been examined. Results suggest that age and proximity to the coast are influencing factors in the response to climate change. For instance, younger people appear to acknowledge the shared personal responsibility of tackling such a challenge, while wishing for a broader community involvement. At the same time, the sense of belonging appears to support long-term, adaptation actions. Overall, the outcomes give hints about the key elements to consider when planning for improving local climate resilience.

Keywords SDG13 Climate action · Climate resilience · Mitigation · Adaptation · Scale · Italy

---

C. Casareale · N. Marchetti · A. Colocci (✉)
Università Politecnica Delle Marche, Ancona, Italy
e-mail: a.colocci@staff.univpm.it

B. Dahiya et al. (eds.), *Disaster Resilience and Human Settlements*,
Advances in 21st Century Human Settlements,
https://doi.org/10.1007/978-981-99-2248-2_13

1 Introduction

Understanding how the geographic scale affects the willingness of people to act to fight climate change and, consequently, the way governments can use the sense of belonging to manage environmental problems is not a new topic for researchers (see Feitelson 1991). Research on people–place relationships, in particular attachment to place and place identity, is beginning to make an important contribution to understand human responses to climate change (Devine-Wright et al. 2015). Nowadays, in fact, it is well known that although climate change is a global phenomenon, the impacts are most acutely experienced at a local scale (Cunsolo Willox et al. 2012). Many parts of the world will need to adapt to an increasingly unstable climate (Bowden et al. 2019), considering that according to the recent reports from the Intergovernmental Panel on Climate Change (IPCC 2021, 2022a, b) the window to avoid dangerous climate change is rapidly closing. Consequently, mitigation seems to be less and less feasible, although still essential in order not to worsen the already very serious condition of the earth system (https://sdgs.un.org/goals/goal13). Despite the growing concern about the ability to adapt both globally and locally (Biesbroek et al. 2013; Porter et al. 2015), much of the adaptation literature remains under-theorized in relation to the social mechanisms involved (Eriksen et al. 2015). Barriers to climate adaptation are often described as unrelated to each other, with a limited explanation of how these barriers are socially constructed and why they persist (Biesbroek et al. 2013; Brulle and Norgaard 2019). Since adaptation often occurs in a more localized context than mitigation, understanding the connections between the place and the people who inhabit it has the potential to break down these barriers (Bowden et al. 2019). Climate and environmental changes are also causing various place-related impacts: alterations in biophysical, geographic and climatic processes (Adger et al. 2007); changes in the level of significant link between people and their environments; disruptions to place-specific socio-cultural activities, such as hunting or fishing, which contribute to health and well-being; transformations in the mental conceptions of place, or rather of the socio-cultural and psycho-social meanings attributed to particular places; alterations to place-based identities held individually and collectively (Cunsolo Willox et al. 2012; Scannell and Gifford 2010). Climate change could affect the sense of belonging in such a way that it can prevail over the attachment of an individual to his/her "place", contributing to the phenomenon of environmental migration as possible adaptation strategy (Tacoli 2009; Warner et al. 2010) instead of encouraging people to adapt locally and increase local climate resilience.

From this perspective, understanding the elements that shape individual and collective sense of place, the relative levels of attachment to or bond with a particular area, and the ways in which individuals create identities connected to and within the place, becomes an essential component of qualitative and quantitative (Devine-Wright 2013) climate resilience research (Denton et al. 2014) and can help reduce climate-related disaster risks (UNDRR 2015).

This study has the following objectives: (a) to analyze the perceived impact that climate change will have on the citizens of Veneto Region (Italy); (b) to identify possible factors that influence their sense of place; (c) to analyze the influence that these factors have on the willingness to counteract climate change and its effects. This work ultimately aims at helping decision-makers to consider the elements that influence the sense of belonging when planning mitigation and adaptation strategies. Such local strategies would be intended to be part of the local development agenda (United Nations 2017), thus improving local climate resilience.

2 Methodology

2.1 Study Area

The research is part of the Interreg Italy-Croatia Project "RESPONSe" (Strategies to adapt to climate change) that aims at empowering local policy-makers to enable climate-smart governance approaches and promote sustainable living in Adriatic marine and coastal areas. Veneto Region (Fig. 1a) is one of the pilot areas involved in the project. The region is the fourth most densely populated region of Italy, with 4,852,453 residents and a density of about 264 people per km^2. About 41% of the population lives in coastal provinces (Metropolitan City of Venice, Rovigo and Padua), where the population density is particularly high. The remaining 59% of the population lives in hilly and mountainous provinces (Verona, Treviso, Vicenza and Belluno) (Istituto Nazionale di Statistica 2021). In the coastal area, there are widespread farms of small size with intensive cultivation, intensive farming, inhabited areas and small-very small industries. Even outside the large, inhabited centers, there is an articulated mix of productive activities, residential settlements, road structures and services. Veneto also has a strong tourist flow linked to attractions such as the city of Venice and its lagoon (Fig. 1b), the Arena of Verona, the Colli Euganei Regional Park (Fig. 1c), the wine cellars spread throughout the region, just to name a few. Despite the strong weight that the region has on the national economy, the enormous potential for further development and the high sense of belonging, at least from an economic point of view (Regione Veneto 2007), Veneto has one of the highest rates of emigration of Italy. Emigrants from Veneto are mainly educated young people, moving to other European countries to find better job opportunities (Osservatorio Veneti nel Mondo 2021). According to data from the Veneto Region, in fact, in 2019 the registrations to AIRE (Registry of Italians Residing Abroad) of Venetian were just under 457 thousand (+5.7% compared to the previous year) with a weight on the total number of residents in the region equal to 9.4%, confirming itself as the fifth Italian region for outgoing migratory flows (Osservatorio Veneti nel Mondo 2021).

Fig. 1 The Veneto Region with provincial borders (*Source* Authors) (**a**). Overview of Venice and its Lagoon (*Source* https://www.interno.gov.it) (**b**). Lago della Costa in the Colli Euganei Regional Park (*Source* http://www.parcocollieuganei.com) (**c**)

2.2 Data Analysis

In order to investigate the level of the sense of belonging of the Veneto population and its influence on the willingness to counteract climate change, the responses provided to a questionnaire distributed to the population as part of the RESPONSe project were analyzed.

Due to the ongoing COVID-19 pandemic, the questionnaire was administered to the population by publishing it on the website of the National Network System for Environmental Protection (SNPA). The questionnaires were collected during the period from March 2020 to April 2021.

The sampling method used for the sample sizing is the non-probabilistic per-quota method (Moser 1952), selecting as a variable in the residence in the Veneto Region.

The questionnaire was structured in two parts:

- Climate change-related questions, aimed at collecting information related to the knowledge, understanding and propensity to adapt of the population;
- General part, aimed at outlining the demographic profile of the participants.

The questions included in the questionnaire are of four types:

- Single-answer questions, for which the respondent can express only one choice;
- Multiple choice questions, for which the respondent can express more than one choice;
- Single-answer questions on a psychometric scale, for which the respondent must express an opinion more or less in agreement with a stated assumption on a "Likert" scale;
- Open questions.

The questionnaire administered to the population as part of the RESPONSe Project consists of 54 questions, 5 of which provide information related to the sense of belonging and were therefore selected for the realization of this study. The analyzes were carried out to verify whether the willingness to adapt to climate change is influenced by the (a) sense of belonging and by demographic characteristics, such as (b) age, (c) proximity to the coast, and (d) level of integration in the community.

To verify the above hypotheses, the sense of belonging-related questions and the demographic questions were selected from the original questionnaire. The analyzed questions are shown in Table 1.

To determine whether there is a correlation between the answers provided by the questionnaires (considered categorical variables, i.e. nominal or ordinal variables with less than 5 rankings), the non-parametric chi-square χ^2 test for independence (Mchugh 2013) was chosen. The chi-square test can be considered significant if the level of significance, the p-value, is lower than 0.05. For levels of p lower than 0.001 the significance of the test is extremely high. To carry out the analysis, the questions with answers on a "Likert" scale were combined into three answers: "Strongly disagree"/"Disagree"; "Undecided"; "Completely agree"/"Agree", depending on the formulation of the question analyzed. Likewise, the multiple-choice questions were

Table 1 Overview of the questions of the RESPONSe questionnaire selected for the analysis

Reference to the part of the questionnaire	Questions analyzed
Sense of belonging[1]	Q8: Climate change will impact your lifestyle
	Q9: What do you think will have to change in your lifestyle?
	Q25: The cost of mitigation of, and adaptation to climate change should be exclusively paid by the government
	Q28: What can you do, at the individual level, to prepare for climate-related hazards?
	Q29: Can you list concrete steps that you and your family have taken to face climate change?
General part	Q31: Age
	Q32: Where do you live?
	Q34: Do you feel integrated in your community?
	Q35: How far do you live from the coast?

Source Authors

grouped into smaller categories, depending on the answer options available in the question analyzed.

Once verified the presence of a dependence between the two variables (nominal or nominal and ordinal), the degree of associations between the variables was evaluated using the Cramer's V index, useful for making this evaluation for variables that each have more than two values (i.e., contingency tables bigger than 2×2). The value of Cramer's V is defined based on the degrees of freedom (df).

Finally, for the ordinal variables the direction of the association, positive or negative, was evaluated through the gamma index (γ) of Goodman and Kruskal. The γ index varies from -1 to 1. Values close to an absolute value of 1 indicate a strong relationship between the two variables, negative or positive. Values close to zero indicate scarcity or absence of relationship (Hryniewicz 2006).

Test processing was done with the use of IBM SPSS Statistic 19 software.

As a final analysis, contingency tables were constructed to verify the degree of association between two of the variables under consideration. This methodology made it possible to evaluate the number of respondents observed for all combinations of the categories of the two variables and to determine whether the variables considered are dependent or independent of each other.

[1] The questions are contained in different sections of the questionnaire (Q8 and Q9 – Risk perception; Q25 – Risk acceptance; Q28 and Q29 – Attitude towards risks and, for the scope of the chapter, are collected under the appellation "sense of belonging".

3 Results and Discussion

A total of 1231 random individuals among the adult citizens of the Veneto region participated in the survey, allowing to consider the responses as representative of the overall regional adult population (95% ± 2.8% confidence level).

The general information concerning the respondents (Table 2) reports a majority of adults (Q31, 66.7%), which corresponds to the data registered for 2020 by the Italian Institute of Statistics (Istituto Nazionale di Statistica, n.d.). Most of the respondents appear to reside in the hinterland (Q32, 62.9%) and are mainly far from the coast (Q35, 87.2%). Overall, the sense of belonging was predominant (Q34, 81.6%) over the other options. On the other hand, the information concerning the approach towards issues related to climate change (Table 3) describes a majority of respondents that agrees, even strongly (Q8, 37.5% and 37.1%, respectively), on the impacts of climate change on personal life, as well as that most of the changes will need to happen at the community (Q9, 45.3%) and individual (Q9, 43.6%) levels. Nevertheless, there is a wide uncertainty (Q25, 37.9%) whether governments should be the only ones to bear the costs of such adaptation and mitigation efforts, although a significant disagreement, even strong, emerges (Q25, 20.7% and 19.1%, respectively). At the same time, at an individual level, there is a wide consensus on the need to lower the personal consumption of resources (Q28, 79.2%) and engaging in informative events (Q28, 51.0%), while most of the respondents have already taken action against climate change, mainly working on the residential domain (Q29, 98.1%).

In order to further explore these findings, comparisons were performed among the questions presented above. As mentioned, where possible, cross-correlation and chi-square χ^2 test for independence were carried out. In the following paragraphs, the most significant results will be discussed.

3.1 The Influence of Age in Shaping the Sense of Place and Perception of Climate Change

For the purpose of analyzing the perceived impact that climate change will have on Veneto's inhabitants, question number Q31 was related to questions Q8 (Table 4) and Q9 (Fig. 2). Basically, age details were associated with the perception of climate change.

Table 4 shows that the population agrees about the problem of climate change impacting daily activities, though age still induces some differences among the responses. Overall, there is a consciousness of what is happening in the world and even in the specific area of Veneto region. Particularly, looking through the results, it is possible to recognize that young people disagree the least about the impact of climate change on lifestyle (0.4%), compared to older people, among whom the uncertainty is greater (4.1%). Probably, young people are growing up receiving information about climate change and suggestions about how to behave. Older people understand the

Table 2 Answers of the general part of the questionnaires (sample size = 1231 individuals)

Question	Answers	Frequencies (N)	Percentages (%)
Q31: Age	Young (18–34 years)	240	19.5
	Adults (35–64 years)	821	66.7
	Elderly (> 64 years)	169	13.7
	Total	1231	100.0
Q32: Where do you live?	Coastal area	457	37.1
	Hinterland	774	62.9
	Total	1231	100.0
Q34: Do you feel integrated in your community?	Yes	1005	81.6
	No	83	6.7
	I don't know	70	5.7
	I prefer not to answer	73	5.9
	Total	1231	100.0
Q35: How far do you live from the coast?	<1000 m	67	5.4
	>1000 m	1073	87.2
	I don't know	62	5.0
	I prefer not to answer	29	2.4
	Total	1231	100.0

Source Authors

current problem, but they may not completely understand what to change in their lifestyle, or they may not grasp well the use of new technologies, so they are not sure of what needs to be changed.

Table 5 summarizes the related measures of association (p-value and gamma). The statistical correlation between Q31 and Q8 is very significant (P value = 0.000), with a moderate association and a negative (gamma = −0.251).

About the responsibility of changing the lifestyle, it is possible to observe (Fig. 2) that elderly people would shift at their individual level (10.9%), adults think that changes must come from the community involvement and responsibility (32.1%), and young people support both the options (6.0% for individual level and 5.9% for the community). The alternative of not changing anything seems to be more refused by elderly people (0.9%) compared to adults and young people (7.3% and 2.0% respectively). The aptitude of sharing problems and interventions supported by the adult and young population in contrast with a less collective perspective of elder people stresses the transition from an individualistic community to a more collectivist one. The cause of the differences among the age classes might be in a different attachment to the territory where they live, that is changing and transforming. Furthermore, people do not have the same updated information and adopt different attitudes. The age influence in shaping the sense of place and perception of climate change emerges also from the analysis of questions about age Q31 correlated with questions Q28 and

Table 3 Answers to the sense of belonging part of the questionnaires (sample size = 1231 individuals)

Question	Answer	Frequency (N)	Percentage (%)
Q8: Climate change will impact your lifestyle[a]	Strongly disagree	32	2.6
	Disagree	53	4.3
	Undecided	227	18.4
	Agree	462	37.5
	Strongly agree	457	37.1
	Total	1231	100.0
Q9: What do you think will have to change in your lifestyle?[b]	Nothing	125	10.2
	Political level	6	0.5
	Community level	558	45.3
	Individual level	537	43.6
	Total	1226	99.6
Q25: The cost of mitigation of, and adaptation to climate change should be exclusively paid by the government[a]	Strongly disagree	235	19.1
	Disagree	255	20.7
	Undecided	466	37.9
	Agree	177	14.4
	Strongly agree	98	8.0
	Total	1231	100.0
Q28: What can you do, at the individual level, to prepare for climate-related hazards?[c]	Nothing	50	4.1
	Insurance	236	19.2
	Consumption	975	79.2
	Events	628	51.0
	Exposure	122	9.9
	Habits	84	6.8
	Total	1227	99.7
Q29: Can you list concrete steps that you and your family have taken to face climate change?[b]	Nothing	96	7.8
	Impacts	729	59.2
	Residential	376	98.1
	Consumption	516	41.9
	Soft[d]	105	8.5
	Total	1208	98.1

[a]Likert scale, [b]Open question, [c]Multiple choice question, [d]Personal habits and actions
Source Authors

Table 4 Cross-table of Q31 and Q8 with frequencies (percentages) of the answers

		Q8. Climate change will impact your lifestyle			
		Strongly disagree/disagree	Undecided	Strongly agree/agree	Total
Q31. Age	Young (18–34)	5 (0.4%)	39 (3.2%)	196 (15.9%)	240 (19.5%)
	Adult (35–64)	67 (5.4%)	138 (11.2%)	616 (50.0%)	821 (66.7%)
	Elderly (>64)	13 (1.1%)	50 (4.1%)	106 (8.6%)	169 (13.7%)
	Total	85 (6.9%)	227 (18.4%)	919 (74.7%)	1230 (100.0%)

Source Authors

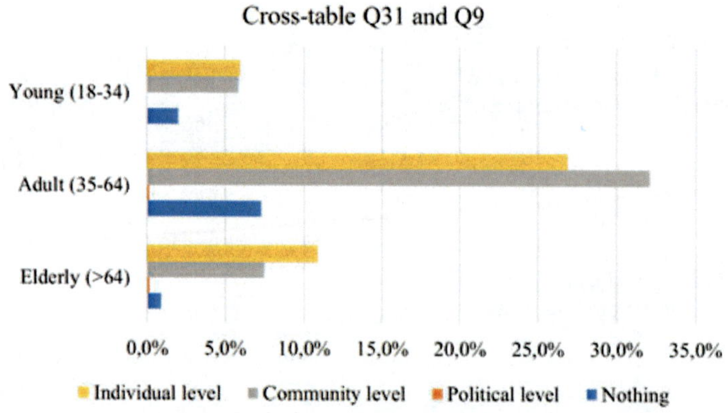

Fig. 2 Graphical representation of the cross-table of Q31 and Q9. *Source* Authors

Table 5 Summary of the measures of the chi-square χ^2 test and the Gamma for the cross-table presented in Table 4

	P value	Gamma
Q31 × Q8	0.000	−0.251

Source Authors

Q29, concerning actions and measures to prepare for climate change hazards, as shown in Fig. 3a and b respectively.

People were asked to choose among a list of actions to do to prepare for climate change hazards (Q28) such as changing habits, moving house to reduce self-exposure, participating in educational and informative events, reducing energy consumption at home, protecting assets with insurance, or doing nothing. All age classes express a common preference for consumption (that has the highest score in all classes), events and insurance (Fig. 3a). Only young people and adults choose exposure, but nobody

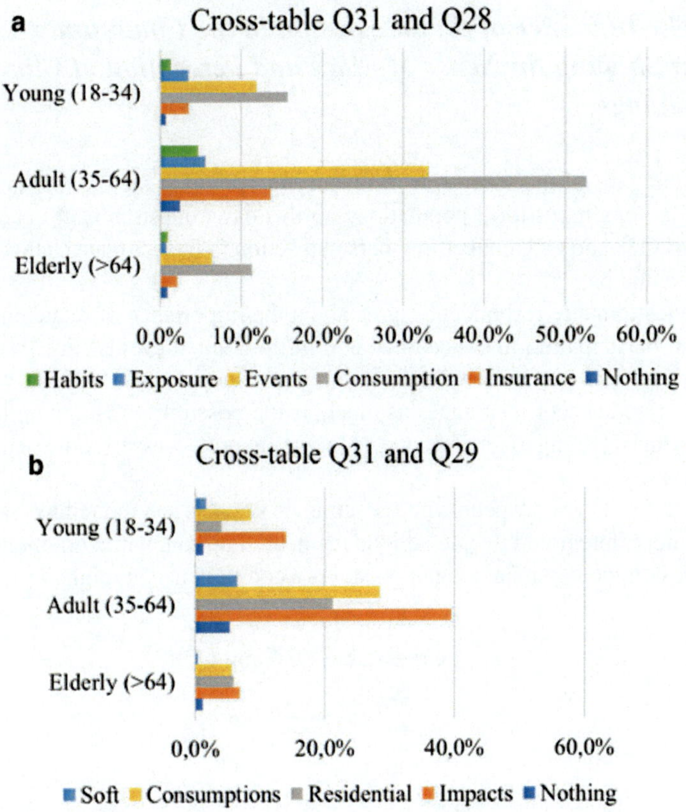

Fig. 3 Graphical representation of the cross-table of Q31 and Q28 (**a**), and Q31 and Q29 (**b**). *Source* Authors

is willing to move house. The interest in increasing self-knowledge about climate change adaptation and mitigation actions is high and the exclusion of the moving home option hints at an integrated population attached to the territory.

When citizens were asked what actions they undertake to cope with climate change, the whole community responded as operating principally on impacts and in reducing consumptions (Fig. 3b). Adults and elderly people added the importance of the residential sector. This option brings out the intention of older people for improving their adapting capacity and the availability of resources instead of totally changing habits and lifestyle. Hence, this implies that belonging to a specific age class actually influences the behavior and the actions to undertake.

3.2 The Influence of the Integration in the Community in Shaping the Sense of Place and Perception of Climate Change

One of the aims of the study was to identify the possible factors that influence the sense of belonging of local populations. In this case, question Q34 is crossed with questions Q28 and Q29 concerning personal action to face climate change, as shown in Fig. 4a and b.

All questionnaire respondents agree about the importance of reducing consumptions and participating in educational and informational events (Fig. 4a). The trust in insurance tools is fundamental for integrated people, while those who do not feel integrated accept insurances as much as the possibility of moving home. This suggests that a feeling of integration encourages to protect assets more than safeguard themselves.

Figure 4b shows how people prefer acting on impacts and reducing consumptions. Furthermore, integrated people add the residential option, while not integrated ones are more confident in managing impacts. As a consequence, it might be assumed that

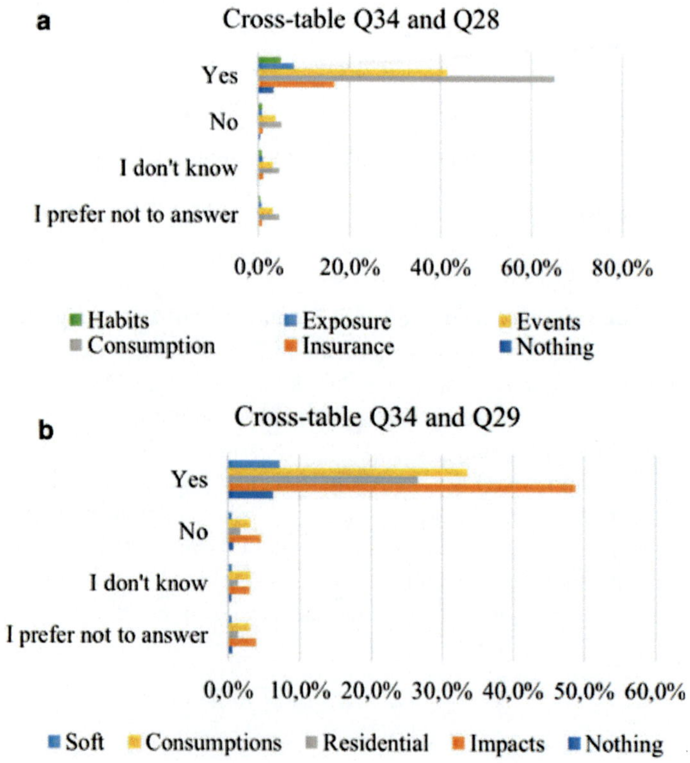

Fig. 4 Graphical representation of the cross-table of Q34 and Q28 (**a**), and Q34 and Q29 (**b**)

the belonging to the territory indeed influences climate hazard perception and thus the preference over actions to cope with climate change.

3.3 The Influence of the Proximity to the Coast in Shaping the Sense of Place and Perception of Climate Change

The relation with strong geographical features, such as the proximity to the coast, was a further element assumed to potentially influence the perception of climate change issues. Hence, this facet was explored through the comparison of question Q35 with questions Q8 (Table 6) and Q25 (Table 7), as well as with the questions Q28 and Q29 (Fig. 5).

Table 6 Cross-table of Q35 and Q8 with frequencies (percentages) of the answer

		Q8. Climate change will impact your lifestyle			
		Strongly disagree/disagree	Undecided	Strongly agree/agree	Total
35. How far do you live from the coast?	<1000 m	3 (0.2%)	5 (0.4%)	59 (4.8%)	67 (5.4%)
	>1000 m	69 (5.6%)	198 (16.1%)	806 (65.5%)	1073 (87.2%)
	I don't know	8 (0.6%)	18 (1.5%)	36 (2.9%)	62 (5.0%)
	I prefer not to answer	5 (0.4%)	6 (0.5%)	18 (1.5%)	29 (2.4%)
	Total	85 (6.9%)	227 (18.4%)	919 (74.7%)	1231 (100.0%)

Source Authors

Table 7 Cross-table of Q35 and Q25 with frequencies (percentages) of the answer

		25. The cost of mitigation of, and adaptation to climate change should be exclusively paid by the government			
		Strongly disagree/disagree	Undecided	Strongly agree/agree	Total
35. How far do you live from the coast?	<1000 m	22 (1.8%)	24 (1.9%)	21 (1.7%)	67 (5.4%)
	>1000 m	440 (35.7%)	410 (33.3%)	223 (18.1%)	1073 (87.2%)
	I don't know	17 (1.4%)	23 (1.9%)	22 (1.8%)	62 (5.0%)
	I prefer not to answer	11 (0.9%)	9 (0.7%)	9 (0.7%)	29 (2.4%)
	Total	490 (39.8%)	466 (37.9%)	275 (22.3%)	1231 (100.0%)

Source Authors

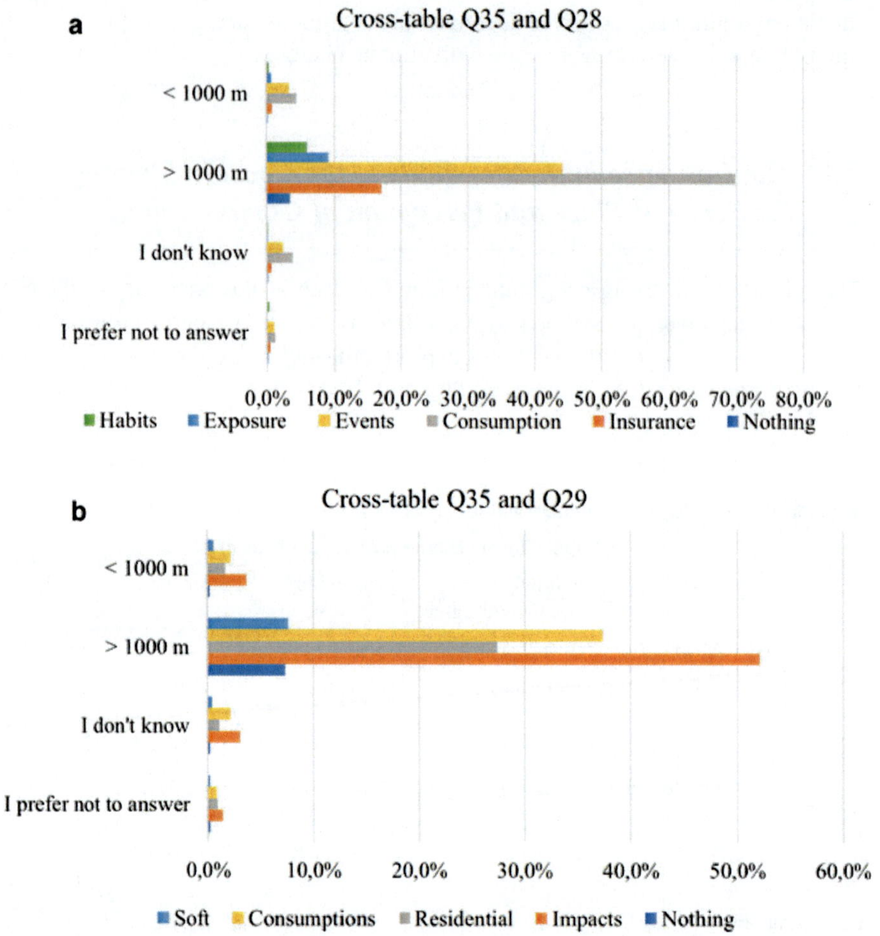

Fig. 5 Graphical representation of the cross-table of Q35 and Q28 (**a**), and Q35 and Q29 (**b**). *Source* Authors

In particular, results show that there is a significant agreement (Table 6, 74.7%) over the impact of climate change on everyday routines independently of the proximity to the coastline, with a strong statistical correlation (Table 8, p-value = 0.002), though the association is not particularly defined (Table 8, Cramer's V = 0.093). Nevertheless, it is noteworthy that respondents residing farther from the coast (<1000 m) appear to be more hesitant than those living nearer to the coasts (>1000 m) to acknowledge such a scenario (Table 6). Although slight, this difference suggests that people living closer to the coast are more aware of the impact that the climate can have on the lifestyle. This result could be related to the fact that people living closer to the coast are more exposed to direct effects of climate change such as sea level rise or coastal flooding and may therefore have a greater awareness of the effects these events could have on the territory. Such exposure increases local

Table 8 Summary of the measures of the chi-square χ^2 test and the Cramer's V (degree of freedom) for the cross-tables presented in Table 7

	P value	Cramer's *V*
Q35 × Q8	0.002	0.093 (2)
Q35 × Q25	0.036	0.074 (2)

Source Authors

awareness on these issues, especially with respect to other types of phenomena that also involve internal areas but could be perceived as common.

When questioning whether the government should sustain the overall costs of mitigation and adaptation efforts, a more varied picture emerged (Table 7). Indeed, respondents residing near the coast assigned almost an equal number of preferences for agreement or disagreement concerning this theme (21 and 22, respectively), although uncertainty showed a modest predominance (24 preferences). On the other hand, respondents farther from the coast were more consistently uncertain and even more in disagreement with such statement, thus suggesting a higher acknowledgement of the need to contribute to the common economic endeavors to cope with climate change. Overall, also in this case the explored geographical factor appears relevant, as the analysis resulted once more statistically significant (p-value $= 0.036$), though the association was not particularly strong (Cramer's $V = 0.074$).

When confronted with potential actions to tackle the effects of climate change, preferences exhibit a strong convergence towards the reduction of home consumptions and the attendance to informative events (Fig. 5a). The high share of agreement received by the reduction of consumptions, especially for answers coming farther from the coast, denotes a significant awareness on the relevance of personal choices in daily activities in the general efforts to cope with climate change. It might be noteworthy that the following emerging domains were insurances and reduction of exposed assets, where the respondents nearer the coast demanded more vividly the limitation of the susceptibility (that is, in terms of exposed assets) rather than the compensation for damages, as if previous events had left a higher sensibility towards this issue. It is also interesting to observe that a rather low share of respondents affirmed that nothing should be done to deal with the issues related to climate change, hence it might be assumed that there is a widespread consensus on a shared as well as personal responsibility towards this theme.

This perspective appears confirmed by the question on the actual actions undertaken, although respondents reported a higher tendency to reduce the impacts of everyday activities, rather than lowering overall consumptions, followed by additional interventions endowed to the residential sector (Fig. 5b). At the same time, there is also a significant share of admissions of having done nothing yet to tackle climate change, especially when considering the respondents farther from the coast. This might suggest that, in spite of the willingness to effectively act on the challenges presented by climate change, there might be some serious barriers to the actual capitalization and implementation of that commitment, which might demand further investigations. Indeed, it might also be possible that the hesitation in recognizing the impending effects of climate change on personal lifestyles (Q8, Table 7) might also delay the engagement of residents living far from the coastline in actual countermeasures (Q29, Fig. 5b). On the contrary, the perception of climate impacts as an urgent issue might foster the prompt response of those residing nearer the coast.

3.4 The Influence of the Geographic Position in Shaping the Sense of Place and Perception of Climate Change

A further geographical factor was explored, here in terms of area of residence defined in the related province. Hence, the dichotomy between coastal area and hinterland was confronted with the preferences over potential (Q28) and implemented (Q29) actions relevant to cope with climate change.

The results confirm the picture provided by the previous analysis based on the proximity to the coast (see Fig. 5). Indeed, the overall priority is assigned to actions devoted to the reduction of consumptions as well as to the involvement in informative events, followed from afar by the increase of insurance coverage (Fig. 6a). Nevertheless, the actually enacted actions only partially correspond to these preferences, as in this case most of the efforts have been devoted to limiting the personal impacts on the overall natural system, especially for coastal residents, though the reduction of consumptions follows closely (Fig. 6b). Once more, the share of respondents that have yet to take action against climate change significantly emerges, especially for the hinterland, thus suggesting the need of further investigations on the root causes of the divergence between expected actions (Fig. 6a) and implemented efforts (Fig. 6b).

Nevertheless, in light of the common trends in the analysis concerning the physical features, it appears reasonable to assume that indeed geographical attributes, in terms of either proximity to the coast or area of residence, significantly influence the perception of the effects of climate change on the local territory, as well as the willingness and the engagement in actual actions against climate change. In particular, it seems that people residing in the coastal area and especially nearer the sea tend to perceive a higher urgency and thus to act more consistently on personal choices to tackle climate impacts. On the contrary, residents of the inner areas appear more hesitant to engage in actual actions, possibly perceiving climate impacts as threats that will affect their territory in the distant rather than in the near future.

4 Conclusions

This study, conducted as part of the Interreg Italy-Croatia Project "RESPONSe", investigated the elements that shape individual and collective sense of place, the level of attachment to or bond with a particular area, and the ways in which individuals create identities connected to place. A total of 1231 adult citizens were reached by administering an online questionnaire, therefore it has been possible to consider

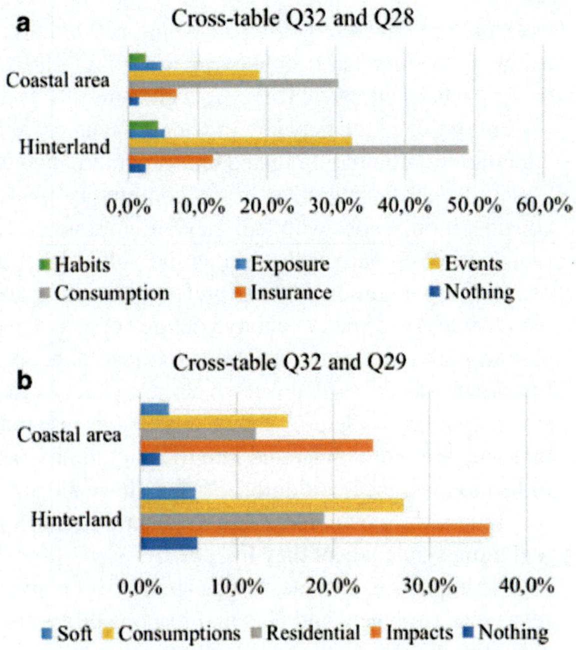

Fig. 6 Graphical representation of the cross-table of Q32 and Q28 (**a**), and Q32 and Q29 (**b**). *Source* Authors

the answers as representative for the overall regional adult population with a 95% (±2.80%) confidence level.

Results showed that young people are more aware of the effects that climate change has on everyday life, show a sense of responsibility for coping with climate change based on personal efforts in the context of a broader community engagement and deny inaction. Adults and elderly people show more uncertainty and consider the support of authorities more significantly. Eventually, elderly people are the least trustful about the efficacy of collective actions. As a matter of fact, the importance of the involvement of the entire community in counteracting climate change is more desired by younger people. Young people feel the urgency of counteracting and express more initiative in dealing with climate change, possibly because they have a greater environmental culture.

Young people show that they prefer more structural actions that reduce the exposure of the physical elements of the community, while adults and elderly people pay greater attention to their homes. This suggests a greater openness of young people to change and to take personal action to shape the community in order to improve their own future. On the contrary, the older generations are less inclined to change and focus their attention more on the safety of their present conditions.

Coastal and inland areas do not seem to show great differences in terms of willingness to act against climate change. Nevertheless, those who live in the innermost areas seem to be a little more skeptical about the impacts of climate change but are a little more inclined to commit personal funds to counteract them, even if they are not convinced of the importance of community involvement. Those who live on the coast have already, concretely acted on the impacts of climate change. This difference may be due to the fact that people living in the hinterland tend to perceive and suffer less the impacts of climate change, as well as to rely mainly on individual initiatives. A possible explanation may be a greater propensity for isolation typical of more closed communities commonly linked to more internal contexts (mountain communities).

The level of integration in the community is also an element to be taken into consideration. Those who feel they are not integrated expect to have to support the reduction of climate change on an individual level, are less sure of the importance of involving communities and prefer actions that are immediately effective. Those who feel less integrated, mainly young people and people living along the coast, also tend to prefer mitigation solutions to those of adaptation, meaning that they prefer limiting the activities that worsen climate change rather than adjusting to the changes that cannot be avoided. As shown in previous research (Simms 2017), those who feel that they are more integrated into the community prefer adaptation actions (mainly linked to insurance). Additionally, they have already put in place mitigation actions focused on their own home and in general prefer long-term solutions because they feel more stable where they live.

The impacts of climate change are more perceived along the coast, where people feel greater urgency and personal responsibility towards climate change as well as expect more actions to reduce adverse events, while they have already acted on the impacts. Those who live in the innermost areas perceive less the impacts of climate change, so much that they also consider that they do not have to do anything to counter the consequences of climate change. In addition, they expect community intervention, rather than individual action, even if they would accept to contribute financially.

These results suggest that the sense of belonging can be used as a means for local authorities to implement efficient adaptation strategies to enhance climate resilience. Local authorities should make use of the openness and progressive spirit of young people, as well as of the sense of "home" of more adult people to define tailored and appropriate adaptation strategies to climate change, while supporting mitigation activities. Authorities should focus on the feeling of belonging to the community to plan long-term adaptation and mitigation measures. At the same time, it would be pivotal to take advantage of the possible positive results to increase, in turn, the sense of belonging among those who feel less attached to the community to enhance the climate resilience of the local community as a whole.

Acknowledgements This research was funded by the EU Italy-Croatia Interreg program through the RESPONSe project (ID 10046849).

References

Adger W, Agrawala NS, Mirza MMQ, Conde C, O'Brien K, Pulhin J, Pulwarty R, Smit B, Takahashi K (2007) Assessment of adaptation practices, options, constraints and capacity. In: Parry ML, Canziani OF, Palutikof JP, van der Linden PJ, Hanson CE (eds) Climate change 2007: impacts, adaptation and vulnerability. contribution of working group ii to the fourth assessment report of the intergovernmental panel on climate change. Cambridge University Press, Cambridge, UK, pp 717–743. https://doi.org/10.1007/978-1-349-02250-2_5

Biesbroek GR, Klostermann JEM, Termeer CJAM, Kabat P (2013) On the nature of barriers to climate change adaptation. Reg Environ Change 13(5):1119–1129. https://doi.org/10.1007/s10113-013-0421-y

Bowden V, Nyberg D, Wright C (2019) Planning for the past: local temporality and the construction of denial in climate change adaptation. Glob Environ Change 57:101939. https://doi.org/10.1016/j.gloenvcha.2019.101939

Brulle RJ, Norgaard KM (2019) Avoiding cultural trauma: climate change and social inertia. Environ Politics 28(5):886–908. https://doi.org/10.1080/09644016.2018.1562138

Cunsolo Willox A, Harper SL, Ford JD, Landman K, Houle K, Edge VL (2012) "From this place and of this place:" climate change, sense of place, and health in Nunatsiavut, canada. Soc Sci Med 75(3):538–547. https://doi.org/10.1016/j.socscimed.2012.03.043

Denton F, Wilbanks TJ, Abeysinghe AC, Burton I, Gao Q, Lemos MC, Masui T, O'Brien KL, Warner K (2014) Climate-resilient pathways: adaptation, mitigation, and sustainable development. In: Field CB, Barros VR, Dokken DJ, Mach KJ, Mastrandrea MD, Bilir TE, Chatterjee M, Ebi KL, Estrada YO, Genova RC, Girma B, Kissel ES, Levy AN, MacCracken S, Mastrandrea PR, White LL (eds) Climate change 2014: impacts, adaptation, and vulnerability. Part A: global and sectoral aspects. Contribution of working group II to the fifth assessment report of the intergovernmental panel on climate change. Cambridge University Press, Cambridge, United Kingdom and New York, NY, USA, pp 1101–1131

Devine-Wright P (2013) Think global, act local? The relevance of place attachments and place identities in a climate changed world. Glob Environ Chang 23(1):61–69. https://doi.org/10.1016/j.gloenvcha.2012.08.003

Devine-Wright P, Price J, Leviston Z (2015) My country or my planet? Exploring the influence of multiple place attachments and ideological beliefs upon climate change attitudes and opinions. Glob Environ Chang 30:68–79. https://doi.org/10.1016/j.gloenvcha.2014.10.012

Eriksen SH, Nightingale AJ, Eakin H (2015) Reframing adaptation: the political nature of climate change adaptation. Glob Environ Chang 35:523–533. https://doi.org/10.1016/j.gloenvcha.2015.09.014

Feitelson E (1991) Sharing the globe. The role of attachment to place. Glob Environ Change 1(5):396–406. https://doi.org/10.1016/0959-3780(91)90005-E

Hryniewicz O (2006) Goodman-Kruskal γ measure of dependence for fuzzy ordered categorical data. Comput Stat Data Anal 51(1):323–334. https://doi.org/10.1016/j.csda.2006.04.014

IPCC (2021) Summary for Policymakers. In: Climate Change 2021: The Physical Science Basis. Contribution of Working Group I to the Sixth Assessment Report of the Intergovernmental Panel on Climate Change [Masson-Delmotte, V., P. Zhai, A. Pirani, S.L. Connors, C. Péan, S. Berger, N. Caud, Y. Chen, L. Goldfarb, M.I. Gomis, M. Huang, K. Leitzell, E. Lonnoy, J.B.R. Matthews, T.K. Maycock, T. Waterfield, O. Yelekçi, R. Yu, and B. Zhou (eds.)]. Cambridge University Press, Cambridge, United Kingdom and New York, NY, USA, pp 3−32. https://doi.org/10.1017/9781009157896.001

IPCC (2022) Summary for Policymakers [H.-O. Pörtner, D.C. Roberts, E.S. Poloczanska, K. Mintenbeck, M. Tignor, A. Alegría, M. Craig, S. Langsdorf, S. Löschke, V. Möller, A. Okem (eds.)]. In: Climate Change 2022: Impacts, Adaptation, and Vulnerability. Contribution of Working Group II to the Sixth Assessment Report of the Intergovernmental Panel on Climate Change [H.-O. Pörtner, D.C. Roberts, M. Tignor, E.S. Poloczanska, K. Mintenbeck, A. Alegría, M. Craig, S. Langsdorf, S. Löschke, V. Möller, A. Okem, B. Rama (eds.)]. Cambridge University

Press, Cambridge, UK and New York, NY, USA, pp 3–33. https://doi.org/10.1017/9781009325844.001

IPCC (2022) Summary for Policymakers. In: Climate Change 2022: Mitigation of Climate Change. Contribution of Working Group III to the Sixth Assessment Report of the Intergovernmental Panel on Climate Change [P.R. Shukla, J. Skea, R. Slade, A. Al Khourdajie, R. van Diemen, D. McCollum, M. Pathak, S. Some, P. Vyas, R. Fradera, M. Belkacemi, A. Hasija, G. Lisboa, S. Luz, J. Malley, (eds.)]. Cambridge University Press, Cambridge, UK and New York, NY, USA. https://doi.org/10.1017/9781009157926.001

Istat. (n.d.). Demografia in cifre

Istituto Nazionale di Statistica (2021) Popolazione residente al 1° gennaio. Retrieved 4 June 2021, from http://dati.istat.it/Index.aspx?DataSetCode=DCIS_POPRES1

Mchugh ML (2013) The chi-square test of independence lessons in biostatistics. Biochemia Medica 23(2):143–149. Retrieved from https://doi.org/10.11613/BM.2013.018

Moser A (1952) Quota sampling. Soc Roy Stat 115(3):411–423

Osservatorio Veneti nel Mondo (2021) Comprendere il fenomeno guardando al futuro Percorsi di approfondimento sul le relazioni familiari transnazionali, il ruolo delle imprese e gli effetti della Brexit Report di ricerca. Venezia-Mestre

Porter JJ, Demeritt D, Dessai S (2015) The right stuff? Informing adaptation to climate change in British Local Government. Glob Environ Chang 35:411–422. https://doi.org/10.1016/j.gloenvcha.2015.10.004

Regione Veneto (2007) P.R.S. Programma regionale di Sviluppo

Scannell L, Gifford R (2010) Defining place attachment: a tripartite organizing framework. J Environ Psychol 30(1):1–10. https://doi.org/10.1016/j.jenvp.2009.09.006

Simms JRZ (2017) "Why would I live anyplace else?": resilience sense of place and possibilities of migration in Coastal Louisiana. J Coastal Res 33(2):408–420. https://doi.org/10.2112/JCOASTRES-D-15-00193.1

Tacoli C (2009) Crisis or adaptation? Migration and climate change in a context of high mobility. Environ Urban 21(2):513–525. https://doi.org/10.1177/0956247809342182

UNDRR (2015) Sendai Framework for Disaster Risk Reduction 2015-2030. Geneva, Switzerland. Available at: https://www.undrr.org/publication/sendai-framework-disaster-risk-reduction-2015-2030

Warner K, Hamza M, Oliver-Smith A, Renaud F, Julca A (2010) Climate change, environmental degradation and migration. Nat Hazards 55(3):689–715. https://doi.org/10.1007/s11069-009-9419-7

From Disasters to the Pandemic. A Study on the EU Solidarity Fund

Giovanni Messina

Abstract The purpose of this contribution, which is to be included in the debate on risk mitigation policies and instruments or disaster support, is to describe the European Union Solidarity Fund (EUSF) intervention framework. This fund, created in 2002 and renewed in 2014, is an instrument capable of intervening retrospectively and on different levels on the territories affected by a natural event at high intensity. Since 2020, the operativity of the Fund has also extended to actions to combat the COVID-19 pandemic, representing, well before the composition of the so-called Recovery Fund, the first form of support against the economic and health disaster generated by the pandemic. Recalling the theoretical context, the contribution focuses on the description of the Fund and, with reference to Italy, identifies the resources allocated and maps the areas of intervention.

Keywords Risk mitigation · Policy · EUSF · Resilience · Best practices

1 Introduction

Although inside the scientific debate the term 'extreme natural event' is most commonly used to identify precisely the paroxysmal situations that impact spaces and communities, 'catastrophe' is the lemma that, in my view, maintains more fully its semantic power. I want then to recall two reflections advanced by two scholars from Palermo, who from the school of cultural anthropology of Antonino Buttitta have articulated different research paths.

Vincenzo Guarrasi, in an essay written with Anna Maria La Monica on the effects of the 1968 Belìce earthquake, identifies in the dimension of the catastrophe a form of decomposition and recomposition of the territorial interweaving and processes that sinks its structure in the matrix of the Greek myth (Guarrasi and La Monica 1995). The catastrophe is the paroxysm that, in its dramatic articulation, calls into

G. Messina (✉)
University of Messina, Messina, Italy
e-mail: giovanni.messina@unime.it

B. Dahiya et al. (eds.), *Disaster Resilience and Human Settlements*, Advances in 21st Century Human Settlements,
https://doi.org/10.1007/978-981-99-2248-2_14

question the previous arrangements and opens to new territorial projects (Dematteis 2002) and new relationships (Raffestin 2012).

From a more exquisitely anthropological point of view, in his work of critical reinterpretation of the thought of Lévi-Strauss, Salvatore D'Onofrio dwells, with a foresight and lucid concern, on the contemporary declination of catastrophe[1]:

> [La] riduzione dello spazio vitale ha contribuito in larga misura all'emergere del temuto effetto noto come globalizzazione. Per Lévi-Strauss, questa è in gran parte il risultato dell'esplosione demografica, in cui dovremmo vedere [...] la "vera catastrofe". In effetti, rivolgendo verso noi stessi gli eccessi inflitti agli altri, gli stili di vita dell'Occidente hanno innescato una crescita demografica il cui risultato, la globalizzazione, cancella le particolarità dei diversi gruppi etnici [...]. Ad essere minacciata è finalmente la pluralità delle forme culturali attraverso cui pensiamo l'unità della condizione umana. [...] Applicato all'agricoltura e all'allevamento, il passaggio alla rivoluzione industriale ha causato la rottura [...] della primitiva armonia tra animali ed esseri umani di cui narrano i miti d'origine. [...] Le infezioni animali sono diventate i sintomi di malattie che possono essere trasmesse direttamente all'uomo e richiedono quindi una doppia riflessione. Da un lato, mostrano quanto abbiamo soggiogato altre specie viventi [...]. D'altra parte, queste infezioni sono strettamente legate alla crescita iperbolica della popolazione umana. (D'Onofrio 2019, pp. 34–36).

Lucid concerns, it was said, in the light of the proliferation, for over a year now, of the COVID-19 pandemic (Korinman 2020; Turco 2021) that has in the so-called zoonotic jump its own etiology. These reflections, regarding the current health situation, have found in the geographical debate an immediate response (Messina 2020): coronavirus and anthropocene are strongly linked (Borruso et al. 2020).

As closely intertwined is the debate between human pressure, climate change and exposure to the risk of natural paroxysms. Geographical research, which has always been focused on detecting the connections between spaces and communities, allows then to activate a holistic approach to the issue; illuminating in this regard is the notation of Montz and Tobin in the journal Applied Geography:

> While the geophysical setting describes the physical processes to which a population may be exposed, it is not sufficient to understand just the degree to which people at a location are threatened by that exposure. In this regard, risk is a simple probability statement about the frequency of extreme events and the numbers exposed; however to fully appreciate hazards, dynamic human factors must also be addressed. Social and demographic characteristics of the population at risk combine with the prevailing political-economic system to determine a community's vulnerability. Through hazards research, much of which is geographically

[1] '[The] reduction in living space has largely contributed to the Mergence of the feared effect known as globalization. For Lévi-Strauss, this is largely the result of the population explosion, in which we should see [...] the "real catastrophe". In fact, by turning the excesses inflicted on others towards ourselves, the lifestyles of the West have triggered a demographic growth whose result, globalization, erases the particularities of the different ethnic groups [...]. What is finally threatened is the plurality of cultural forms through which we think of the unity of the human condition. [...] Applied to agriculture and livestock, the transition to the industrial revolution caused the break [...] of the primitive harmony between animals and humans whose origin myths tell. [...] Animal infections have become symptoms of diseases that can be transmitted Directly to humans and therefore require double reflection. On the one hand, they show how much we have subjugated other living species [...]. On the other hand, these infections are closely linked to the hyperbolic growth of the human population'. Translated by the Author.

based, our conception of vulnerability and the factors that contribute to it have progressed significantly (Montz and Tobin 2011, p. 2).

It is Serenella Iovino who then, strongly, puts the theme of the critical approach to the reading of the increasingly problematic relationships between man and ecosystems. The concept of an 'eco-criticism' then moves on a crucial epistemological track. Taking up the powerful image of the landscape as a text (Cosgrove and Jackson 1987), Iovino proposes a holistic critical reading of the dynamics of power and consumption that the environmental criticality deliver us:

> A text is something that can be read: a book, an inscription on a wall, a musical score, a poem, a picture, a film, a theater play. But "text" can also be something else: for example, the *material texture* of meanings, experiences, processes, and substances that make the life of places and beings. A text, in this sense, emerges from the encounter of actions, discourses, imagination, and physical forces that congeal in material forms. Landscapes are texts, and so are bodies. They are texts, because through them we read embodied narratives of social and power relations, biological balances and imbalances, and the concrete shaping of spaces, territories, human, and nonhuman life. (Iovino 2016, p. 3)

In the Anthropocene (Crutzen and Stoermer 2000), an era in which criticality and potentiality of the intense relationships that connect communities to spaces are highlighted (Giorda 2019), the size of the risk from extreme natural events then becomes the indicator of individual and collective perceptions and, together, guidance for policies called to intervene in areas particularly exposed to the destructive action of natural elements (Kasperson and Dow 1993; Gaillard and Dibben 2008; Shi 2019). Multidisciplinary is then scientific debate on risk containment policies with respect to extreme natural events and, in particular, on the tools of intervention available to communities (Thomalla et al. 2006; Mercer 2010; Kousky 2012; Newman et al. 2017; Antronico and Maricioni 2018).

There is a wide national and international bibliography on the many aspects involving study, monitoring, and risk management policies from natural events. In the multitude of sources, of particular interest are the two United Nations documents (Poljanšek et al. 2019) which, adopted in 2015, have focused on reducing the risk of a natural event from the strong intensity and on combating climate change: the *Sendai* Framework and the *2030 Agenda for Sustainable Development*. It seemed interesting to me then to focus on the EU Solidarity Fund (EUSF), a financial instrument[2] of the European Commission in support of the territories affected by the cataclysms.

> The European Union Solidarity Fund (EUSF) was set up to respond to major natural disasters and express European solidarity to disaster-stricken regions within Europe. The Fund was created as a reaction to the severe floods in Central Europe in the summer of 2002. Since then, it has been used for 80 disasters covering a range of different catastrophic events including floods, forest fires, earthquakes, storms and drought. 24 different European countries have been supported so far for an amount of over 5 billion €

[2] Look at https://ec.europa.eu/regional_policy/en/funding/solidarity-fund/.

From 1 April 2020, the Fund was the first response to the economic and health emergency generated by the pandemic: considered as an extreme natural event or, better, a disaster, has seen its scope extended to health crises.[3]

As part of the above-mentioned debate on risk mitigation and disaster relief policies and instruments, this contribution aims to describe the EUSF intervention framework and, with reference to Italy, identify more precisely the resources allocated and the areas of intervention. For the analysis, reference will be made to the official data and to the reference scientific literature.

2 The EU Solidarity Fund

The EUSF, in accordance with the EU Regulation 2012 of 2002, which was subsequently renewed with the EU Regulation 661 of 2014, was created to partially share among the Member States the expenditure to be disbursed in the event of disasters. It is therefore a financial instrument available to the European Community to implement concrete solidarity and mutual aid policies to support the recovery in the event of floods, earthquakes or storms. Natural disasters inflict considerable economic damage on territories (Fig. 1). It is a deeply transacting instrument and manages to intervene, ex-post, *at* national, regional and local levels.

As pointed out by a group of Romanian researchers,[4] the EUSF was born out of the need to integrate national government funds with international resources that, if necessary, are disbursed with a precise specification (Ciumas et al. 2017). As Stefan Hochrainer-Stigler, Anna Lorant and Eva-Cristina Petrescu note:

> Until 2014 the fund operated with an annual budget of €1 billion. However, the latest Multiannual Financial Framework (MFF 2014–2020) has halved its budget to €500 million (2011 prices) and added a temporal risk-spreading dimension. The primary aim of the EUSF is to finance emergency operations undertaken by public authorities to alleviate non insurable damages. Hence, it covers only a fraction of the total damages: compensation has averaged about 3% of total direct losses since 2002. In addition, it should be noted that the EUSF is a 'virtual' fund – in the event of disaster, the money is raised above and beyond the normal EU budgeting procedure. (Hochrainer-Stigler et al. 2016, p. 252)

The principle of solidarity set out in the name of the Fund really is the inspiration of the instrument. It began in the period between 2002 and 2008, when Europe, which had been hit by more than 100 floods, began to develop, symmetrically with the objectives of *Agenda 21* (UN 1993), a strong sensitivity to the issue of environmental disaster risk management. In the survey of policies and instruments to combat the

---

[3] Look at https://ec.europa.eu/regional_policy/en/funding/solidarity-fund/covid-19.

[4] 'Romania is ranked 82 from a total of 173 countries as regards the natural catastrophic risk (cat risk). When this ranking was realized the structural susceptibility was considered and also the management capacity of a natural catastrophe and the exposure to natural hazards. If we take into account only Europe, Romania is the Fourth country being outrun only by Albania, Serbia and Greece'. (Ciumas et al. 2017 p. 59).

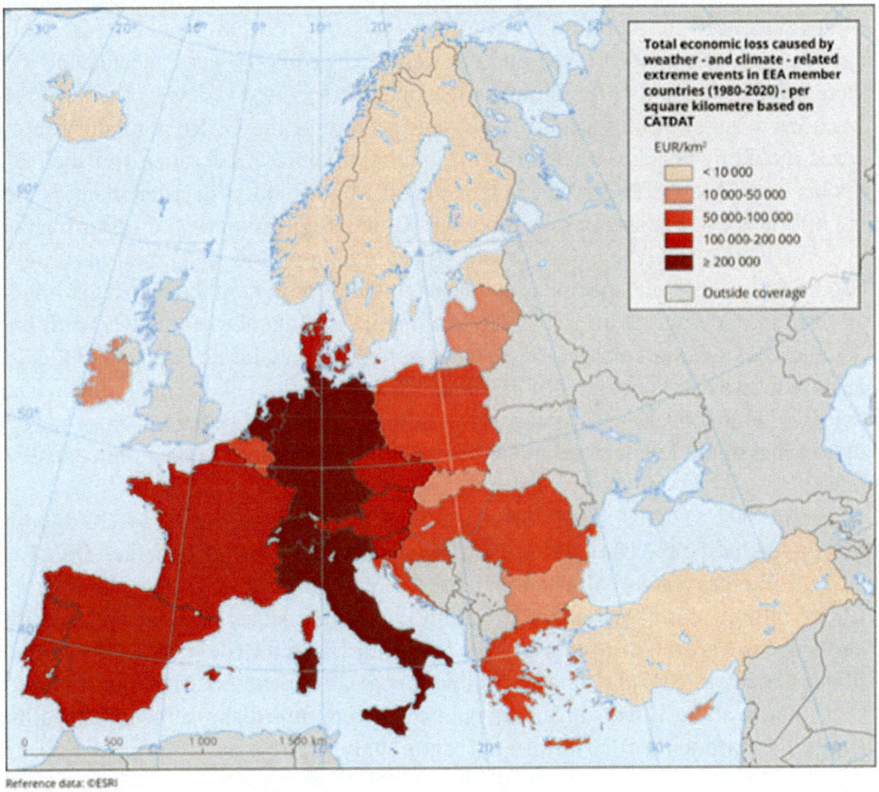

Fig. 1 Total economic loss caused by weather- and climate-related extreme events in EEA member countries (1980–2020). *Source* https://www.eea.europa.eu/data-and-maps/figures/total-economic-loss-caused-by-1

effects of disasters (Aakre et al. 2010), I report the notes of Slavko Bogdanovic on the EUSF:

> [From 2002] the EU Solidarity Fund (EUSF) was established, with the aim to enable Community to respond rapidly, efficiently, and in a flexible manner to emergency situations [Article 1], which are inter *alia major* natural disasters (*i.e.* such disasters resulting in at least one of the Member States in damage estimated either over three billion in 2002 prices, or more than 0.6% of its GNI). EUSF may be mainly mobilized in case of serious repercussions on living conditions, natural environment, and economy [Articles 2.1, 2.2]. The aim of EUSF is to complement the efforts of the affected State (or the region) to carry out emergency operations, which could comprise: Immediate restoration of works of infrastructure and plants in the fields of [*inter alia*] water and waste water; Provide temporary accommodation and funding rescue services to satisfy the immediate needs of population; Immediate securing of preventive infrastructures and measures for immediate protection of cultural heritage; Immediate cleaning-up of disaster-stricken areas, including natural zones [Article 3.2]. EUSF may only intervene for emergency operations, and not for phases preceding an emergency. (Bogdanovic 2011, p. 51)

As evidenced by the Fund's activity report, from 2002 to the end of 2020, 142 applications were made to the Fund; 96 were successful, 43 were rejected and three were withdrawn.[5] Also important is the document edited, among others by John Bachtler, which outlines a more up-to-date picture of the EUSF regulatory framework, including changes made in 2014. In relation to the 'Categories, thresholds for applications' the EUSF may be activated for: Major natural disaster (damage over €3 billion at 2011 prices or over 0.6% MS GNI); Regional natural disaster (damage > 1.5% NUTS 2 GDP (1%, for outermost regions); and Neighbouring eligible country (threshold not specified, but based on damage in neighbour). Regarding the 'Budgetary Ceiling', from 2014, 'Up to a maximum annual total of €500 million at 2011 prices since start of 2014–2020 multi-annual financial framework. On 1 October each year, at least one quarter of the annual amount of the Fund should remain available in order to cover needs arising until the end of the year', whereas 'Prior to 2014, up to a maximum annual total of €1 billion (at current prices)' (Bachtler et al. 2018, p. 6).

From 2002 to 2018, 88 EUSF activations were carried out in 24 EU Member States—including the United Kingdom, which in 2007 and 2015 benefited from EUR 222.6 million as a result of a national and regional flood that caused damage of over EUR 6 billion—for a total of 5, 536 billion Euro. Fires, storms, floods and especially earthquakes were the main catastrophic events for which the EUSF was activated. The floods that hit, in 2012, Lavamünd (Schwarz et al. 2018), the Austrian municipality of Carinthia, near the Slovenian border, recorded a minimum disbursement, 0.24 million Euros; the earthquakes that devastated central Italy in 2016 and 2017 (Calderoni et al. 2017), with an estimated damage of almost 22 billion Euro, have benefited from the support of the largest EUSF, almost 1.2 billion Euro.[6]

Italy is the country that has drawn the most from the resources of the EUSF. In 2019, the Fund has allocated more than 2.8 billion Euros to the Italian territory (Bergianti 2019). The synoptic table (Table 1) summarizes the picture of the interventions. It emerges the great importance, in terms of impact and damage detected, of seismic phenomena, the only ones to have a supra-regional scale.

To these funds should be added the €211.7 millions allocated to support the damage that caused between the end of October and November 2019 were inflicted on all of Italy, affected by extreme climatic events that have caused serious damage with floods and landslides, culminated in the flooding of Venice.[7]

---

[5] Look at https://cohesiondata.ec.europa.eu/stories/s/An-overview-of-the-EU-Solidarity-Fund-2002-2020/qpif-qzyn/.

[6] Look at https://ec.europa.eu/regional_policy/sources/thefunds/doc/interventions_since_2002.pdf.

[7] Look at https://ec.europa.eu/commission/presscorner/detail/it/ip_20_778.

Table 1 Natural disaster and EUSF aid in Italy from 2002 to 2019

Occurrence	Natural disaster	Scale	Damage (millions of Euro)	EUSF aid (millions of Euro)	Total EUSF aid (millions of Euro)
October 2002	Earthquake (Molise)	Regional	1.558	30.8	2.792.9
October 2002	Eruption of Volcano Etna (Sicily)	Regional	894	16.8	
April 2009	Earthquake (Abruzzo)	Major	10.212	493.8	
October 2010	Floods (Veneto)	Regional	676	16.9	
October 2011	Floods (Liguria and Tuscany)	Regional	723	18.1	
May 2012	Earthquakes (Emilia-Romagna)	Major	13.274	670.2	
November 2013	Floods (Sardinia)	Regional	652	16.3	
October 2014	Floods	Regional	2.241	56	
August 2016 to January 2017	Earthquake (Central Italy)	Major	21.879	1.196,8	
October 2018	Floods	Major	6.630.3	277,2	

Source https://ec.europa.eu/regional_policy/sources/thefunds/doc/interventions_since_2002.pdf

3 Conclusions

On 11 March 2020, the WHO proclaimed the pandemic dimension of COVID-19 infection. Since[8] March 31, the EUSF has expanded its scope for intervention in the health crisis. In the premises of the document with which the European Parliament updates the EUSF Regulation[9] we read:

> In the event of major public health emergencies, the Union should show its solidarity with Member States and the population concerned by providing financial assistance to help the population affected, to contribute to a rapid return to normal living conditions in the affected regions and to contain the spreading of infectious diseases.

The EUSF, in relation to the propagation of COVID-19, was called upon to intervene, well before the reflection on the Recovery *Plan for Europe was*[10] launched, to support Member States in assisting quickly, including from a medical point of view, the citizens affected by the disease and in setting up structures for[11] the prevention,

[8] Look at https://www.who.int/emergencies/diseases/novel-coronavirus-2019/events-as-they-happen.

[9] Look at https://eur-lex.europa.eu/legal-content/EN/TXT/?uri=uriserv:OJ.L_.2020.099.01.0009.01.eng&toc=OJ:L:2020:099:TOC.

[10] Look at https://ec.europa.eu/info/strategy/recovery-plan-europe_en.

[11] Look at https://ec.europa.eu/regional_policy/sources/thefunds/solidarity/pdf/20200401_EUSF_thresholds_health_emergencies_2020.pdf.

monitoring and control of the spread of the contagion, aimed at mitigating the risk with respect to public health. As for the budget, it was expected that, for major public health crises, total public Expenditure related to the crisis must exceed 0.3% of GNI or €1.5 billion in 2011 prices. For Italy, the maximum aid ceiling is approximately EUR 1.8 billion.

As for us, to return to the premises of the contribution, it is very interesting the convergence that leads the instrument of intervention to mitigate the impacts of natural disasters to address also health. This is the most empirical evidence of the systematic relationship between man and the environment in the Anthropocene. Catastrophes, in their being both destruction and palingenesis, require holistic vision and common policies, strategic planning and operational tools.

References

Aakre S, Banaszak I, Mechler R, Rübbelke D, Wreford A, Kalirai H (2010) Financial adaptation to disaster risk in the European Union Identifying roles for the public sector. Mitig Adapt Strat Glob Change 15:721–736

Antronico L, Maricioni M (ed) (2018) Natural hazards and disaster risk reduction policies. Il Sileno

Bachtler J, Begg I, Ferry M, Ogilvie J (2018) Ex post evaluation of the European Union solidarity fund 2002–2016. Final Report. General Directorate for Regional and Urban Policy-European Commission

Bergianti E (2019) I costi dei disastri naturali in Italia. Oggi Scienza. https://oggiscienza.it/2019/09/16/costi-disastri-naturali-italia/

Bogdanovic S (2011) Solidarity principle as a system tool for mitigation of flood impacts. In: Hefnawy M (Ed) Advances in food protection focus on food safety and defence. Springer, pp 49–55

Borruso G, Balletto G, Murgante B, Castiglia P, Dettori M (2020) Covid-19. Diffusione spaziale e aspetti ambientali del caso italiano. Semestrale Di Studi e Ricerche Di Geografia 2:39–56

Calderoni G, Rovelli A, Di Giovambattista R (2017) rupture directivity of the strongest 2016–2017 central Italy earthquakes. J Geophys Res Solid Earth 122:9118–9131

Ciumas C, Coca RA, Chis DM (2017) The role of public sector in catastrophic risk management. Knowl Horiz Econ 2:59–63

Cosgrove D, Jackson P (1987) New directions in cultural geography. Area 19(2):95–101

Crutzen PJ, Stoermer EF (2000) The 'Anthropocene'. Int Geosph–Biosph Program Newsl 17–18. http://www.igbp.net/download/18.316f18321323470177580001401/1376383088452/NL41.pdf

Dematteis G (2002) Progetto implicito. Il contributo della geografia umana alle scienze del territorio, Franco Angeli

D'Onofrio S (2019) Lévi-Strauss e la catastrofe: Nulla è perduto, possiamo riprenderci tutto, Mimesis

European Commission (2022a) COVID-19-EU solidarity fund. https://ec.europa.eu/regional_policy/en/funding/solidarity-fund/covid-19

European Commission (2022b) EUSF thresholds 2020 for major health emergencies. https://ec.europa.eu/regional_policy/sources/thefunds/solidarity/pdf/20200401_EUSF_thresholds_health_emergencies_2020.pdf

European Commission (2022b) EU solidarity fund. https://ec.europa.eu/regional_policy/en/funding/solidarity-fund/